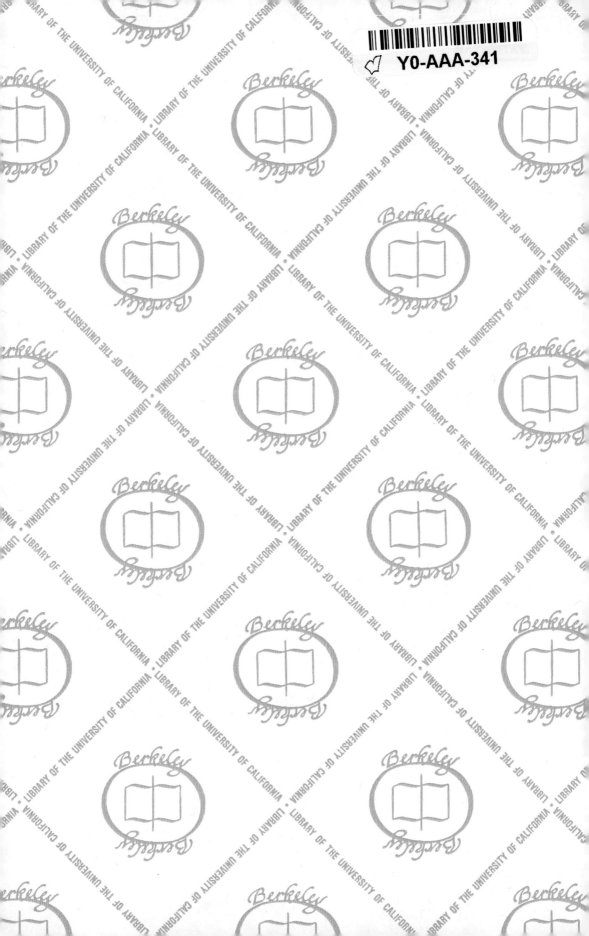

CONTEMPORARY MATHEMATICS

Titles in This Series

Titles in This Series

Volume

27 **Microlocal analysis,** M. Salah Baouendi, Richard Beals, and Linda Preiss Rothschild, Editors

28 **Fluids and plasmas: geometry and dynamics,** Jerrold E. Marsden, Editor

29 **Automated theorem proving,** W. W. Bledsoe and Donald Loveland, Editors

30 **Mathematical applications of category theory,** J. W. Gray, Editor

31 **Axiomatic set theory,** James E. Baumgartner, Donald A. Martin, and Saharon Shelah, Editors

32 **Proceedings of the conference on Banach algebras and several complex variables,** F. Greenleaf and D. Gulick, Editors

33 **Contributions to group theory,** Kenneth I. Appel, John G. Ratcliffe, and Paul E. Schupp, Editors

34 **Combinatorics and algebra,** Curtis Greene, Editor

35 **Four-manifold theory,** Cameron Gordon and Robion Kirby, Editors

36 **Group actions on manifolds,** Reinhard Schultz, Editor

37 **Conference on algebraic topology in honor of Peter Hilton,** Renzo Piccinini and Denis Sjerve, Editors

38 **Topics in complex analysis,** Dorothy Browne Shaffer, Editor

39 **Errett Bishop: Reflections on him and his research,** Murray Rosenblatt, Editor

40 **Integral bases for affine Lie algebras and their universal enveloping algebras,** David Mitzman

41 **Particle systems, random media and large deviations,** Richard Durrett, Editor

42 **Classical real analysis,** Daniel Waterman, Editor

43 **Group actions on rings,** Susan Montgomery, Editor

44 **Combinatorial methods in topology and algebraic geometry,** John R. Harper and Richard Mandelbaum, Editors

45 **Finite groups—coming of age,** John McKay, Editor

46 **Structure of the standard modules for the affine Lie algebra $A_1^{(1)}$,** James Lepowsky and Mirko Primc

47 **Linear algebra and its role in systems theory,** Richard A. Brualdi, David H. Carlson, Biswa Nath Datta, Charles R. Johnson, and Robert J. Plemmons, Editors

48 **Analytic functions of one complex variable,** Chung-chun Yang and Chi-tai Chuang, Editors

49 **Complex differential geometry and nonlinear differential equations,** Yum-Tong Siu, Editor

50 **Random matrices and their applications,** Joel E. Cohen, Harry Kesten, and Charles M. Newman, Editors

51 **Nonlinear problems in geometry,** Dennis M. DeTurck, Editor

52 **Geometry of normed linear spaces,** R. G. Bartle, N. T. Peck, A. L. Peressini, and J. J. Uhl, Editors

53 **The Selberg trace formula and related topics,** Dennis A. Hejhal, Peter Sarnak, and Audrey Anne Terras, Editors

54 **Differential analysis and infinite dimensional spaces,** Kondagunta Sundaresan and Srinivasa Swaminathan, Editors

55 **Applications of algebraic K-theory to algebraic geometry and number theory,** Spencer J. Bloch, R. Keith Dennis, Eric M. Friedlander, and Michael R. Stein, Editors

56 **Multiparameter bifurcation theory,** Martin Golubitsky and John Guckenheimer, Editors

57 **Combinatorics and ordered sets,** Ivan Rival, Editor

Combinatorics and Ordered Sets

CONTEMPORARY MATHEMATICS

Volume 57

Combinatorics and Ordered Sets

Proceedings of the AMS-IMS-SIAM Joint Summer Research Conference held August 11–17, 1985, with support from the National Science Foundation

Ivan Rival, Editor

AMERICAN MATHEMATICAL SOCIETY

Providence • Rhode Island

The AMS-IMS-SIAM Joint Summer Research Conference in the Mathematical Sciences on Combinatorics was held at Humboldt State University, Arcata, California on August 11–17, 1985, with support from the National Science Foundation, Grant DMS-8415201.

1980 *Mathematics Subject Classifications*. 06A10, 06A5, 06C05, 90B35, 0504, 05C20, 03D15.

Library of Congress Cataloging-in-Publication Data

Combinatorics and ordered sets.

(Contemporary Mathematics; v. 57)

Bibliography: p.

1. Ordered sets–Congresses. 2. Combinatorial set theory–Congresses. I. Rival, Ivan, 1947– . II. American Mathematical Society. III. Institute of Mathematical Statistics. IV. Society for Industrial and Applied Mathematics. V. Series: Contemporary mathematics (American Mathematical Society); v. 57.

QA171.48.C65 1986 511.3′2 86-8006

ISBN 0-8218-5051-2, ISSN 0271-4132

CONTENTS

PREFACE

Ordered sets abound in mathematics, for instance in algebra, combinatorics, geometry, model theory, set theory and topology. The theory of ordered sets has applications throughout mathematics, and beyond in operations research, computer science and the physical and social sciences. The link with modern combinatorial theory is a major source of the current vitality in ordered sets itself.

This volume is a collection of the principal expository lectures presented at COMBINATORICS and ORDERED SETS a conference held at Humboldt State University, Arcata, California from August 11 to August 17, 1985. This weeklong conference was one of the 1985 Joint Summer Research Conferences in the Mathematical Sciences under the direction of the AMS-IMS-SIAM. It was supported by a grant from the National Science Foundation. The aim of this conference was to highlight several of the leading combinatorial themes in ordered sets today. As a conference featuring surveys on ordered sets it is a continuation of two recent NATO Advanced Study Institutes. The first, SYMPOSIUM on ORDERED SETS, was held in Banff, Canada, from August 28 to September 12, 1981 and was intended as an introduction to all current topics in the theory of ordered sets[1]. The second, GRAPHS and ORDER, was held in Banff too, from May 18 to May 31, 1984 and was to document the role of graphs in the theory of ordered sets and its applications[2].

The nine articles presented here cover a wide range of combinatorial themes in ordered sets. In *Order-theoretic aspects of scheduling*, Werner Poguntke focusses on the famous three-machine problem to illustrate order-theoretic aspects of scheduling theory. In *Radon transforms in combinatorics and lattice thoery*, Joseph P.S. Kung, surveys techniques he used to settle the longstanding 'matching conjecture' between the join irreducible and the meet irreducible elements in a finite modular lattice. In *Recursive ordered sets*, Henry Kierstead elaborates on the problem to 'effectively'

---

[1] *Ordered Sets* (I. Rival, ed.), D. Reidel Publishing Co., Dordrecht, 1982.

[2] *Graphs and Order* (I. Rival, ed.), D. Reidel Publishing Co., Dordrecht, 1985.

decompose a (countable) ordered set into few chains. In *Orientations and reorientations of graphs*, Oliver Pretzel surveys uses of the 'pushdown' operation which reorients a diagram to produce another order with the same covering graph. In *Abstract convexity and meet-distributive lattices*, Paul Edelman synthesizes the many and varied occurrences of the meet-distributive property a frequently recurring concept. In *Correlation and order*, Peter Winkler surveys techniques that settled the conjecture that it is more likely, in a binary sorting problem, that $a < b$ if it becomes known that a is less than some other element c. In *Retracts: graphs and ordered sets from the metric point of view*, E.M. Jawhari, D. Misane and Maurice Pouzet formulate and develop an intriguing general view point for retraction, unifying ideas in ordered sets and in graphs by placing them both within a general theory of metric spaces. In *Antichains and cutsets*, Mohamed El-Zahar and Nejib Zaguia survey cutsets (subsets which meet every maximal chain) a theme which has seen remarkable activity in just the past few years. Finally, in *Stories about order and the letter* **N** (en), Ivan Rival considers the role played by the notion of subdiagram especially in structure and optimization themes in ordered sets.

The participation and presence of R.P. Dilworth was considerable - just as it was at the earlier SYMPOSIUM on ORDERED SETS. At that earlier meeting in 1981 there was, rather unexpectedly, one single result that seemed to play a motivating role in many of the lectures. It was the well known 'Chain Decomposition Theorem' that, *in an ordered set the minimum number of chains whose union is all of the set equals the maximum number of pairwise noncomparable elements*[4]. At this meeting in Arcata his influence again seemed to be ubiquitous. The themes in one day's lectures alone, spanned almost two decades of his fundamental results[3,4,5]. The theory of ordered sets is a lively area today and it has a rich legacy.

COMBINATORICS and ORDERED SETS was a relatively short meeting (five full working days). To avoid the prospect that some, among the many participants would perhaps, not become aware - until it was too late - of the presence of others with whom they would like to consult, we organized two opening sessions (*Introductory sketches*) on the first morning. These were explicitly intended as an opportunity for all participants to introduce themselves - publicly. Over forty did according to these suggested guidelines: write your name on the blackboard and discuss your current research interests - all within three minutes! It seemed to work. On the basis of these introductory sketches we were able to monitor common interests and therefore, to design *Special Sessions*, spontaneously and following the initiatives of the participants. The value of these conferences is often in the magic and spontaneity that

---

[3]R.P. Dilworth (1940) Lattices with unique irreducible decompositions, *Ann. Math.* 41, 771-777.

[4]R.P. Dilworth (1950) A decomposition theorem for partially ordered sets, *Annals of Math.* 51, 161-166.

[5]R.P. Dilworth (1954) Proof of a conjecture on finite modular lattices, *Annals of Math.* 60, 359-364.

occurs between the main lectures; the lectures themselves constitute the formal structure.

These weeklong Summer Research Conferences are inspired by the American Mathematical Society. Although they are held at different locations each summer their format and overall administration is uniform and smooth. There is little doubt that we all owe much to the tireless efforts of Carole Kohanski, the Summer Research Conference Coordinator of the American Mathematical Society. Besides all of her administrative work, her comments and suggestions helped in designing an effective, yet spontaneous, scientific programme.

I am particularly grateful for the consistent patience, encouragement and enthusiasm of my wife Hetje.

Calgary, Canada, February 20, 1986 Ivan Rival

PARTICIPANTS

Michael Albertson (U.S.A.)
Brian Alspach (Canada)
Margaret Bayer (U.S.A.)
Mary Bennett (U.S.A.)
Gary Bloom (U.S.A.)
Kenneth Bogart (U.S.A.)
Graham Brightwell (England)
Gerard Chang (Taiwan)
Stephen Comer (U.S.A.)
Julien Constantin (Canada)
Maria Contessa (Italy)
Hong Dang (U.S.A.)
Elias David (England)
Walter A. Deuber (Germany)
Robert P. Dilworth (U.S.A.)
Dwight Duffus (U.S.A.)
Paul Edelman (U.S.A.)
Paul Erdös (Hungary)
Peter Frankl (France)
Stephen Grantham (U.S.A.)
George Grätzer (Canada)
Michel Habib (France)
Mark Halsey (U.S.A.)
Katherine Heinrich (Canada)
Roland Jégou (France)
Jeffry Kahn (U.S.A.)
Henry Kierstead (U.S.A.)
David Klarner (U.S.A.)
Daniel Kleitman (U.S.A.)
V. Krishnamurthy (U.S.A)
Joseph Kung (U.S.A.)
Renu Laskar (U.S.A.)
Klaus Leeb (Germany)
Ko-Wei Lih (Taiwan)
Frank R. McMorris (U.S.A.)
Robert Melter (U.S.A.)
Shawkwei Moh (China)
Joseph Neggers (U.S.A.)
Heinrich Niederhausen (U.S.A.)
Richard Nowakowski (Canada)
Sergei Ovchinnikova (U.S.A.)
Werner Poguntke (Germany)
Maurice Pouzet (France)
Oliver Pretzel (England)
Hans Prömel (Germany)
Yuyuan Qin (China)
Robert Quackenbush (Canada)
Ivan Rival (Canada)
Vojtech Rödl (Czechoslovakia)
Ivo G. Rosenberg (Canada)
Michael Saks (U.S.A.)
Rainer Schrader (Germany)
Steven Simpson (U.S.A.)
Michael Stone (Canada)
Willaim Trotter (U.S.A.)
Miroslaw Truszczynski (Poland)
Bernd Voigt (Germany)
Edward Wang (Canada)
N. Zaguia (Tunisia)

SCIENTIFIC PROGRAMME

Monday, August 12, 1985

Chair: I. Rival

All: *Introductory sketches*

W. Poguntke, *Order-theoretic aspects of scheduling.* (I)

All: *Introductory sketches*

J. Kung, *Radon transforms in combinatorics and lattice theory.* (I)

H. Kierstead, *Recursive combinatorics of ordered sets.* (I)

Tuesday, August 13, 1985

I. Rival, *Stories about order and the letter* **N** (*en*).

M. Habib, *Linear extensions and depth-first search.*

O. Pretzel, *Orientations and reorientations of graphs.*

P. Edelman, *Abstract convexity and meet-distributive lattices.*

J. Kung, *Radon transforms in combinatorics and lattice theory.* (II)

Special Session I:

Chair: W. Deuber

P. Erdős, *Problems and results.*

M. Albertson, *Homomorphisms and independence in triangle-free graphs.*

S. Moh, *On chessmen moving.*

D. Kleitman, *Combinatorial enumeration.*

Wednesday, August 14, 1985

M. Pouzet, *Retracts: graphs and ordered sets from the metric point of view.*

W. Poguntke, *Order-theoretic aspects of scheduling.* (II)

Thursday, August 15, 1985

N. Zaguia, *Antichains and cutsets.*

W.T. Trotter, *Constructions of linear extensions.*

G. Grätzer, *Partition rank functions for general lattices.*

Special Session II:

Chair: R.P. Dilworth

E. David, *On the distributive property of quasi-ordered sets.*

I.G. Rosenberg, *Orders admitting a majority or near-unanimity isotone operation.*

R.W. Quackenbush, *Gaps, holes and near-unanimity functions.*

J. Constantin, *The fixed point property in posets (and graphs).*

Special Session III:

Chair: D. Duffus

G. Brightwell, *A universal correlation inequality for finite posets.*

A. Rucinski, *Balanced extensions of graphs.*

G. Bloom, *Labelled graphs.*

Friday, August 16, 1985

H. Kierstead, *Recursive combinatorics of ordered sets.* (II)

Special Session IV:

Chair: S. Comer

D. Klarner, *The number of graded posets is alternately congruent to* 1 *and* 3 (*mod* 6).

Y. Qin, *On a jar-metric principle in optimization.*

Special Session V:

Chair: B. Alspach

M. Truszczynski, *Matroids and jump number in* N*-free posets.*

N. Zaguia, *Interchanging chains.*

Contemporary Mathematics
Volume 57, 1986

ORDER-THEORETIC ASPECTS OF SCHEDULING

Werner Poguntke

ABSTRACT. Deterministic scheduling concerns the allocation over time of scarce resources in the form of *machines* or *processors* to different activities. An order on the activity set, due to technological or other constraints, may dictate that of two given activities one must be performed before the other so that only certain allocations are feasible. The problem is to devise efficient algorithms in order to determine a feasible solution that is optimal with respect to some criterion.
This article mainly concentrates on the *m-machine problem*. The *jump number problem* is also considered and used to illustrate order-theoretic aspects of scheduling theory. It is the aim of this article to work out some of these common principles. In particular, the main topics treated are *the role of order ideals, the role of linear extensions, min-max results*, and *transformation techniques*.

§ 1. INTRODUCTION

Deterministic scheduling concerns the allocation over time of scarce resources in the form of *machines* or *processors* to different activities ("jobs" or "tasks"). An order on the activity set, due to technological or other constraints, may dictate that of two given activities one must be performed before the other so that only certain allocations are feasible. The problem is to devise efficient algorithms in order to determine a feasible solution that is optimal with respect to some criterion.

Various techniques have been developed for the solution of scheduling problems. The reader is referred to the survey articles [Graham et al. 1979; Lawler & Lenstra 1982; Lenstra & Rinnooy Kan 1984; Rival 1984] or the book [Coffman 1976] which presents an attractive collection of articles written by several experts.

As is common in computer science, we consider a problem *well-solved* if it is solvable by an algorithm whose running time is bounded by a polynomial function

---

1980 Mathematics Subject Classification. 06A10, 68C15, 90B35.

of the problem size. In practical applications, it is of course very important *which* polynomial can be taken as a bound. This will not be in the center of interest in this article.

The theory of computational complexity makes it possible to distinguish formally between well-solved problems and NP-*hard problems* for which the existence of a polynomial algorithm is very unlikely. We refer to [Garey & Johnson 1979] for a comprehensive treatment of complexity theory.

It is the aim of this article to work out some of the basic principles of scheduling problems from the order-theoretic point of view. This could help to bring to the attention of order theorists many interesting and challenging problems. The basic ideas that will be treated are *order ideals, linear extensions, min-max results*, and *transformation techniques*. We will mainly concentrate on the m-*machine problem*. The other problem that will be considered throughout is the *jump number problem*. Both problems can be formulated in merely order-theoretic terms.

All ordered sets occurring in this article will be finite. We try not to be too formal in out notation. In particular, an ordered set $(P,<)$ will be denoted by "P", and if different orderings on the same set are considered, we will just use different letters.

Here is the *jump number problem*.
Suppose a single machine is to perform a set of jobs, one at a time; certain precedence constraints are given for the jobs. Any job which is performed immediately after a job which is not constrained to precede it requires a "setup" or "jump" producing some fixed additional cost. The problem is: *Schedule the jobs to minimize the number of jumps*.
In order-theoretic terms, we are dealing with an ordered set and its *linear extensions*. One way to define a linear extension is as an isotone bijection

$$\alpha: P \to \{1,2,\ldots,|P|\}.$$

Another way of viewing a linear extension is as a total ordering L on the same set as P such that $x<y$ in L whenever $x<y$ in P.
For a linear extension α of P, set

$$s(P,\alpha) = |\{(x,y) \in P \times P \mid x \nless y \text{ in } P \text{ and } \alpha(x)+1 = \alpha(y)\}|,$$

and let $\quad s(P) = \min\{s(P,\alpha) \mid \alpha$ is a linear extension of $P\}$.

By the *jump number problem*, we will in the sequel understand the following:

> *Evaluate* $s(P)$ *and construct a linear extension with* $s(P)$ *many jumps.*

We give an example in Figure 1. As usual, *diagrams* are used to represent ordered sets pictorially in the plane.

FIGURE 1

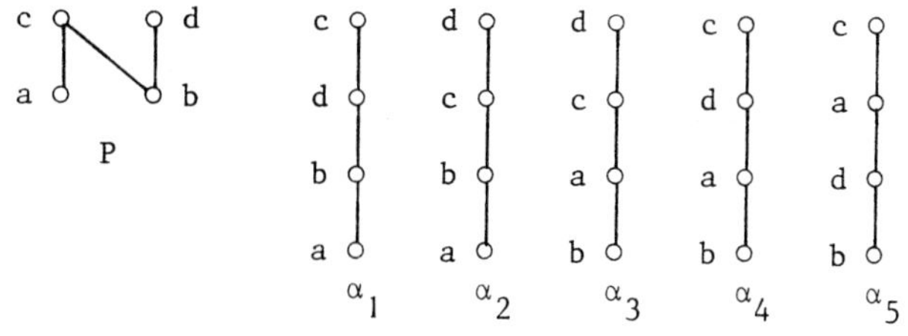

$s(P,\alpha_1) = s(P,\alpha_2) = s(P,\alpha_3) = 2,$
$s(P,\alpha_4) = 3, \quad s(P,\alpha_5) = 1 = s(P)$

The jump number problem has been shown to be NP-hard in [Pulleyblank 1981]. Good algorithms have been found for several restricted classes of ordered sets. The reader is referred to the survey paper [Rival 1984].

Here is the m-*machine problem: Schedule unit-time jobs subject to precedence constraints on* m *identical, parallel machines to minimize the maximum completion time.*

Again, the jobs are the elements of an ordered set P. An m-*machine schedule* can be defined as a one-to-one map σ of P to the direct product $\{1,2,\ldots,m\} \times \mathbb{N}$ whose second projection σ_2 is strictly order-preserving. Thus, to each job $x \in P$, the schedule assigns an ordered pair (i,n). The i-th machine, $1 \le i \le m$, processes the job x during the n-th time unit, $n \in \mathbb{N}$. A job cannot be processed until after all of its predecessors are processed; that is, if $x < y$ in P then $\sigma_2(x) < \sigma_2(y)$ in $\mathbb{N}$. The m-*machine-problem* can now be understood in the following way:

Find a schedule which attains

$$\min_{\substack{\sigma \\ \text{schedule}}} \max_{x \in P} \sigma_2(x) .$$

We call *optimal* any schedule which does.

Figure 2 shows an ordered set P and a 2-machine schedule σ for P. σ is indicated by plotting two horizontal time axes, one for each machine. Blocks are placed above the axes to indicate when a job starts and when its processing ends. Cross-hatch shading is commonly used to indicate that a machine is idle. Such a diagram for a schedule is called a *Gantt chart*.

FIGURE 2

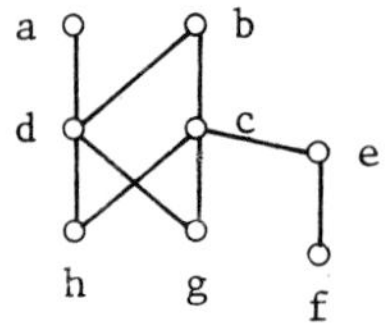

An ordered set P.

g	f	e	c	b
h	d	a		

A 2-machine schedule for P.

When asking for the computational complexity of the m-machine problem (measured in the size of P), one has to decide whether m ought to be fixed or not. In fact, it has been shown by [Ullman 1975] that the problem is NP-hard if m is arbitrary. Good algorithms have been found only for very restricted classes. On the other hand, there are polynomial algorithms in case $m = 2$ (see [Fujii, Kasami & Ninomiya 1969; Coffman & Graham 1972; Sethi 1976; Gabow 1982]). The reader is referred to the survey articles [Lawler & Lenstra 1982; Lenstra & Rinnooy Kan 1984].

We now get to an outstanding open problem:

> *For any fixed* $m \geq 3$, *no polynomial algorithm has been found for the* m*-machine problem, nor has* NP-*hardness been shown.*

We will get to special cases later in the discussion.

Linear extensions and m-machine schedules can also be viewed as *partial extensions*. It will be convenient for us also to be able to use this terminology.

Let P be an ordered set. An ordered set E with the same ground set is called a *partial extension* of P if $x < y$ in E whenever $x < y$ in P; thinking of an ordering as a set of pairs, this just means that the order of P is contained in the order of E. E is a linear extension if it is a maximal partial extension, i.e. a total order (or *chain*). An m-machine schedule σ for P induces via the "time order" a partial extension P^t of P: $x <^t y$ *if and only if* $\sigma_2(x) < \sigma_2(y)$.

Figure 3 shows the time order P^t for the schedule σ of Figure 2.

The *linear sum* $P_1 \oplus \ldots \oplus P_n$ of ordered sets $P_1, \ldots, P_n$ is the ordered set obtained by putting the P_i's in a tower with P_1 at the bottom and P_n at the top (see Figure 3 for an example).

It is now clear that the m-machine problem can also be understood in the following way:

Given an ordered set P, *find a partial extension* E *such that*

(i) E *is a linear sum of antichains,* $A_1, \ldots, A_k$;

(ii) $|A_i| \le m$ *for each* $1 \le i \le k$;

(iii) k *is minimal with respect to* (i) *and* (ii).

FIGURE 3

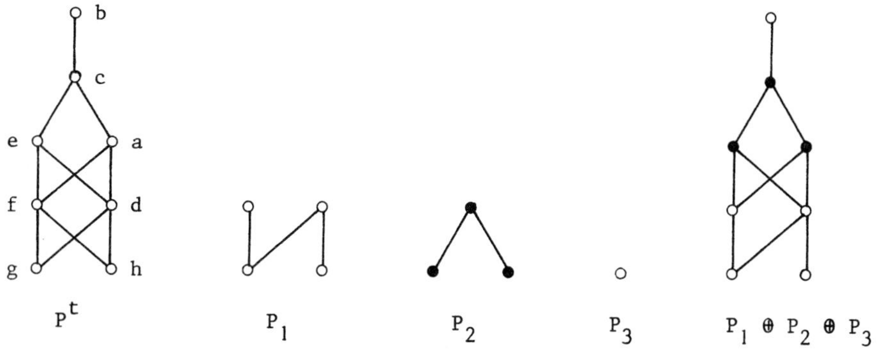

§ 2. THE ROLE OF ORDER IDEALS

An *order ideal* of an ordered set P is a subset $A \subseteq P$ with the property that $a \in A$ and $b < a$ implies $b \in A$.

If E is a partial extension of P corresponding to an m-machine schedule, i.e. E is a linear sum of antichains, $A_1, \ldots, A_k$, with $|A_i| \le m$ for all $i \le k$, then

$$A_1 \subseteq A_1 \cup A_2 \subseteq \ldots \subseteq A_1 \cup \ldots \cup A_k$$

is an ascending chain of order ideals of P.

On the other hand, if

$$\emptyset = 0_o \subseteq 0_1 \subseteq \ldots \subseteq 0_k = P$$

is an ascending chain of order ideals such that for each $1 \le i \le k$, $A_i := 0_i - 0_{i-1}$ consists of at most m maximal elements of 0_i, then the linear sum $A_1 \oplus \ldots \oplus A_k$ defines an m-machine schedule for P.

In other words: an m-machine schedule can also be viewed as an ascending chain of order ideals.

Let $G_m(P) = (\mathcal{O}(P), \to)$ be the following directed graph: $\mathcal{O}(P)$ is the set of order ideals of P; for $A,B \in \mathcal{O}(P)$, there is an edge, $A \to B$, if $A \subsetneq B$ and $B - A$ consists of at most m maximal elements of B. An optimal m-machine schedule for P now corresponds to a shortest path connecting $\emptyset$ to P in $G_m(P)$.
The following conclusion can be drawn:

> *Let* L *be a class of finite ordered sets,* m *an integer. If* $G_m(P)$ *can be constructed in polynomial time for* $P \in L$*, then the m-machine problem can be solved in polynomial time for* L.

This conclusion holds since there are well-known good algorithms to find shortest paths in graphs.

This can be applied to ordered sets of bounded width. The *width* of P, denoted by $w(P)$, is the largest cardinality of an antichain in P. For an integer q, let L_q be the class of finite ordered sets P with $w(P) \leq q$. The above conclusion shows:

> *The m-machine problem can be solved for* L_q *in polynomial time.*

In fact, in this statement, one does not even have to fix m, since $m \leq q$ can be assumed anyway. It has to be observed that an ordered set P with n elements and $w(P) \leq q$ has at most n^q many order ideals, and $G_m(P)$ can be constructed in polynomial time.
The only place we found these arguments written down is in [Pelzer 1984]. In fact, the definition of $G_m(P)$ given there is slightly more effective using the fact that not *all* order ideals of P have to be constructed: if $O \subseteq P$ is already constructed and $P-O$ has at least m minimal elements, then at the next recursion step only those ideals $O' = O \cup S$ with $|S| = m$ are constructed as successors of O.

Figure 4 shows an example for the construction of a sufficiently large subgraph of $G_2(P)$.

FIGURE 4

step 1 $\emptyset \begin{array}{c} \nearrow \{f,g\} \\ \rightarrow \{f,h\} \\ \searrow \{g,h\} \end{array}$

step 2 $\emptyset \begin{array}{c} \nearrow \{f,g\} \searrow \{e,f,g,h\} \\ \rightarrow \{f,h\} \nearrow \\ \searrow \{g,h\} \rightarrow \{d,f,g,h\} \end{array}$

step 3 $\emptyset \begin{array}{c} \nearrow \{f,g\} \rightarrow \{e,f,h,g\} \rightarrow \{c,d,e,f,g,h\} \\ \rightarrow \{f,h\} \nearrow \\ \searrow \{g,h\} \rightarrow \{d,f,g,h\} \rightarrow \{a,d,e,f,g,h\} \end{array}$

step 4 $\emptyset \begin{array}{c} \nearrow \{f,g\} \searrow \{e,f,g,h\} \Rightarrow \{c,d,e,f,g,h\} \\ \Rightarrow \{f,h\} \nearrow \end{array}$ $\Longrightarrow$ $\begin{array}{c} P \\ \nearrow \end{array}$

$\searrow \{g,h\} \rightarrow \{d,f,g,h\} \rightarrow \{a,d,e,f,g,h\} \rightarrow \{a,c,d,e,f,g,h\}$

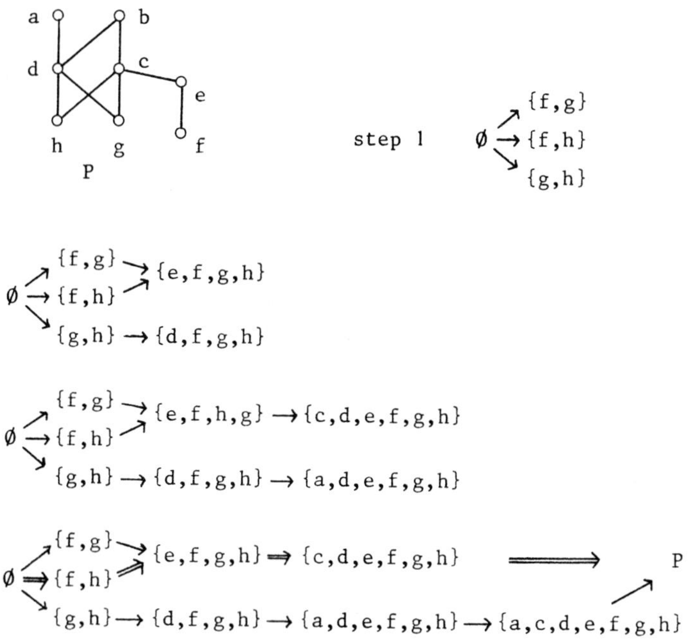

$G_2(P)$ with a shortest path from $\emptyset$ to P

The above considerations can be slightly generalized.

For x,y ϵ P, let x be *equivalent* to y, x $\sim$ y, if x and y have the same upper and lower covers. P/$\sim$ is the factor structure defined in the obvious way identifying equivalent elements. For an integer q, let $\tilde{L}_q$ be the class of all finite ordered sets P with $w(P/\sim) \le q$.

We claim:

> The m-*machine problem can be solved in polynomial time for* $\tilde{L}_q$.

The idea of proof of this claim is the following.

Let m and q be fixed, and let P be given. We now construct an ordered set P^* on the same set as P: if a $\sim$-class C has more than m elements, the order is strengthened by changing the antichain C into m chains (each of them having $\lceil \frac{|C|}{m} \rceil$ or $\lfloor \frac{|C|}{m} \rfloor$ many elements). It is now easy to see that every optimal schedule for P^* is optimal for P as well. Since $P^* \epsilon L_{mq}$, the claim follows.

Figure 5 shows an example for P and P^* (with m = 3, q = 2).

FIGURE 5

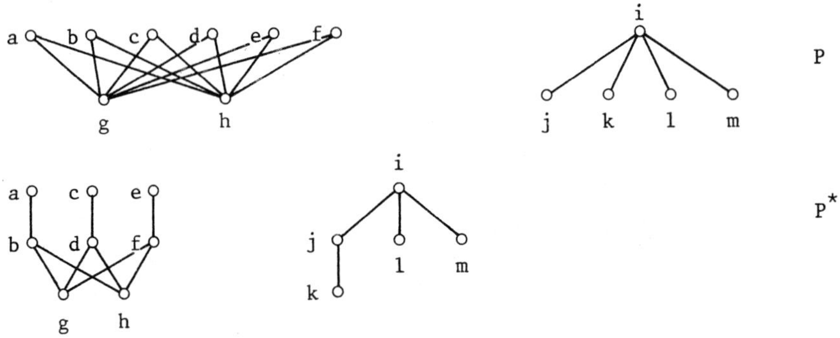

A disjoint union of q ordered sets each of which is a linear sum of antichains, is called a *level order* (with q components). P of Figure 5 is a level order with two components. Observe that a level order with q or less components belongs to $\tilde{L}_q$.
A polynomial algorithm to produce an optimal m-machine schedule for level orders (and fixed m) is given in [Dolev & Warmuth 1982b] without the restriction to q components. We will come back to this topic in the fifth paragraph.

Similar methods can be used for the jump number problem. This time, one has to consider *pointed ideals*. A *pointed ideal* of P is a pair (O,a) such that O is an order ideal of P and a is a maximal element of O.

Let $\mathcal{P}(P)$ be the set of all pointed ideals of P. For $(I,a),(J,b) \in \mathcal{P}(P)$, let there be an edge

$$(I,a) \rightarrow (J,b)$$

if $I \cup \{b\} = J$ and $b \notin I$.
We complete the construction by setting

$$\mathcal{P}_c(P) = \mathcal{P}(P) \cup \{\emptyset, P\}$$

and adding all the arrows $\emptyset \rightarrow (\{a\},a)$ (a minimal in P)
and $(P,b) \rightarrow P$ (b maximal in P).
The arrows are assigned a *cost* which is always 0 except in the cases

$$(I,a) \rightarrow (J,b) \quad \text{with} \quad a \nless b$$

when the cost is 1.

It is now obvious that a jump-optimal linear extension of P corresponds to a shortest path in $(\mathcal{P}_c(P), \rightarrow)$ from $\emptyset$ to P.

This allows the following conclusion (cf. [Colbourn & Pulleyblank 1984]):

The jump number problem can be solved for L_q *in polynomial time.*

Figure 6 shows an example of $(P_c(P), \to)$ with a shortest path.

FIGURE 6

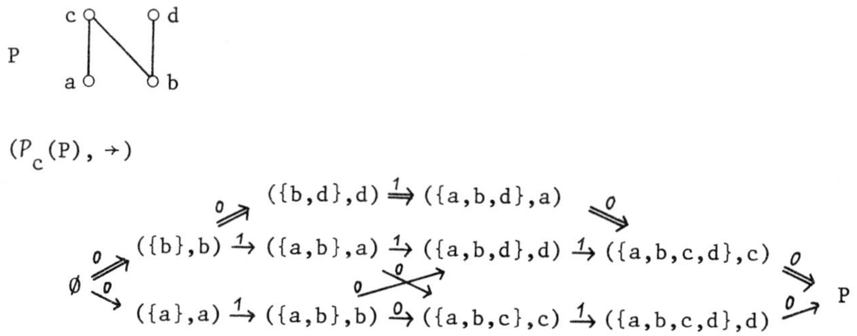

§ 3. THE ROLE OF LINEAR EXTENSIONS

The jump number problem is a one-machine scheduling problem where the objective is to find a certain (optimal) linear extension. For the m-machine problem, linear extensions are closely related to *priority lists*. Before we come to this, we will describe the solution of the 2-*machine flow-shop problem*. It has the remarkable property that an order on the set of jobs can be defined which is not present in the statement of the problem, and any linear extension then gives an optimal solution.

The ideas follow the lines of [Johnson 1954] which was one of the first papers dealing with scheduling problems.

A set of jobs $P = \{p_1, \dots, p_n\}$ is given. Each of them has to be executed on machine 1 first, then on machine 2. The execution times are $a_1, \dots, a_n$ (on machine 1) and $b_1, \dots, b_n$ (on machine 2). As usual, at most one job can be on a machine at a given time, and once a job is started on a machine, it has to be executed without interruption. The problem is to find a schedule that minimizes the overall execution time.

The idea in the solution of this problem is to schedule p_i early if a_i is small and late if b_i is small.

Figure 7 shows an example for illustration.

FIGURE 7

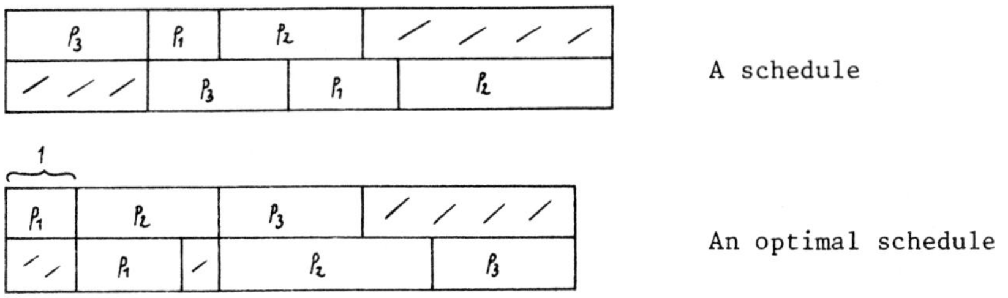

$P = \{p_1, p_2, p_3\}, \quad a_1 = 1, \quad b_1 = 1.5, \quad a_2 = a_3 = b_3 = 2, \quad b_2 = 3$

The first thing that has to be observed is: *There is an optimal schedule with the same order of the jobs on both machines.* This can easily be seen: If $p_1, \ldots, p_n$ is the sequence on machine 1, and $p_1, \ldots, p_i, p_j, \ldots$ $(j \neq i+1)$ is the sequence on machine 2, then putting p_{i+1} between p_i and p_j on machine 2 and schedule the jobs on machine 2 as early as possible does not increase the execution time.

The problem now is to find a certain linear ordering on the set P. If such an ordering L is given, a unique schedule can be associated by scheduling in this order without idle times between jobs on machine 1 and, in the same order, start jobs on machine 2 as early as possible.

The second important observation is the following:
If $L = p_1 \ldots p_{i-1} p_i \ldots p_j \ldots p_n$, *read as a schedule in the above sense, has execution time* T, *and if* $a_j \le a_i, b_j$, *then the execution time* T' *of* $L' = p_1 \ldots p_{i-1} p_j p_i p_{i+1} \ldots p_{j-1} p_{j+1} \ldots p_n$ *satisfies* $T' \le T$.
Seeing this is not too hard.
By symmetry, the condition $b_i \le a_i, b_j$ leads to a similar conclusion.

This gives the idea to define a relation $<$ on P by

$$p_i < p_j \text{ if and only if } \min\{a_i, b_j\} < \min\{a_j, b_i\}.$$

Obviously, $<$ is anti-reflexive.

The next claim is that $<$ is also transitive, i.e. $(P,<)$ is an ordered set. To see this, assume $p_i < p_j < p_k$. It is not possible that $b_j \le a_i$ and $a_j \le b_k$, since this would imply $b_j < a_j$ and $a_j < b_j$. Assume $a_i \le b_j$ and $b_k \le a_j$. This means that $a_i < a_j, b_i$ and $b_k < a_k, b_j$, hence $\min\{a_i, b_k\} < \min\{b_i, a_k\}$. The other cases can be treated similarly.

Now finally it follows by the above remarks:

Any linear extension L *of* (P,<) *induces an optimal schedule.*

This is true since starting from *any* linear ordering, one can get to *any* linear extension of (P,<) by a series of switches as described above without increasing the execution time.

As a consequence, the following algorithm produces an optimal ordering: Look for the smallest number among $a_1,\ldots,a_n,b_1,\ldots,b_n$. If this is an a_i, put p_i first in the order; if it is a b_i, put p_i last. After that, remove p_i (as well as a_i and b_i), and apply the same procedure, thus working to the middle from both ends. (In the case of ties, the choice is free.)

The solution of the 2-machine flow-shop problem is also our first example of the use of transformation methods to get other schedules from a given schedule. We will come back to this in the fifth paragraph.

Now what is the role of linear extensions for the m-machine problem? As announced above, they can be used to get *priority lists*.

Let P be an ordered set and $\pi: P \to \mathbb{N}$ a mapping. For given m, a partial extension $E = A_1 \oplus \ldots \oplus A_k$ of P corresponding to an m-schedule for P can be constructed recursively using π as a priority list in the following way:

> *Set* $A_o = \emptyset$.
> *Assume* $A_o,\ldots,A_i$ *have already been defined. If* $A_o \cup \ldots \cup A_i = P$, *then* $k = i$. *Otherwise, let* $M_i = \min (P - (A_o \cup \ldots \cup A_i))$. *If* $|M_i| \le m$, *let* $A_{i+1} = M_i$.
> *Otherwise,* A_{i+1} *is chosen to consist of* m *elements of* M_i *such that for no* $x \in M_i - A_{i+1}$ *and* $y \in A_{i+1}$ *is* $\pi(x) > \pi(y)$.

A schedule obtained in this way is called a *list schedule*. Since each optimal schedule without unforced idle times is obviously a list schedule, it is enough to look at list schedules. Because of this, the m-machine problem could be reduced to the question:

> *What are the right priorities?*

Observe that if π is not one-to-one, then the schedule constructed from π in general is not unique.

Figure 8 gives an example; the labels indicate the π-values.

FIGURE 8

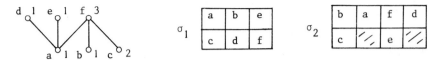

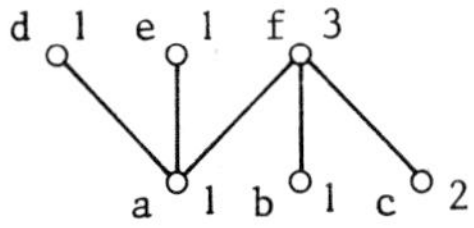

σ_1

a	b	e
c	d	f

σ_2

b	a	f	d
c	//	e	//

σ_1 and σ_2 are induced by π.

What one would like is, of course, to find in an effective way a priority list π such that *every* induced schedule is optimal.

A priority list should be easily produceable in the algorithmic sense.
Most of the priority lists commonly used are *antitone*, i.e. $x < y$ in P implies $\pi(x) > \pi(y)$.
An antitone priority list π which is one-to-one can be "normalized" so that $\{1,2,\ldots,|P|\}$ is the image set. In that case,

$$\alpha = |P| + 1 - \pi$$

is a linear extension of P. This justifies the statement that the m-machine problem consists of finding the "right" linear extension.

We would like to introduce three of the most common priority lists. Figure 9 illustrates them using examples.
Two notations are needed. For $x \in P$, let $\uparrow x = \{y \in P \mid y > x\}$ be the *upset* of x; $\lambda(x)$ is the *level* of x, i.e. the length of a longest chain in $\uparrow x \cup \{x\}$.

$\mu: P \to \mathbb{N}$ defined by $\mu(x) = |\uparrow x|$ is called the *upset list*. Any induced schedule is an *upset schedule*. Upset schedules need not be optimal, not even for two machines.

$\lambda: P \to \mathbb{N}$, taking levels as priorities, is called the *level list*. Level lists produce *level schedules*. We prove in the next paragraph Hu's Theorem saying that for trees, every level schedule (which in this case is also upset) is optimal.
In general, level schedules need not be optimal.

A CG-*list* (meaning "Coffman-Graham") is defined in the following way. The maximal elements of P, say $|\max(P)| = l$, are labelled arbitrarily with 1 to l. Now, doing induction, assume labels 1 to k have been assigned. Each $x \in P$ with the property that all its upper covers have already been labelled gets assigned the decreasing sequence of the labellings of its upper covers; a lexicographically smallest with respect to this assignment then gets the label k+1. The outcoming CG-list γ is one-to-one and depends on the labelling of the maximal elements. For $m = 2$, every CG-schedule is optimal; a proof will be outlined in the next paragraph. It is easy to see that every CG-schedule is a level schedule.

FIGURE 9

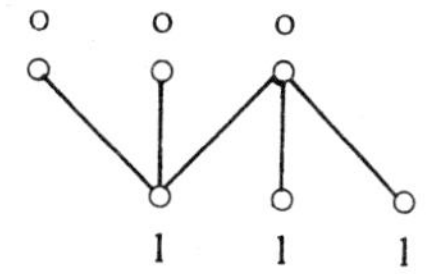

The level list. Not every level schedule is optimal.

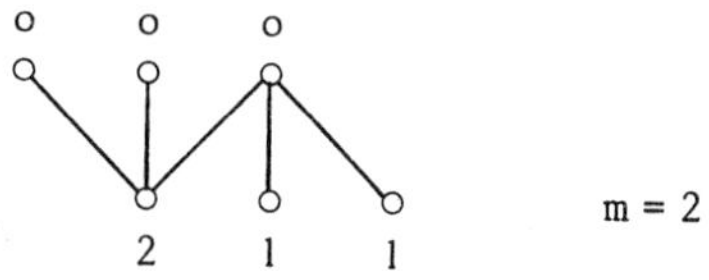

The upset list. Every upset schedule is optimal.

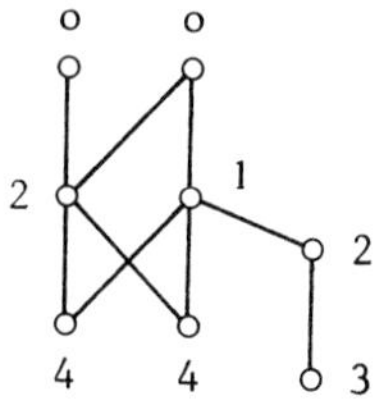

The upset list. No upset schedule is optimal.

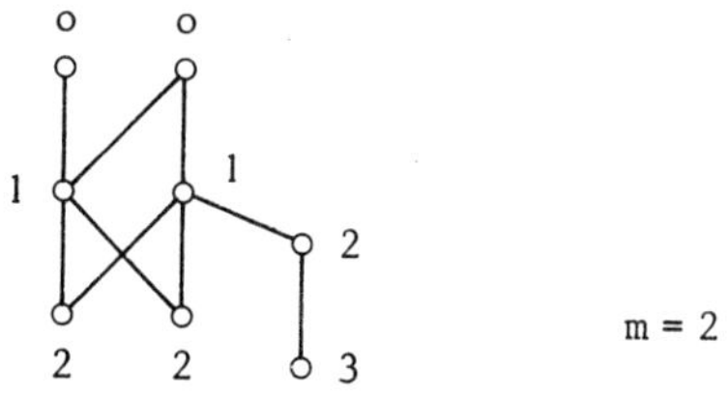

The level list. Every level schedule is optimal.

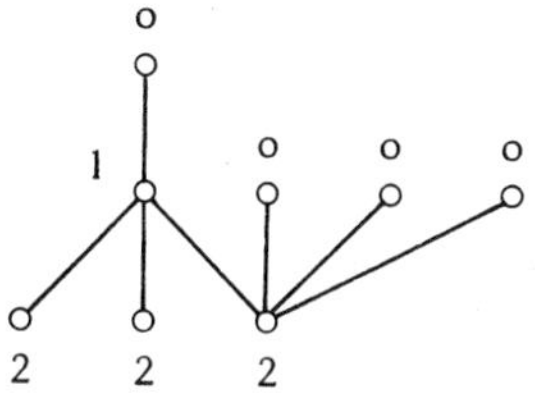

The level list. Not every level schedule is optimal.

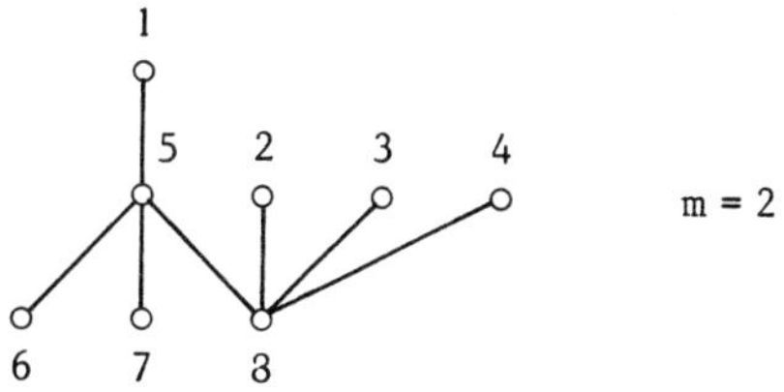

A CG-list. Every CG-schedule is optimal.

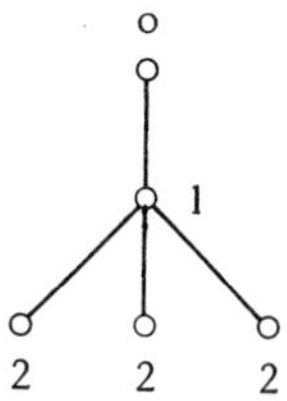

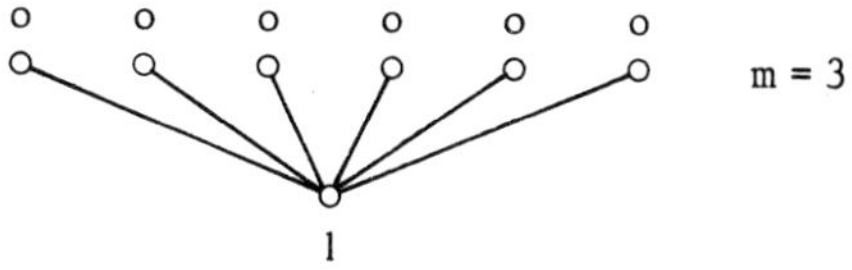

The level list. No level schedule is optimal.

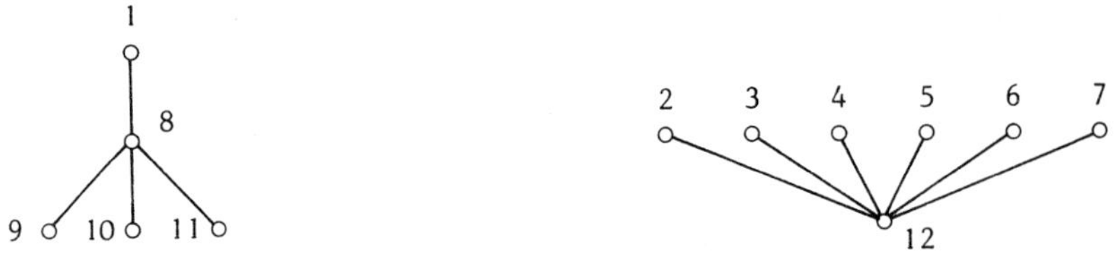

A priority list inducing an optimal schedule.

The priorities (or *labellings*) described above - as well as combinations of these - have been considered by many authors. But, as already mentioned earlier, for no fixed $m \geq 3$ has there been found a polynomially produceable labelling which leads to optimal schedules for all ordered sets. Also, the use of *dynamic lists* did not lead to large improvements; the idea of a dynamic list is that new priorities are assigned at each time stage in the schedule when it is being constructed.

Since there is this basic open problem, similar as in other areas of combinatorial optimization a lot of work done concerning the m-machine problem deals with the following questions:

> *What is the worst-case behaviour of upset, level,*
> CG- *or other lists when used as heuristics?*
> *What is their average behaviour?*
> *What are large classes of ordered sets for which*
> *these lists produce optimal schedules?*

We will not go deep into these questions. Here are just a few remarks.

As far as we know, not too much has been done concerning the second question. As to the third question, we already mentioned Hu's result for trees. More recent results are in [Nett 1979; Papadimitriou & Yannakakis 1979; Simon 1984]; see also [Ecker 1977]. Many articles, old and new, deal with the first question. We would like to mention the result of [Chen & Liu 1975] that the execution time of level schedules is at most $2 - \frac{1}{m-1}$ times the optimum (provided $m \geq 3$). Another interesting result is the one by [Lam & Sethi 1977] saying that the corresponding bound for CG-schedules is $2 - \frac{2}{m}$ $(m \geq 2)$. We also refer to the survey article of R.L. Graham in [Coffman 1976].

At the end of this paragraph, we would like to point out that most of the recent progress in devising effective algorithms to find optimal m-machine schedules for restricted classes of ordered sets does not use the idea of list scheduling, but is based on other techniques (cf., e.g., [Garey et al. 1983; Dolev & Warmuth 1980, 1982a, 1982b, 1984]). This will be taken up in the fifth paragraph.

§ 4. MIN-MAX THEOREMS

To work out the central points of this and the next paragraph, let us write down the general setting for a combinatorial optimization problem:

> *Given a finite set* S *and a function* $f: S \to \mathbb{R}$, *find an* $s^* \in S$ *such that* $f(s^*) \le f(s)$ *for all* $s \in S$.

In the cases we are considering, e.g. S is the set of m-machine schedules for an ordered set P, and f is the schedule length, algorithms to find an optimal solution are usually not measured using $|S|$ as the "size" of the problem. Instead, some other number defined by a structure from which S can be constructed figures as the size; for the m-machine problem and the jump number problem, this is just $|P|$. Since $|S|$ usually grows exponentially as a function of this other number, the aim is to find an optimal member of S *without* constructing *all* of S.

Apart from the problem to find a good algorithm, the other central question that comes up now is the following:

> *How can one show that a certain* $s^* \in S$ *is optimal, without constructing all of* S?

There seem to be two standard ways to solve this problem:

1. *One can prove a min-max result.*
2. *One can use transformation techniques.*

Transformation techniques will be treated in the next paragraph.

The lucky case is when a min-max result holds. The situation typically is the following:

There is some other set, T, also derived from the structure underlying S, and a function $g: T \to \mathbb{R}$ such that $g(t) \le f(s)$ for all $t \in T$ and $s \in S$. Hence, all the values $g(t)$ are *a priori* lower bounds for the optimal value of $f(s)$.
In this situation, if one can show that for a certain $s^* \in S$ (hopefully *constructed* by a good algorithm),

$$f(s^*) = g(t')$$

for some $t' \in T$, then s^* must be optimal, and the equality

$$\min_{s \in S} f(s) = \max_{t \in T} g(t)$$

holds.

Min-max results are very attractive goals in any mathematical field dealing with optimization. One reason is that they always seem to go with good algorithms. The second reason is they allow elegant proofs of optimality. One well-known example from combinatorial optimization is the max-flow min-cut Theorem for graphs; Dilworth's decomposition Theorem of ordered sets into chains can be viewed as a variation of this (see [Dilworth 1950]).

As our first example, we will sketch the proof of a min-max result for the jump number problem. The following Theorem is shown in [Duffus, Rival & Winkler 1982]:

If P *is a cycle-free ordered set, then* $s(P) = w(P) - 1$.

By *cycle-free*, as usual, it is meant that no cycle (see Figure 10) is isomorphic to a subset of P.

FIGURE 10

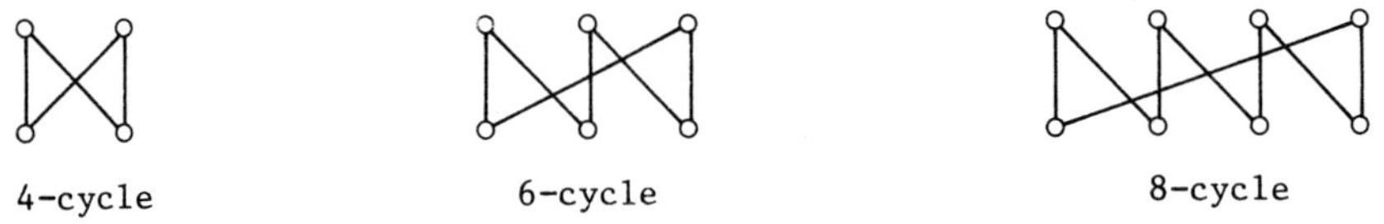

This is a min-max result, since the following is obviously true for any ordered set P:

If $A \subseteq P$ *is an antichain and* α *is a linear extension of* P, *then* $s(P,\alpha) \geq |A| - 1$.

In the general setting outlined above, S is the set of linear extensions of P, f counts the jumps, T is the set of antichains in P, and $g(A) = |A| - 1$ for $A \in T$.

The proof of the Theorem goes by constructing a linear extension with $w(P) - 1$ many jumps:

Let $k = w(P)$ and $C_1,\ldots,C_k$ a covering of P by k maximal chains. Such a covering exists by Dilworth's Theorem and can be obtained in an effective way. The next step is to show that, because there are no cycles in P, subchains $C'_1,\ldots,C'_k$ of $C_1,\ldots,C_k$ still covering P can be constructed such that $C'_1 \oplus \ldots \oplus C'_k$ is a linear extension of P.

As it stands, the two best known algorithms for the m-machine problem, the algorithm of Hu (for trees and arbitrary m) and of Coffman and Graham (for

$m = 2$ and arbitrary P) lead to min-max results and hence to nice optimality proofs. The solution of the 2-machine problem in [Fujii, Kasami & Ninomiya 1969] uses a maximal 2-matching in the incomparability graph of P and is thus based on the max-flow algorithm. The question that remains, of course, is if more results of that kind can be deduced. To state it differently:

Is there more "hidden structure" in the m*-machine problem to get further min-max results?*

In the terminology introduced, this hidden structure would have to be found in the form of a set T and a function g.

We now give a proof of Hu's Theorem following the arguments of [McHugh 1984].

Let P be a tree, i.e. a connected ordered set with greatest element e such that each $x \neq e$ has precisely one upper cover. The claim is:

For any m, *every level schedule for* P *is optimal.*

So let m be arbitrary, and let a level schedule $\bar{\sigma}$ for P be given with $\omega_{\bar{\sigma}}$ its execution time. We denote the *height* of P, $\max\{\lambda(x) \mid x \in P\}$, by $\lambda(P)$. For $0 \le i \le \lambda(P)$, let

$$P_i = \{x \in P \mid \lambda(x) \ge i\}.$$

The important (a priori) observation now is that for *any* m-machine schedule σ for P and *any* $0 \le i \le \lambda(P)$,

$$\omega_\sigma \ge i + \left\lceil \frac{|P_i|}{m} \right\rceil .$$

The optimality of $\bar{\sigma}$ now will simply follow from the fact that there is a k between 0 and $\lambda(P)$ such that

$$\omega_{\bar{\sigma}} = k + \left\lceil \frac{|P_k|}{m} \right\rceil .$$

Let $P^t = A_1 \oplus \ldots \oplus A_n$ be the partial extension (time order) of P associated with the schedule $\bar{\sigma}$.

Let $1 \le k \le n$ be largest with the property that A_k consists of m elements with the same λ-value v. ($k = 0$ if no such A_i exists.)

If in A_{k+1} there are still elements with λ-value v, then

$$\omega_{\bar{\sigma}} = v + 1 + k = v + \left\lceil \frac{|P_v|}{m} \right\rceil .$$

If this is not the case, then

$$\omega_{\bar{\sigma}} = v + \frac{|P_v|}{m} = v + \left\lceil \frac{|P_v|}{m} \right\rceil .$$

This proves the claim.

Figure 11 shows an example of a tree and an optimal level schedule (for $m = 3$).

FIGURE 11

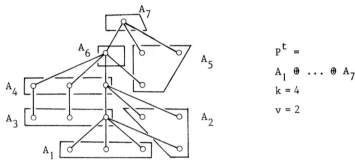

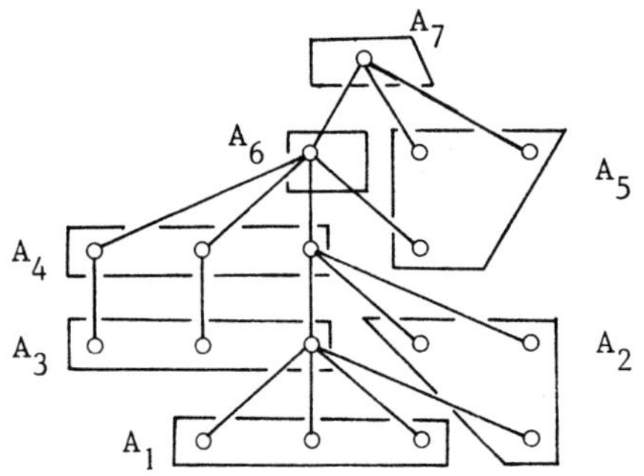

$P^t =$

$A_1 \oplus \ldots \oplus A_7$

$k = 4$

$v = 2$

What has indeed been shown is: *If* P *is a tree and* m *is arbitrary, then*

$$\max_{0 \le i \le \lambda(P)} \left(i + \left\lceil \frac{|P_i|}{m} \right\rceil\right) = \min_{\sigma \text{ schedule}} \omega_\sigma ;$$

in fact, any level schedule acutally "finds" a k such that $k + \left\lceil \frac{|P_k|}{m} \right\rceil$ is maximum.

We next outline a proof of the Coffman-Graham Theorem:

> *If* $m = 2$, *then every* CG-*schedule is optimal for arbitrary* P.

The "hidden structure" here is in the *towers*.

A sequence of subsets $S_1, \ldots, S_k$ of an ordered set P is called a *tower* in P if for all $1 \le i \le k-1$, $x \in S_i$ and $y \in S_{i+1}$, $x < y$ holds. The key observation giving an a priori lower bound for the execution time is: if $S_1, \ldots, S_k$ is *any* tower in P and σ is *any* 2-machine schedule for P, then

$$\omega_\sigma \ge \left\lceil \frac{|S_1|}{2} \right\rceil + \ldots + \left\lceil \frac{|S_k|}{2} \right\rceil .$$

Now let $\bar{\sigma}$ be a 2-machine CG-schedule for P deduced from a CG-labelling $\gamma: P \to \{1, 2, \ldots, |P|\}$. The proof goes by constructing a sequence $S_1, \ldots, S_k$ of subsets of P using the schedule $\bar{\sigma}$ and by showing that

(i) $S_1, \dots, S_k$ is a tower in P;

(ii) $n = \omega_{\bar{\sigma}} = \left\lceil \frac{|S_1|}{2} \right\rceil + \dots + \left\lceil \frac{|S_k|}{2} \right\rceil$.

Let $P^t = A_1 \oplus \dots \oplus A_n$ be the time order of P with respect to $\bar{\sigma}$. To simplify formulations, we use the ordered set P^* deduced from P by putting a "dummy" element into each A_i that is a singleton. Dummies represent idle times in the schedule and are labelled 0. (As an example, Figure 12 shows P^* for P^t of Figure 3.)

FIGURE 12

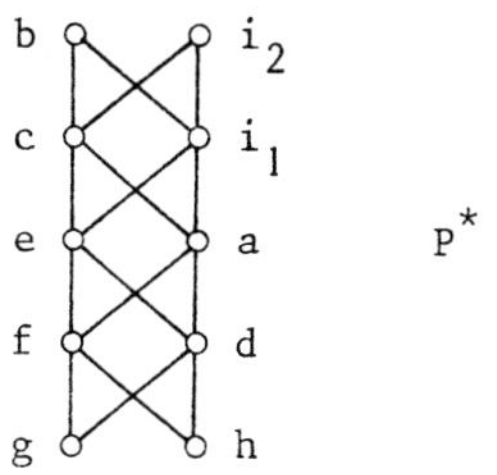

Elements $v_1, w_1, \dots, v_k, w_k$ of P (the w_i's might be dummies) are now defined inductively in the following way:

- $v_1 \in A_n$ is the element of A_n with higher label, w_1 (possibly dummy) is the other element.
- Assume $v_1, w_1, \dots, v_i, w_i$ have been defined, and let $A_j = \{v_i, w_i\}$. Now look for the largest $l < j$ such that A_l contains an x with $\gamma(x) < \gamma(v_i)$; define w_{i+1} to be this x and v_{i+1} the other element of A_l. If such an l does not exist, the procedure stops with $k = i$.

We now assume $v_1, w_1, \dots, v_k, w_k$ have been defined.
T_i, for $1 \le i \le k$, is the set of elements x of P scheduled between v_{i+1} and v_i, i.e. $v_{i+1} <^t x <^t v_i$, plus the element v_i itself. (An example is given in Figure 13.)

Using several inductive arguments, one can now show that $S_1, \dots, S_k$ with $S_i = T_{k+1-i}$ is a tower in P. The proof of this, of course, is the hard part and uses the CG-labelling of the elements (see [Coffman & Graham 1972]).

The min-max result obtained reads as follows: *If* P *is any ordered set and* $m = 2$, *then*

$$\max_{\substack{S_1, \dots, S_k \\ \text{tower in } P}} \left(\left\lceil \frac{|S_1|}{2} \right\rceil + \dots + \left\lceil \frac{|S_k|}{2} \right\rceil \right) = \min_{\sigma \text{ schedule}} \omega_\sigma;$$

in fact, any CG-schedule "detects" an optimal tower.

FIGURE 13

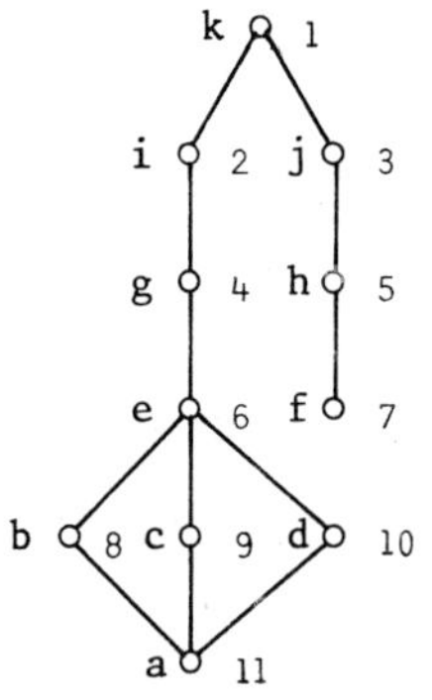

P with CG-labelling

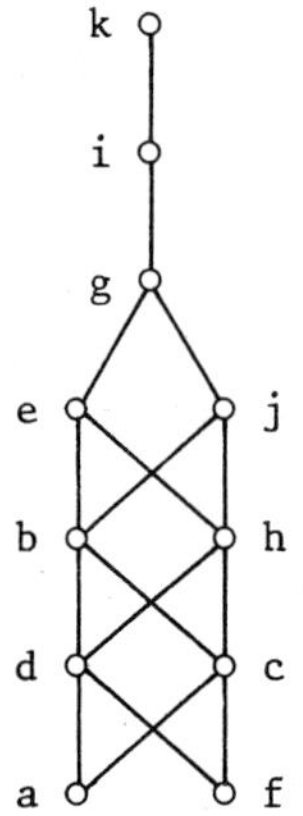

P^t (time order) associated with CG-schedule

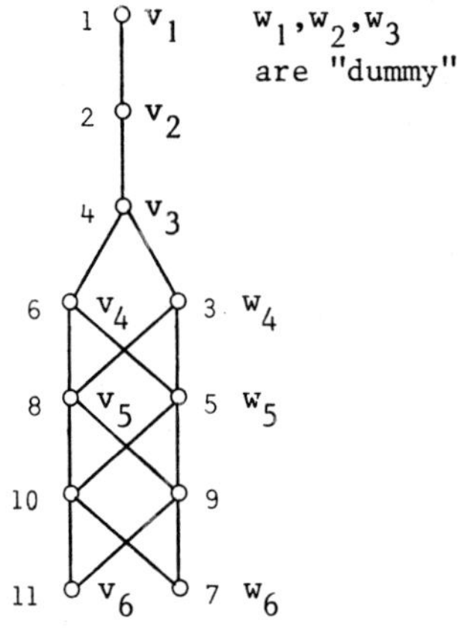

Finding the tower

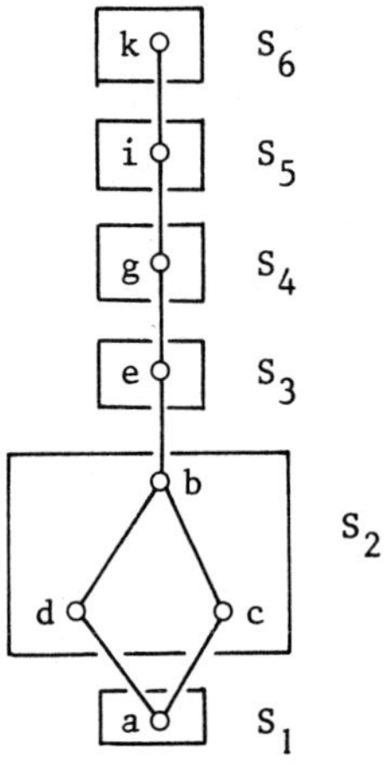

An optimal tower

CG-schedules are not optimal for $m \geq 3$, as was already mentioned in the preceding paragraph: In the last example of Figure 9 (with $m = 3$), *no* level schedule is optimal; a CG-list for this example and the corresponding schedule are shown in Figure 14.

FIGURE 14

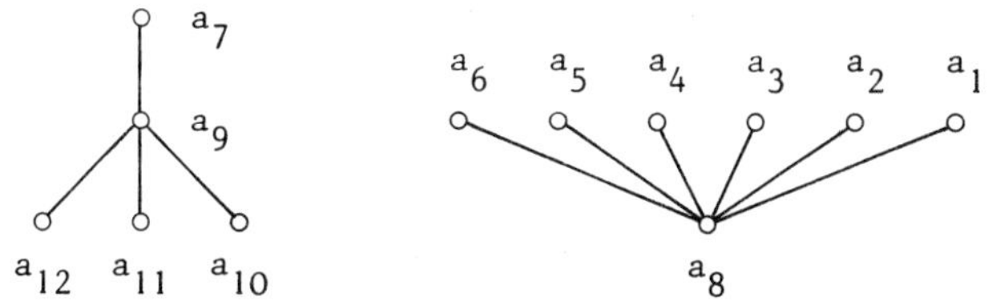

The indexing describes a CG-list

a_{12}	a_9	a_7	a_4	a_1
a_{11}	a_8	a_6	a_3	
a_{10}		a_5	a_2	

The corresponding CG-schedule

Also, for $m = 3$, an analogous min-max result to the Coffman-Graham Theorem does not hold. This is not surprising, since we do not have a good algorithm, but this is what seems to be needed to establish a min-max result.

Figure 15 gives an example: An optimal 3-machine schedule has execution time 5, but using towers (and $\left\lceil\frac{|S_1|}{3}\right\rceil + \ldots + \left\lceil\frac{|S_k|}{3}\right\rceil$ as a measure) only gets one to the maximum value of 4.

FIGURE 15

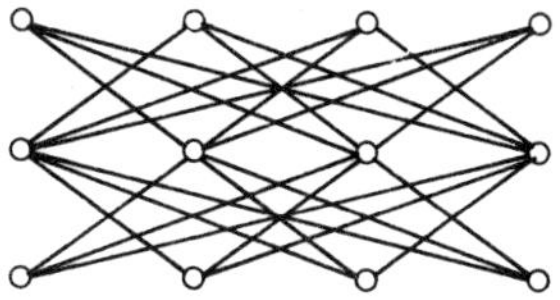

Some work was done to generalize the tower idea. One result obtained along these lines is the one in [Lam & Sethi 1977] mentioned above which gives bounds on the performance of CG-schedules for $m \geq 3$.

§ 5. TRANSFORMATION TECHNIQUES

We already applied transformation techniques in the second paragraph to solve the 2-machine flow-shop problem.
Such techniques can serve different purposes. Transformations can be part of an algorithm to find optimal solutions. They can also be used to show that certain proposed algorithms do what they ought to. And finally, they can be used to prove general statements on the behaviour of the objective function or the optimal value $f(s^*)$, respectively.

What are some of the typical situations?

The most direct application of transformations is when there is an algorithm transforming any given $s \in S$ into an optimal element. The algorithm increasing flows in graphs is of that sort. (We remind the reader of our general setting: a finite set S and a function $f: S \to \mathbb{R}$ are given, and the problem is to find an $s^* \in S$ for which $f(s^*)$ is minimum.)

Now suppose a subset S' of S has the property that it is substantially smaller than S and is guaranteed to contain an optimal element. This reduces the cases to look at and can be very valuable if the problem considered is known to be NP-complete. In this case, the proof that S' contains an optimal element will typically use some transformation to show that "without loss of generality", it is enough to look at S'.

Another situation is given when one wants to show that every element of S' is optimal. Typically, S' will consist of the elements of S that are outcomes of a proposed algorithm. This situation can be handled by devising an algorithm transforming *any* optimal element of S into *any* given element of S' without increasing the objective function. (The proof of Johnson's Theorem in the second paragraph went like this.)

The reverse method is used in [Rival 1983] where it is shown that *any* $s' \in S'$ can be transformed, without changing the objective function, into *any* optimal element of S.

Let us first give a brief example for the use of transformations in the jump number problem.
The claim simply is:

> *If* P *and* Q *are ordered sets, then*
> $s(P+Q) = s(P) + s(Q) + 1.$

(As usual, $P+Q$ denotes the disjoint union or "cardinal sum".) This seems trivial, but how can one *prove* it?

One way is by observing the following: if

$$A_1 \oplus B_1 \oplus \dots \oplus A_i \oplus B_i \oplus \dots \oplus A_n + B_n$$

(with $A_i \subseteq P$, $B_j \subseteq Q$, A_1 possibly empty and $n \geq 2$) is an optimal linear extension of $P+Q$, then interchanging A_n and B_n to get

$$A_1 \oplus B_1 \oplus \dots \oplus B_{n-1} \oplus B_n \oplus A_n$$

results in a linear extension with *not more jumps*. Doing induction and using symmetry finally gives a linear extension with all of P and Q in one block, and this cannot result in a smaller number of jumps than $s(P) + s(Q) + 1$.

In the above terminology, the situation here is the following: S is the set of all linear extensions of $P+Q$, f counts the jumps. S' is the set of linear extensions of the form $L_P \oplus L_Q$ or $L_Q \oplus L_P$, where L_P and L_Q are linear extensions of P and Q, respectively. The claim is that each element optimal in S' is also optimal in S. A transformation algorithm not increasing the jump number gradually changes any element of S to one of S'. Since the optimal elements of S' obviously have $s(P) + s(Q) + 1$ many jumps, this proves what has been claimed.

We now want to look at some transformation techniques that were used for the m-machine problem. Of course, this will not be a complete survey.

The first has the form of a general Theorem and reduces the set of schedules that have to be considered.
For $p,q \in P$, we say that p *dominates* q if $\downarrow p \subseteq \downarrow q$ and $\uparrow q \subseteq \uparrow p$.
The following is true:

> *If* $p,q \in P$ *and* p *dominates* q, *then there is an optimal* m*-machine schedule* σ *for* P *with* $p \nsucc^t q$.

The proof goes like this: Let σ' be an m-machine schedule for P that treats q earlier than p, $q <^t p$. The assumptions assure that just interchanging p and q yields a schedule (of the same length). Hence, if σ' is optimal, then the new schedule is.

Figure 16 shows an example (for $m = 2$).

FIGURE 16

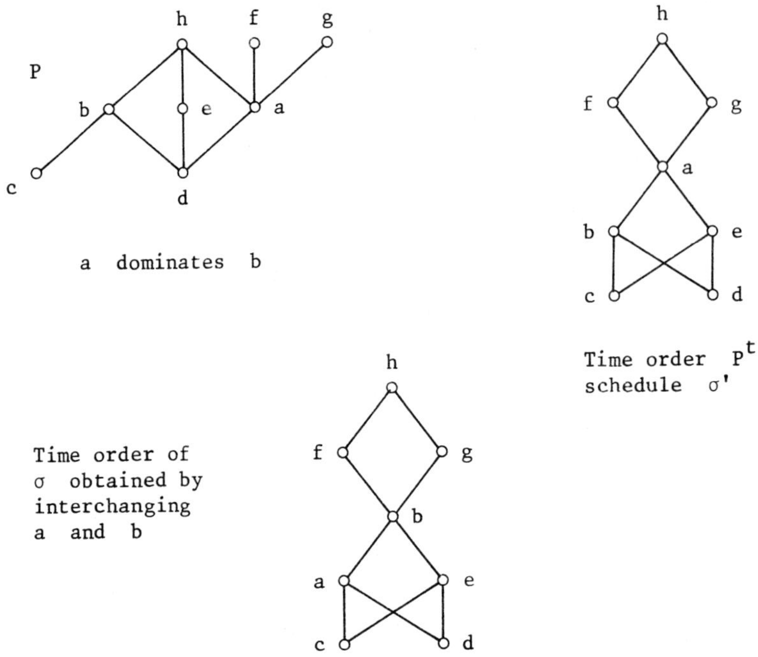

a dominates b

Time order P^t of a schedule σ'

Time order of σ obtained by interchanging a and b

This, of course, can be generalized in the following way. For subsets A and B of P, we say that A *dominates* B if the following are true:

(i) There is an order-isomorphism $\eta: A \to B$, where A and B carry the structure inherited from P;

(ii) If $c < a \in A$ with $c \notin A$, then $c < \eta(a)$ and $c \notin B$;

(iii) If $c > \eta(a)$ with $c \notin B$, then $c > a$ and $c \notin A$.

One now can show similarly as above:

> *If* A *dominates* B *in* P, *then there is an optimal* m-*machine schedule* σ *with* $a \not\succ^t \eta(a)$ *for each* $a \in A$.

If there is much domination in P, then the above can reduce the problem of finding an optimal m-machine schedule to a certain extent. An algorithm based on this (being still exponential) was proposed in [Ramamoorthy et al. 1972].

Of a similar character as the above techniques is *interchanging matched subsets* as considered in [Poguntke & Rival 1984]. It concerns transformation of a given schedule into one which is more like a level schedule - without increasing the execution time. Here is the general technical setting.

Let P be an ordered set and let σ be an m-machine schedule for P. Call disjoint, nonempty subsets X^* of P^* and Y^t of P^t *matched subsets* of the schedule σ provided that the following conditions are satisfied:

M 1. There is an isomorphism ϕ of X^* onto Y^t such that for each $x \in X^*$,

(i) $x <^t \phi(x)$,

(ii) $\lambda(x) < \lambda(\phi(x))$,

and for $u,v \in X^*$,

(iii) if $u <^t v$ then $\phi(u) \not>^t v$;

M 2. For each $u,v \in X^*$, if $\sigma_2(u) = \sigma_2(v)$ then $|\lambda(u) - \lambda(v)| \leq 1$;

M 3. For each $s,t \in Y^t$, if $s \in Y_i$ and $t \in Y_{i+1}$, then $\lambda(t) \geq \lambda(s) - 1$.

In M 3, we use the decompositions

$$X^* = X_1 \oplus X_2 \oplus \ldots \quad \text{and}$$

$$Y^t = Y_1 \oplus Y_2 \oplus \ldots$$

induced by the time order of σ. Obviously the isomorphism ϕ maps X_i onto Y_i. Also, if $X^* \cap P$ and $Y^t \cap P$ are both chains we call the subsets X^*, Y^t *matched chains*.

In the example shown in Figure 17, $X^* = \{h,d,a\}$ and $Y^t = \{f,e,c\}$ are matched chains in the schedule σ.

FIGURE 17

P

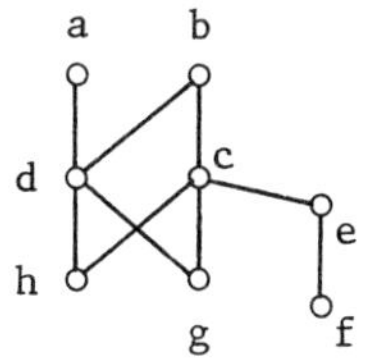

σ

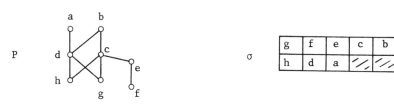

g	f	e	c	b
h	d	a		

P^t

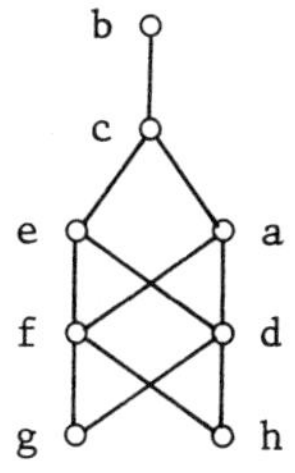

P^*

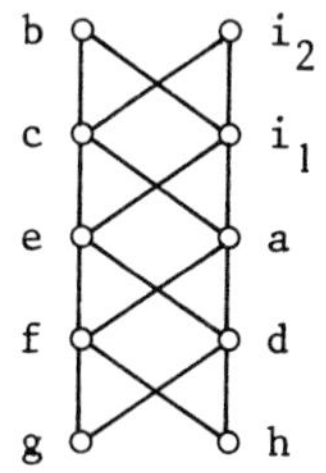

We say that τ is *obtained from* σ *by interchanging* the matched subsets X^* and Y^t provided that

$$\tau(z) = \begin{cases} \sigma(z), & \text{if } z \notin X^* \cup Y^t \\ \sigma(\phi(z)), & \text{if } z \in X^* \cap P^t \\ \sigma(\phi^{-1}(z)), & \text{if } z \in Y^t. \end{cases}$$

For example, from the above schedule $\sigma = \sigma^1$ in Figure 17, we get an optimal schedule σ^3 by a sequence of two matched chain interchanges (see Figure 18).

FIGURE 18

Interchanging $X^* = \{h,d,a\}$ and $Y^t = \{f,e,c\}$ results in the schedule σ^2:

g	h	d	a	b
f	e	c	i_1	i_2

Interchanging $X^* = \{i_1\}$ and $Y^t = \{b\}$ finally produces σ^3 which is an optimal level schedule:

g	h	d	a
f	e	c	b

What can be done with this?
The first result in [Poguntke & Rival 1984] gives one more technical condition under which τ is again a schedule. In this case, obviously, the execution time is not increased.
The main result then shows that any schedule obtained by interchanging matched subsets is "measurably" more like a level schedule:

Let σ *be an* m-*machine schedule for* P *and let* τ *be a schedule for* P *obtained from* σ *by interchanging matched subsets. Then* $|D_\tau| < |D_\sigma|$.

For a schedule σ, $D_\sigma = \{(x,y) \mid x,y \in P, \ \lambda(x) < \lambda(y) \text{ and } x <^t y\}$.

C_σ is defined to be the subset of D_σ with $(x,y) \in C_\sigma$ if in addition y is "available at x", i.e. $z < y$ implies $z <^t x$. Interestingly, it may actually be that for a schedule τ obtained from σ by interchanging matched subsets, $|C_\tau| > |C_\sigma|$.

Interchanging matched subsets does not seem to have obvious consequences as to devise efficient algorithms. The novelty is in the transformation which can be shown to produce big reductions in certain instances.

Another kind of reduction result which is also proved using transformations is the *Elite Theorem* of [Dolev & Warmuth 1980, 1982b, 1984]. It can be applied to get statements about optimal m-machine schedules in case P is not connected. Very roughly, it says that there always is an optimal schedule that starts with elements of components having large height.

It is worthwhile to be more precise.

Let P and m be given. The *median* of P with respect to m, $\mu(P,m)$, is defined to be the smallest integer μ with the property that there are less than m components having height at least μ. The *elite* of P is the set

$$E_m(P) = \{p \in P \mid p \text{ minimal}, \ \lambda(p) > \mu(P,m)\}.$$

Figure 19 shows an example.

FIGURE 19

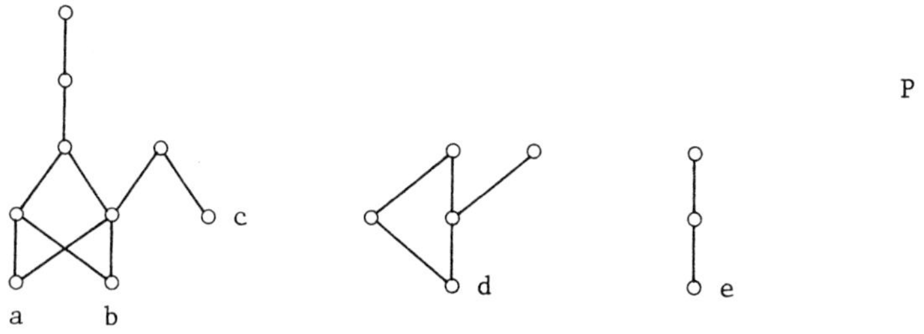

$\mu(P,2) = 3$, hence $E_2(P) = \{a,b\}$.

$\mu(P,3) = 3$, hence $E_3(P) = E_2(P)$.

$\mu(P,4) = 0$, hence $E_4(P) = \{a,b,c,d,e\}$.

The *Elite Theorem* says the following:

> *Let* P *and* m *be given. If* $E_m(P)$ *has more than* m *elements, then there is an optimal* m*-machine schedule for* P *that starts with* m *elements of* $E_m(P)$. *If* $|E_m(P)| \le m$, *then there is an optimal schedule starting with* m *elements of greatest level. If* $E_m(P) = \emptyset$, *then every level schedule is optimal.*

The examples of Figure 20 show the power, but also the restrictedness of the Elite Theorem.

FIGURE 20

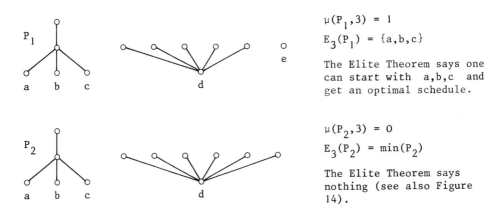

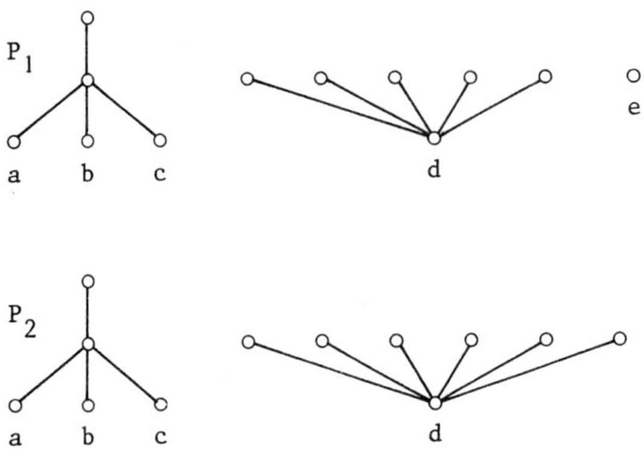

$\mu(P_1,3) = 1$

$E_3(P_1) = \{a,b,c\}$

The Elite Theorem says one can start with a,b,c and get an optimal schedule.

$\mu(P_2,3) = 0$

$E_3(P_2) = \min(P_2)$

The Elite Theorem says nothing (see also Figure 14).

When applying the Elite Theorem to get an optimal schedule, one has of course to be aware that the median $\mu(P,m)$ is a "dynamic line" in that it changes after m elements (for the first time interval of the schedule) have been chosen and removed from P.

The Elite Theorem together with "profile scheduling" and certain "merging" techniques are then used in [Dolev & Warmuth 1982b] to get polynomial algorithms (with m fixed) if P is an opposing forest or a level order. In the case of level orders, these concepts help to get rid of the restriction as to the number of components that we made in the second paragraph.
Algorithms for opposing forests are also given in [Garey et al. 1983]. Further quite recent results using transformations are in [Nett 1979; Papadimitriou & Yannakakis 1979; Dolev & Warmuth 1982a; Simon 1984].

We should mention at the end the result of [Mayr 1981] saying that if m is *not* fixed, then the m-machine problem is NP-hard even for the restricted classes of opposing forests with two components or level orders.

§ 6. CONCLUDING REMARKS

1. What are promising ways to make further progress in the m-machine problem? Since it might be too optimistic to hope to find many more powerful min-max results, new transformation techniques together with methods like "profile scheduling" and "merging" seem to be needed.

2. The outstanding problem still is to find, for fixed $m \geq 3$, a polynomial algorithm for the m-machine problem. But even the following question concern-

ing approximation algorithms has not been answered.

For an approximation algorithm A_m for the m-machine problem, let $\omega(A_m)$ be the supremum of

$$\omega_{A_m}(P) \big/ \omega_{opt}(P) ,$$

where P ranges over all finite ordered sets and $\omega_{A_m}(P)$ is the execution time of the schedule found by A_m, $\omega_{opt}(P)$ is the optimal execution time. $\omega(A_m)$ measures the worst-case behaviour of A_m. Here is the question:

> *Is there a sequence* $A_m^1, A_m^2, \ldots$ *of polynomial approximation algorithms such that* $\lim_{i\to\infty} \omega(A_m^i) = 1$?

3. The third remark concerns schedules viewed as partial extensions. Let Q be any set, and let $(\mathcal{O}(Q),\subseteq)$ be the ordered set of orderings on Q. The smallest element is just the antichain order $\{(q,q) \mid q \in Q\}$ on Q. If a greatest element 1 is adjoined, then $\tilde{\mathcal{O}}(Q) = \mathcal{O}(Q) \cup \{1\}$ becomes a lattice with the linear orderings being the coatoms. There are several interesting algebraic and combinatorial questions concerning this lattice that have been looked at. [Dean & Keller 1968] and [Avann 1972] consider properties of the interval of $\mathcal{O}(Q)$ below some linear order.

   For $O \in \mathcal{O}(Q)$, let $c(O)$ denote the number of covering pairs in O. The problem we would like to address is:

   > *Given* $O \in \mathcal{O}(Q)$, *find an extension* $O' \supseteq O$ *with* $c(O')$ *maximal.*

   There are some connections to schedules. For example, it is clear that if O is given and O' is an optimal extension relative to the condition that $w(O') \leq m$, then O' is an m-machine schedule for O. Is it necessarily optimal?

   As is easy to see, it is *not* true that *every* optimal m-machine schedule is also c-optimal.

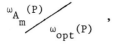

REFERENCES

S.P. AVANN (1972): The lattice of natural partial orders. Aequationes Math. 8, 95-102.

N.F. CHEN, C.L. LIU (1975): On a class of scheduling algorithms for multiprocessor computing systems. Lecture Notes in Computer Science, vol. 24, Springer-Verlag, New York, pp. 1-16.

E.G. COFFMAN, jr., ed. (1976): Computer and job-shop scheduling theory. New York: John Wiley & Sons.

E.G. COFFMAN, jr., R.L. GRAHAM (1972): Optimal scheduling for two-processor systems. Acta Informat. 1, 200-213.

C.L. COLBOURN, W.R. PULLEYBLANK (1984): Minimizing setups in ordered sets of fixed width. Order 1, 225-228.

R.A. DEAN, G. KELLER (1968): Natural partial orders. Can. J. Math. 20, 535-554.

R.P. DILWORTH (1950): A decomposition theorem for partially ordered sets. Ann. of Math. 51, 161-166.

D. DOLEV, M.K. WARMUTH (1980): Scheduling wide graphs. Preprint, Stanford.

D. DOLEV, M.K. WARMUTH (1982a): Scheduling precedence graphs of bounded height. IBM Research Report.

D. DOLEV, M.K. WARMUTH (1982b): Profile scheduling of opposing forests and level orders. IBM Research Report.

D. DOLEV, M.K. WARMUTH (1984): Scheduling flat graphs. Preprint, Jerusalem.

D. DUFFUS, I. RIVAL, P. WINKLER (1982): Minimizing setups for cycle-free ordered sets. Proc. Amer. Math. Soc. 85, 509-513.

K. ECKER (1977): Organisation von parallelen Prozessen. Mannheim: Bibliographisches Institut.

M. FUJII, T. KASAMI, K. NINOMIYA (1969): Optimal sequencing of two equivalent processors. SIAM J. Appl. Math. 17, 784-789.

H.N. GABOW (1982): An almost-linear algorithm for two-processor scheduling. J. Assoc. Comp. Mach. 29, 766-780.

M.R. GAREY, D.S. JOHNSON (1979): Computers and intractability: a guide to the theory of NP-completeness. San Francisco: Freeman.

M.R. GAREY, D.S. JOHNSON, R.E. TARJAN, M. YANNAKAKIS (1983): Scheduling opposing forests. SIAM J. Alg. Disc. Meth. 4, 72-93.

R.L. GRAHAM, E.L. LAWLER, J.K. LENSTRA, A.H.G. RINNOOY KAN (1979): Optimization and approximation in deterministic sequencing and scheduling: a survey. Ann. Disc. Math. 5, 287-326.

T.C. HU (1961): Parallel sequencing and assembly line problems. Oper. Res. 9, 941-948.

S.M. JOHNSON (1954): Optimal two and three stage production schedules with setup times included. Naval Res. Logist. Quart. 1, 61-68.

S. LAM, R. SETHI (1977): Worst-case analysis of two scheduling algorithms. SIAM J. Comput. 6, 518-536.

E.L. LAWLER, J.K. LENSTRA (1982): Machine scheduling with precedence constraints. In: Ordered Sets (ed. I. Rival), Dordrecht: D. Reidel, pp. 655-675.

J.K. LENSTRA, A.H.G. RINNOOY KAN (1984): Two open problems in precedence constrained scheduling. In: Orders: Description and Roles (ed. M. Pouzet / D. Richard), Ann. Disc. Math. 23, 509-521.

E. MAYR (1981): Well structured programs are not easier to schedule. Preprint, Stanford.

J.A.M. McHUGH (1984): Hu's precedence tree algorithm: a simple proof. Naval Res. Logist. Quart. 31, 409-411.

E. NETT (1979): Deterministische Strategien zum Scheduling von Taskabhängigkeitsstrukturen für Mehrprozessorsysteme. Report Nr. 122, GMD, Bonn.

C.H. PAPADIMITRIOU, M. YANNAKAKIS (1979): Scheduling interval-ordered tasks. SIAM J. Comput. 8, 405-409.

H. PELZER (1984): Zur Komplexität des 3-Maschinen-Scheduling-Problems. Diplomarbeit, TH Aachen.

W. POGUNTKE, I. RIVAL (1984): Level schedules by interchanging matched subsets. Preprint, Calgary.

W.R. PULLEYBLANK (1981): On minimizing setups in precedence constrained scheduling. Preprint, Bonn.

C.V. RAMAMOORTHY, K.M. CHANDY, M.J. GONZALES (1972): Optimal scheduling strategies in a multiprocessor system. IEEE Trans. Comput. C-21, 137-146.

I. RIVAL (1983): Optimal linear extensions by interchanging chains. Proc. Amer. Math. Soc. 89, 387-394.

I. RIVAL (1984): Linear extensions of finite ordered sets. In: Orders: Description and Roles (ed. M. Pouzet / D. Richard), Ann. Disc. Math. 23, 355-370.

R. SETHI (1976): Scheduling graphs on two processors. SIAM J. Comput. 5, 73-82.

H.U. SIMON (1984): On the performance of critical path schedules for well structured programs. Preprint, Saarbrücken.

J.D. ULLMAN (1975): NP-complete scheduling problems. J. Comput. Syst. Sci. 10, 384-393.

Fachbereich Mathematik
Technische Hochschule Darmstadt
D - 6100 Darmstadt
Federal Republic of Germany

Contemporary Mathematics
Volume 57, 1986

RADON TRANSFORMS IN COMBINATORICS AND LATTICE THEORY

Joseph P. S. Kung *

ABSTRACT. This paper is intended to be a survey of the applications of the finite Radon transform in combinatorics and lattice theory. In the first section, we introduce the basic notions, and, more importantly, the point of view, of the theory of finite Radon transforms. Several applications to combinatorics and statistics are given. The remaining sections are devoted to the applications of Radon transforms to finding matchings and proving rank or covering inequalities in finite lattices.

1. THE FINITE RADON TRANSFORM

1.1. ORIGINS

Let $f:\mathbb{R}^n \to \mathbb{R}$ be a real-valued function defined on n-dimensional Euclidean space satisfying certain integrability properties which need not concern us here. The *Radon transform* Tf of f is the real-valued function defined on the manifold of affine hyperplanes in $\mathbb{R}^n$ by the formula

$$Tf(H) = \int_H f,$$

where the integral is taken over the hyperplane H relative to the Lebesgue measure. The theory of Radon transforms is an important component of modern analysis and is the cornerstone of computed tomography. (A good reference is Shepp 1982; the survey article by Cormack in this volume is particularly accessible.)

1980 Mathematics Subject Classification. 05-02, 06-02.
* Partially supported by a North Texas State University Faculty Research Grant.

Since a finite analogue of integration is summation, one can propose the following definition.

(1.1.1) DEFINITION. Let $\mathscr{C}$ be a collection of subsets of a finite set S. If $f:S \to \mathbf{k}$ is a function from S to a field $\mathbf{k}$, then the *Radon transform* $Tf:\mathscr{C} \to \mathbf{k}$ is the function from $\mathscr{C}$ to $\mathbf{k}$ defined by the formula: for $A \in \mathscr{C}$,

$$Tf(A) = \sum_{x \in A} f(x) .$$

The Radon transform defines a linear transformation T from the vector space S* of functions from S to $\mathbf{k}$ to the vector space $\mathscr{C}^*$ of functions from $\mathscr{C}$ to $\mathbf{k}$. The central problem in the theory of finite Radon transforms is the following.

(1.1.2) THE RECONSTRUCTION PROBLEM. Can the function f be reconstructed from its Radon transform Tf? That is, is the linear transformation T invertible? More generally, what is the rank of T?

Relative to the standard basis of delta functions,

$$\delta_x(y) = 1 \text{ if } x = y, \text{ and } 0 \text{ otherwise,}$$

the matrix of the linear transformation T is the incidence matrix $I(\mathscr{C}|S)$ with rows indexed by $\mathscr{C}$ and columns indexed by S whose A,x-entry is 1 if $x \in A$ and 0 otherwise. Thus, for reductionists, the study of Radon transforms in "just" the study of incidence matrices. However, as we hope to show, thinking of incidence matrices as Radon transforms suggests reconstruction methods - most of them recursive or inductive - for inverting incidence matrices which would be hard to discover from any other point of view.

The finite Radon transform was discovered by Ethan Bolker around 1976 and his work has been a major influence on the subject. Bolker's work (see Bolker pre) is focused on finite analogues of the central ideas in the theory of Radon transforms in analysis: inversion formulas, relation to the Laplacian and other differential operators, ranges of Radon transforms, group actions and homogeneous spaces, and relation to group representation theory. Although we shall not give an account of this work here, its importance cannot be overemphasized.

Our aim in this paper is to give a survey of the theory of finite Radon transforms. Most of the work in this area is on the applications of the Radon transform to finding matchings in lattices.

1.2. DESIGNS AND FISHER'S INEQUALITY.

Perhaps the earliest proof that an incidence matrix has maximum possible rank occurs in the theory of block designs. Recall that a *t-design* on the point set S is a collection $\mathcal{B}$ of subsets of S called *blocks* satisfying

BD1. Every block has the same size.

BD2. For every t-subset A of S, the number of blocks containing A equals a fixed number independent of A.

For example, the lines of a projective plane form a 2-design on the points of the plane.

A classical result in design theory is Fisher's inequality. This states that if $t \geq 2$, the number of blocks in a t-design is at least the number of points, that is, $|\mathcal{B}| \geq |S|$. The usual proof of Fisher's inequality is to proceed *via* the following stronger result.

(1.2.1) THEOREM. Let $\mathcal{B}$ be the collection of blocks of a t-design on the set S. If $t \geq 2$, then the incidence matrix $I(\mathcal{B}|S)$ has rank $|S|$ over the rational numbers.

We shall prove (1.2.1) using Radon transforms. Every proof of (1.2.1) uses the following standard fact about designs, and our proof is no exception.

(1.2.2) LEMMA. Let $\mathcal{B}$ be a t-design on S with $t \geq 2$. Then for every point x in S, the number of blocks containing x is a fixed number α independent of x. Similarly, if x and y are distinct points in S, the number of blocks containing both x and y is a fixed number β independent of x and y. In addition, $\alpha > \beta$.

SKETCH OF PROOF. First prove that a t-design is also a (t-1)-design. □

Now let $f:S \to \mathbb{Q}$ be a function from S to the rational numbers $\mathbb{Q}$. We shall show how f can be reconstructed from its Radon transform $Tf:\mathcal{B} \to \mathbb{Q}$. Let x be a point in S. Then, by the previous lemma, we have

(1.1) $$\sum_{B:x\in B} Tf(B)$$

$$= [\#\text{blocks containing } x]f(x) + [\#\text{blocks containing } x \text{ and } y] \sum_{x\neq y} f(y)$$

$$= (\alpha - \beta)f(x) + \beta \sum_{y\in S} f(y).$$

Define the *mass* of f by

$$\text{mass}(f) = \sum_{y\in S} f(y).$$

By solving Equation (1.1), we can write f(x) in terms of the Radon transform Tf and the mass of f:

(1.2) $$f(x) = \frac{1}{\alpha - \beta}\Big[\sum_{B:x\in B} Tf(B) - \text{mass }(f)\Big].$$

The mass of f can be found by the following consistency condition:

(1.3) $$\text{Mass}(f) = \sum_{x\in S} f(x)$$

$$= \frac{1}{\alpha - \beta}\Big[\sum_{x\in S}\Big(\sum_{B:x\in B} Tf(B)\Big)\Big] - \frac{\beta|S|}{\alpha - \beta}\,\text{mass}(f).$$

Since $\alpha - \beta > 0$, the coefficient of mass(f) in (1.3) is positive and hence, mass(f) can be calculated from the Radon transform Tf using (1.3). Once mass(f) is calculated, the function f can be reconstructed using (1.2). This completes the proof of (1.2.1).

The idea of using mass(f) "now" even if it is not yet known and calculating it "later" by means of a consistency condition is potentially quite useful. We shall call it the *missing mass method*.

1.3. POINT RECONSTRUCTION AND DATABASE SECURITY

A natural variation on the central problem (1.1.2) is to ask what information about the Radon transform Tf is needed to reconstruct the function value f(x) at a given point x. Usually, this question is posed as follows.

(1.3.1) THE POINT RECONSTRUCTION PROBLEM. Let $\mathcal{C}$ be a collection of subsets of a finite set S, x a given point in S, and $\mathcal{C}_x$ the subcollection in $\mathcal{C}$ consisting of all the subsets in $\mathcal{C}$ containing x. Can the function value f(x) be reconstructed from the Radon transform Tf restricted to the subcollection $\mathcal{C}_x$?

The answer to (1.3.1) for Radon transforms in analysis can be rather surprising. As an example, the Radon transform over hyperplanes in $\mathbb{R}^n$ is point-reconstructible if the dimension n is odd and not point-reconstructible if n is even!

In the finite case, the answer is more predictable. A useful rule-of-thumb is: if the Radon transform is reconstructible and $\mathcal{C}$ contains the entire set S, then the answer is "yes"; the answer is "no" otherwise. If the answer is "no", then adding the extra information of the mass of f would generally suffice to reconstruct f(x). The following result is a simple instance of this phenomenon.

(1.3.2) PROPOSITION. Let $\mathcal{B}$ be a t-design on the set S with $t \geq 2$ and $f: S \to \mathbb{Q}$ a function from S to the rational numbers $\mathbb{Q}$. If x is a point in S, the function value f(x) can be reconstructed from the mass of f and the Radon transform Rf restricted to the subcollection $\mathcal{B}_x$ of blocks containing x.

(1.3.2) follows immediately from Equation (1.2.2). It is easy to show by example that the Radon transform of a t-design is never point-reconstructible without the additional information of the mass.

The point reconstruction problem has an amusing interpretation in terms of security of statistical databases (see DeMillo *et al.* 1983 for background). Statistical databases (for example, the U.S. census data] are usually open to sociologists and other researchers in the following way. A user is allowed to specify a subset and ask for its size and the mean of a random variable X on it [for example, he may ask for the number and mean income of all mathematicians]. If a user is allowed unlimited access, he can, for ulterior motives of his own, deduce the value of X at a specific point p [for example, if he is allowed to ask for the size and mean income of three sets A, B, and C, where $A \cap B = \{p\}$ and $A \cup B = C$, then he can calculate the income of individual p by

$$\text{income}(p) = |A|\text{mean}(A) + |B|\text{mean}(B) - |C|\text{mean}(C)].$$

Since $|A|\text{mean}(A)$ is the Radon transform of the random variable X at the subset A, the problem of protecting privacy is equivalent to the problem of ensuring that for a given user, the collection of subsets for which he is allowed information is not point-reconstructible. Usually, the user is only allowed to ask about "large" subsets. If this is the case, then round-off errors would render most reconstructed values meaningless even when additional restrictions are absent.

1.4. TRANSLATES OF SUBSETS IN GROUPS

It is no surprise that there are applications of the Radon transform in statistics. For example, one version of the classical reconstruction problem is: Can a distribution function be reconstructed from its marginal distributions? (See Cormack 1982.) Another application (Diaconis 1983) is to projection pursuit for discrete data. Let X be a finite set. A *projection base* of X is a collection $\mathscr{C}$ of subsets of X satisfying

PB1. Every subset in $\mathscr{C}$ has the same size

PB2. There exists a partition $\{\mathscr{C}_1, \mathscr{C}_2, \ldots, \mathscr{C}_j\}$ of $\mathscr{C}$ such that each subcollection $\mathscr{C}_i$ is a partition of X.

If $f: X \to \mathbb{R}$ is a real-valued function, the Radon transform $Tf: \mathscr{C}_i \to \mathbb{R}$ is called the *projection of* f *along the direction* $\mathscr{C}_i$. This terminology is motivated by the following example. Let X be a finite-dimensional vector space over a finite field. The collection $\mathscr{C}$ of all affine subspaces of a given dimension in X is a projection base of X. For each subspace U of X, let $\mathcal{C}_U$ be the subcollection of all translates of U. The subcollections $\mathcal{C}_U$ partition $\mathcal{C}$ and $\{\mathcal{C}_U\}$ satisfies PB2. If $\mathcal{C}$ is the collection of lines, then the Radon transform $Tf:\mathcal{C}_\ell \to \mathbb{R}$ are the projections of f along the direction given by ℓ. Projection bases are used to analyse finite data sets: the idea here is to find projections which are not uniform and hence, show a discernible structural property in the data set.

Another example of a projection base is obtained by taking a finite group G and taking $\mathcal{C}$ to be the collection of all conjugates of cosets of a subgroup H in G. More precisely,

$$\mathcal{C} = \{x(Hy)x^{-1}: x,y \in G\}.$$

Let $\mathcal{C}_K$ be the subcollection of cosets of a subgroup K conjugate to H. Then $\{\mathcal{C}_K\}$ partitions $\mathcal{C}$ and satisfies PB2. Because $\mathcal{C}_K$ is a partition of G, the Radon transform on $\mathcal{C}_K$ is not reconstructible unless K is the trivial subgroup of size one. This example motivated the following question:

> Let G be a finite group and S a subset of G. Let $\mathcal{T}(S)$ be the collection of all translates $Sx = \{yx: y \in S\}$ of S. Is the Radon transform over $\mathcal{T}(S)$ reconstructible?

This question was posed and studied in Diaconis and Graham 1985. We shall now give a sampling of their results.

The reconstruction problem is closely related to the representation theory for G.

(1.4.1) LEMMA. The Radon transform on $\mathcal{T}(S)$ is reconstructible if and only if for every irreducible representation $\rho:G \to GL(n)$, the matrix

$$\sum_{s\in S} \rho(s^{-1})$$

is non-singular.

This theorem was used by Diaconis and Graham to study the nearest neighbor transform in S_n, the symmetric group on n elements. If π and σ are permutations in S_n, their *Cayley distance* $d(\pi,\sigma)$ is defined by

$$d(\pi,\sigma) = \text{minimum number } k \text{ of transpositions } \tau_1, \tau_2, \ldots, \tau_k \text{ such that } \tau_1\tau_2\ldots\tau_k\pi = \sigma.$$

The Cayley distance $d(\pi,\sigma)$ is the distance between π and σ in a Cayley graph of the group S_n; it is *bi-invariant*, in the sense that $d(\pi\eta,\sigma\eta) = d(\eta\pi,\eta\sigma) = d(\pi,\sigma)$. Now let

$$B = \{\tau: \tau \text{ is a transposition or the identity}\}.$$

The *translate* $B\pi$ of B is the set of nearest neighbors of π relative to the Cayley distance, that is,

$$B\pi = \{\sigma: d(\pi,\sigma) \le 1\}.$$

(1.4.2) THEOREM. The Radon transform over $\mathcal{T}(B)$ in the symmetric group S_n is reconstructible if and only if $n = 1,3,4,5,6,8,10$ or 12.

The proof uses rather detailed knowledge concerning irreducible representations of S_n.

In addition to the symmetric group, Diaconis and Graham also studied the Radon transform over translates in the additive group Z_2^k of the k-dimensional vector space over the field Z_2 of order two, with the subset taken to be the Hamming ball or sphere of a given radius. As one may expect, the finite Fourier transform and Krawtchouk polynomials were used extensively. In addition, Diaconis and Graham determined the computational complexity of deciding reconstructibility.

(1.4.3) THEOREM. The problem:

Given the input a subset $S \subseteq Z_2^k$, decide whether the Radon transform over $\mathcal{T}(S)$ is reconstructible

is NP-complete.

The proof consists of showing that this problem is polynomially equivalent to the minimum weight codeword problem: Given the input a binary matrix M and an integer t, decide whether the subspace (or code) generated by the row vectors of M contains a vector of Hamming weight t.

The Radon transform over translates of a subset in Z_2^k has also been applied to the problem of reconstructing a graph from its vertex-switchings (Stanley 1985). Let X be a simple graph on the vertex set $\{x_1,\ldots,x_n\}$. To *switch* the graph X *at the vertex* x_i is to delete all the edges in X incident on x_i and insert all the possible edges incident on x_i not in X. The vertex-switching reconstruction problem is

> Can the graph X be reconstructed from the multiset of unlabelled graphs X_i obtained by switching X at each of the vertices x_i?

A partial answer to this problem is the following theorem due to Stanley.

(1.4.4) THEOREM. If the number of vertices is not congruent to 0 modulo 4, then X can be reconstructed from its vertex-switchings.

The idea of the proof is to identify a graph X with the characteristic vector of its edges in $Z_2^{\binom{n}{2}}$, the vector space of $\binom{n}{2}$-tuples over Z_2 with coordinates indexed by two-element subsets of $\{x_1,\ldots,x_n\}$. A switch at the vertex x_i corresponds to adding the characteristic vector C_i of the star $K_{1,n-1}$ with center x_i to the characteristic vector of X. Now let Γ be the set $\{C_1,\ldots,C_n\}$. It turns out that the Radon transform over the translates of Γ is reconstructible if and only if $n \not\equiv 0 \pmod 4$. The theorem can now be proved using this fact.

2. RANK INEQUALITIES IN LATTICES

In the remainder of this paper, we describe several examples of Radon transforms and matchings in finite lattices. We shall follow - more or less - the historical order of development, starting with matchings and rank inequalities in geometric lattices, continuing on to Rival's conjecture on matchings in modular lattices and an explicit matching version of Dilworth's covering theorem, and ending with a general axiomatic sufficient condition for the existence of a matching between two subsets in a finite lattice.

We begin by recalling several useful facts about Möbius functions.

2.1. MÖBIUS FUNCTIONS AND RADON TRANSFORMS

Let L be a finite partially ordered set and $f:L \to \mathbf{k}$ a function from L to a field $\mathbf{k}$. The *Radon transform* $Tf:L \to \mathbf{k}$ is the function defined by:

$$Tf(x) = \sum_{y:y\leq x} f(y).$$

The *Möbius function* of L is the function $\mu:L \times L \to \mathbf{k}$ defined by:

$$\mu(x,y) = 0 \text{ if } x \nleq y,$$
$$\mu(x,x) = 1,$$
$$\mu(x,y) = -\sum_{x<z\leq y} \mu(z,y).$$

One way to invert the Radon transform is to use the Möbius inversion formula.

(2.1.1) THE MÖBIUS INVERSION FORMULA. Let $f,g:L \to \mathbf{k}$ be functions defined from the partially ordered set L to a field $\mathbf{k}$. Then

$$g(x) = \sum_{y:y\leq x} f(y) \iff f(x) = \sum_{y:y\leq x} g(y)\mu(y,x)$$

and, dually,

$$g(x) = \sum_{y:y\geq x} f(y) \iff f(x) = \sum_{y:y\geq x} \mu(x,y)g(y).$$

Using (2.1.1), we obtain the following inversion formula for the Radon transform:

$$f(x) = \sum_{y:y\leq x} \mu(y,x)Tf(y).$$

For our purposes, this inversion formula is only partially useful since it requires that Tf(y) be known for all elements y less than or equal to x. In most of our applications, the reverse is true - usually, we know Tf(y) for elements y "near the top" of a lattice. For such situations, the following Möbius function identity is often helpful.

(2.1.2) LEMMA. Let L be a finite lattice and let $f:L \to \mathbf{k}$ be a function from L to the field $\mathbf{k}$. Then

$$\sum_{y:x\leq y\leq 1} \mu(y,1)Tf(y) = \sum_{z:x\vee z=1} f(z). \tag{2.1}$$

PROOF (cf. Doubilet 1972 and Dowling and Wilson 1975). The left hand side suggests that we do Möbius inversion on the upper interval $[x,1]$. However, the function f is defined on all of L. To sidestep this difficulty, we define a new function $f_x(y)$ from $[x,1]$ to $\mathbf{k}$ by:

$$f_x(y) = \sum_{z:x\vee z=y} f(z).$$

Because (a) the elements in L are partitioned into equivalence classes by the relation $a \sim b$ if and only if $a \vee x = b \vee x$, and (b) $z \leq y$ for $y \in [x,1]$ if and only if $x \vee z \leq y$, we have, for any element y in $[x,1]$,

$$Tf(y) = \sum_{z:z\leq y} f(z) = \sum_{z:x\leq z\leq y} f_x(z).$$

Applying Möbius inversion to f_x in $[x,1]$, we obtain

$$\sum_{y:x\leq y\leq 1} \mu(y,1)Tf(y) = f_x(1) = \sum_{z:z\vee x=1} f(z). \qquad \square$$

2.2. MATCHINGS AND INCIDENCE MATRICES

The significance of Radon transforms for the study of matchings lies in the following general fact from linear algebra. Let $\mathcal{A}$ and $\mathcal{B}$ be finite sets and let $\mathbf{k}$ be a field. Let $\mathcal{A}^*$ be the vector space over $\mathbf{k}$ of functions defined from $\mathcal{A}$ to $\mathbf{k}$. Define $\mathcal{B}^*$

similarly. Now let $T:\mathcal{A}^* \to \mathcal{B}^*$ be a linear transformation from $\mathcal{A}^*$ to $\mathcal{B}^*$ and let $M(T)$ be the matrix of T relative to the standard basis of delta functions δ_x, where $\delta_x(y) = 1$ if $x = y$ and 0 otherwise.

(2.2.1) LEMMA. Suppose the linear transformation T has rank $|\mathcal{A}|$. Then there exists an injection $\sigma:\mathcal{A} \to \mathcal{B}$ such that the $\sigma(x),x$-entry of $M(T)$ is non-zero.

PROOF. The matrix $M(T)$ has rank $|\mathcal{A}|$ and hence there exists a non-singular square matrix $M'(T)$ of size $|\mathcal{A}|$. The determinant of $M'(T)$ is non-zero. Thus, there exists a non-zero term in its expansion. This non-zero term gives the injection σ. □

We now specialize this result to Radon transforms in lattices. Let $\mathcal{J}$ and $\mathcal{M}$ be subsets of a lattice L. The *incidence matrix* $I(\mathcal{M}|\mathcal{J})$ is the matrix with rows indexed by $\mathcal{M}$ and columns indexed by $\mathcal{J}$ with x,y-entry 1 if $x \geq y$ and 0 otherwise. Let **k** be a field and let

L^* = space of all functions $f:L \to \mathbf{k}$,
$\mathcal{M}^*$ = space of all functions $f:\mathcal{M} \to \mathbf{k}$,
$\mathcal{J}^*$ = subspace of L^* consisting of functions f such that $f(x) = 0$ unless $x \in \mathcal{J}$.

(2.2.2) PROPOSITION. Suppose that the Radon transform $Tf:L \to \mathbb{Q}$ of a function f in $\mathcal{J}^*$ can be reconstructed given the restriction $Tf:\mathcal{M} \to \mathbb{Q}$ of the Radon transform to $\mathcal{M}$. Then the incidence matrix $I(\mathcal{M}|\mathcal{J})$ has rank $|\mathcal{J}|$ and there exists an injection $\sigma:\mathcal{J} \to \mathcal{M}$ such that for all x in $\mathcal{J}$, $x \leq \sigma(x)$.

PROOF. To distinguish the restriction $Tf:\mathcal{M} \to \mathbb{Q}$ from the "full" Radon transform Tf, let us denote it by $T^{\#}f$. The situation can now be summarized by the following commutative diagram:

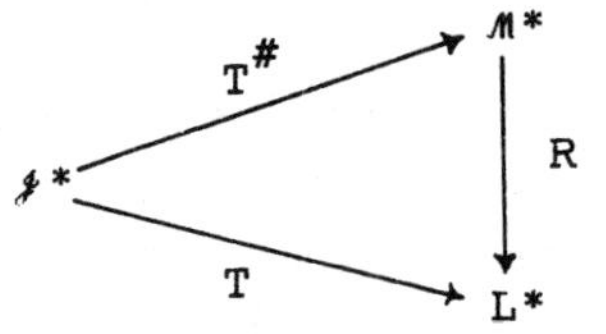

where R is the linear transformation defined as follows: Let $\mathcal{U}$ be a subspace in $\mathcal{M}^*$ complementing the range of $T^{\#}$. If $g \in$ Range $T^{\#}$ and $g = T^{\#}f$, then Rg equals the reconstructed Radon transform Tf. If $g \in \mathcal{U}$, then Rg equals 0.

Now observe that since the Radon transform T is invertible by the Möbius inversion formula (2.1.1), T has rank $|\mathcal{J}|$. Since $T = RT^{\#}$, both $T^{\#}$ and R have rank $|\mathcal{J}|$. Hence, the incidence matrix $I(\mathcal{M}|\mathcal{J})$, which equals the matrix of $T^{\#}$ relative to the standard basis, also has rank $|\mathcal{J}|$. The existence of σ follows from (2.2.1).

□

An injection $\sigma:\mathcal{J} \to \mathcal{M}$ such that $x \le \sigma(x)$ for all x in $\mathcal{J}$ is sometimes referred to as a *matching from $\mathcal{J}$ into $\mathcal{M}$*.

2.3. UPPER COMPLEMENTS

The right hand side of Equation (2.1) suggests another Radon transform on a lattice. Let x be an element of a lattice L. An element y in L is said to be an *upper complement* of x if $x \vee y = 1$. For each element x in L, let C^x be the subset of upper complements of x, i.e.

$$C^x = \{y: y \vee x = 1\}.$$

Now let $f:L \to \mathbf{k}$ be a function. The *upper complement transform* $Cf:L \to \mathbf{k}$ is the function defined by

$$Cf(x) = \sum_{y:y\in C^x} f(y).$$

(2.3.1) PROPOSITION. Let L be a lattice in which $\mu(x,1) \ne 0$ for every x in L. Then the upper complement transform is reconstructible.

FIRST PROOF (Doubilet 1972; Dowling and Wilson 1975). From (2.1.2), we have

$$\sum_{y:x\le y\le 1} \mu(y,1)Tf(y) = Cf(x).$$

By Möbius inversion, we obtain

$$\mu(x,1)Tf(x) = \sum_{y:x\le y\le 1} \mu(x,y)Cf(y).$$

Since $\mu(x,1) \ne 0$,

$$Tf(x) = \sum_{y:x\leq y\leq 1} \frac{\mu(x,y)}{\mu(x,1)} Cf(y).$$

Using this and Möbius inversion, we obtain

$$\begin{aligned} f(x) &= \sum_{z:0\leq z\leq x} \mu(z,x)Tf(z) \\ &= \sum_{\substack{z:0\leq z\leq x \\ y:z\leq y\leq 1}} \left[\frac{\mu(z,x)\mu(z,y)}{\mu(z,1)}\right] Cf(y). \end{aligned}$$

Thus,

$$(2.2) \qquad f(x) = \sum_{y:x\leq y\leq 1} \left[\sum_{z:0\leq z\leq x\wedge y} \left[\frac{\mu(z,x)\mu(z,y)}{\mu(z,1)}\right]\right] Cf(y).$$

This equation provides an explicit inversion formula for Cf(x). [We remark that this inversion formula, applied to the lattice of partitions of a set, yields an effective proof of the fundamental theorem of symmetric functions in the form: the monomial symmetric functions are expressible in terms of the elementary symmetric functions. See Doubilet 1972 for details.] □

SECOND PROOF. We first show that Tf can be reconstructed given Cf by going inductively down the lattice. Since Tf(1) = Cf(1), we are off to a good start. Let x be an element in L. By induction, we may suppose that Tf(y) is already reconstructed for all y > x. Using (2.1.1), we have

$$\mu(x,1)Tf(x) = Cf(x) - \sum_{y:x<y\leq 1} \mu(y,1)Tf(y).$$

The terms on the right hand side are either given or already computed. This, as $\mu(x,1) \neq 0$, Tf(x) can be calculated. The proof can now be completed by observing that f can be reconstructed from Tf by Möbius inversion. □

At first sight, it may seem that the second proof is merely a rephrasing of the first proof - and so it is. However, by being ploddingly inductive, we obtain a strengthening of (2.3.1).

Let L^x be the set of elements x in L such that $\mu(x,1) \neq 0$ and let $J(L^x|L)$ be the matrix with rows indexed by L^x and columns indexed by L whose x,y-entry is 1 if $x \vee y = 1$ and 0 otherwise.

(2.3.2) PROPOSITION. The matrix $J(L^x|L)$ has rank $|L^x|$.

PROOF. Using the inductive algorithm in the second proof, we can reconstruct the restriction $Tf:L^x \to \mathbb{Q}$ of the Radon transform Tf to L^x from the restriction $Cf:L^x \to \mathbb{Q}$. The algorithm still works in this situation because if $\mu(x,1) = 0$, the value Tf(x) always appears with a zero coefficient in Equation (2.1) and can always be ignored. From this, we conclude that the matrix $J(L^x|L)$ and the incidence matrix $I(L^x|L)$ have the same rank. Since $I(L|L)$ is a non-singular matrix, $I(L^x|L)$, and hence $J(L^x|L)$, have rank $|L^x|$. □

We end by stating the matching versions of (2.3.1) and (2.3.2).

(2.3.3) COROLLARY. Let L be a finite lattice. Then there exists an injection γ from L^x to L such that for every x in L^x, $\gamma(x) \vee x = 1$. In particular, if L is a lattice in which $\mu(x,1) \neq 0$ for every x, there exists a bijection γ from L to L such that for every x in L, $\gamma(x) \vee x = 1$.

PROOF. Use (2.2.1). □

The second part of (2.3.3) was first proved in Dowling and Wilson 1975.

An element y is said to be a *lower complement* of an element x if $x \wedge y = 0$. The notion of a lower complement is order-dual to the notion of an upper complement. By looking at the order dual, it is easy to obtain dual versions of all the results in this subsection.

2.4. GEOMETRIC AND SEMIMODULAR LATTICES

A lattice L is said to satisfy the *Jordan-Dedekind chain condition* if for any pair of elements x and y in L, every maximal chain between x and y has the same length. In such a lattice, the *rank* r(x) of an element x is the length of a maximal chain from 0 to x. The *corank* of x is the length of a maximal chain from x to 1 and equals r(1) - r(x). An *atom* or *point* is an element of rank one and a *coatom* or *copoint* is an element of a corank one. A lattice is said to be *atomic* [*coatomic*] if every element is a join [meet] of atoms [coatoms].

Let L be a lattice satisfying the Jordan-Dedekind chain condition. The lattice L is said to be *semimodular* if its rank function r satisfies the *semimodular inequality*: for $x,y \in L$,

$$r(x) + r(y) \geq r(x \vee y) + r(x \wedge y).$$

If, in addition, L is atomic, we say that L is a *geometric* lattice.

In a semimodular lattice L, two elements are said to form a *modular pair* if

$$r(x) + r(y) = r(x \vee y) + r(x \wedge y).$$

An element x is said to be *modular* if for every element y, x and y form a modular pair. If every element in L is modular, we say that L is a *modular* lattice. A useful fact about modular elements is *Dedekind's transposition principle* (Dedekind 1900):

(2.4.1) PROPOSITION. An element x is modular if and only if for every element y, the map $u \to u \vee y$, $[x \wedge y, x] \to [y, x \vee y]$ is an isomorphism with inverse $u \to u \wedge x$.
PROOF. See Birkhoff 1967, p. 82. □

For studying Radon tranforms, the most significant fact about geometric lattices is the following theorem due to Rota.

(2.4.2) PROPOSITION. Let L be a geometric lattice and let $x \leq y$ in L. Then $\mu(x,y)$ is non-zero and has sign $(-1)^{r(y)-r(x)}$.
PROOF. See Rota 1964 or Aigner 1979, p. 171. □

Rota's theorem is false for semimodular lattices in general. For many applications, the following result is an adequate substitute.

(2.4.3) PROPOSITION. Let L be a semimodular lattice and let $x \leq y$ in L. Then, y is a join of atoms in the upper interval $[x,1]$ if and only if the Möbius function $\mu(x,y)$ is non-zero.
PROOF. The proposition follows easily from the following two facts.

(2.4.4) PROPOSITION. Let L be a semimodular lattice and let $\mathcal{A}$ be a subset of atoms of L. Then the sublattice $\langle\mathcal{A}\rangle$ in L generated by $\mathcal{A}$ is a geometric lattice. The rank function of $\langle\mathcal{A}\rangle$ equals the restriction of the rank function of L to $\langle\mathcal{A}\rangle$.

PROOF. Let x be an element in $\langle\mathcal{A}\rangle$ and let

$$0 = x_0 < x_1 < \ldots < x_r = x$$

be a maximal chain in $\langle\mathcal{A}\rangle$ from 0 to x. Since $\langle\mathcal{A}\rangle$ is an atomic lattice and x_i covers x_{i-1} in $\langle\mathcal{A}\rangle$, $x_i = x_{i-1} \vee p_i$ for some atom $p_i \not\leq x_i$. Considering x_i, x_{i-1}, and p_i as elements of L and applying the semimodular inequality, we conclude that x_i also covers x_{i-1} in L. Thus, the rank of x in $\langle\mathcal{A}\rangle$ equals the rank in L and the rank function in $\langle\mathcal{A}\rangle$ also satisfies the semimodular inequality. As $\langle\mathcal{A}\rangle$ is atomic, $\langle\mathcal{A}\rangle$ is a geometric lattice. □

(2.4.5) PROPOSITION. Let L be a lattice. Then

$$\mu(0,1) = \sum_n (-1)^n q_n$$

where q_n is the number of n-element subsets of atoms whose join is 1.

PROOF. This is an easy instance of Rota's crosscut theorem (see Rota 1964 or Aigner 1979, p. 168). □

By (2.4.5), if y is a join of atoms in [x,1], the Möbius function $\mu(x,y)$ in [x,1] equals the Möbius function $\mu(x,y)$ in L^x, the sublattice in L generated by the atom in [x,1]. As L^x is a geometric lattice, $\mu(x,y)$ is non-zero. On the other hand, if y is not a join of atoms in [x,1], the sum in (2.4.5) is empty and $\mu(x,y) = 0$.

□

2.5. WHITNEY NUMBER INEQUALITIES

Let L be a finite lattice satisfying the Jordan-Dedekind chain condition. The *Whitney numbers* W_k *(of the second kind)* are defined by

$$W_k = \text{number of elements in L of rank } k.$$

There are many interesting and - at the present state of knowledge - difficult conjectures about Whitney numbers of geometric lattices. The foremost among them is the *unimodality conjecture* (first posed in Harper and Rota 1971, p. 209).

(2.5.1) CONJECTURE. Let L be a geometric lattice of rank n. Then there exists a positive integer k such that

$$W_0 \leq W_1 \leq W_2 \leq \ldots \leq W_{k-1} \leq W_k$$

and

$$W_k \geq W_{k+1} \geq \ldots \geq W_{n-1} \geq W_n.$$

Another conjecture is that geometric lattices are "top-heavy."

(2.5.2) CONJECTURE. Let L be a geometric lattice of rank n. Then for $0 \leq k \leq n/2$, $W_k \leq W_{n-k}$.

Although there has been almost no progress on the unimodality conjecture, the "top-heaviness" conjecture has been partially verified by a set of inequalities due to Dowling and Wilson (1975). [The special case $W_1 \leq W_{n-1}$ has been proved in various versions in Motzkin 1951, Basterfield and Kelly 1968, Greene 1970, 1975, Heron 1973, Woodall 1976, and Kung 1979.]

(2.5.3) THEOREM. Let L be a geometric lattice of rank n. Then for $0 \leq k \leq n/2$,

$$W_0 + W_1 + \ldots + W_k \leq W_{n-k} + W_{n-k+1} + W_{n-1} + \ldots + W_n.$$

To illustrate the available techniques, we shall give three subtly different proofs of this theorem. The following notation will be used in all three proofs. Let L be a geometric lattice of rank n. The subsets $\mathcal{L}_k$ and $\mathcal{U}_k$ of "lower" and "upper" elements are defined as follows:

$$\mathcal{L}_k = \{x: r(x) \leq k\},$$

$$\mathcal{U}_k = \{x: r(x) \geq n - k\}.$$

Let L* be the vector space of functions defined from L to $\mathbb{Q}$, the rational numbers, and $\mathcal{L}_k{}^*$ the subspace of functions f in L* such that $f(x) = 0$ unless x is in $\mathcal{L}_k$.

All three proofs depend on the following lemma (Dowling and Wilson 1975).

(2.5.4) LEMMA. Let f be a function in $\mathcal{L}_k^*$ and Cf its upper complement transform. If $r(x) < n - k$, then $Cf(x) = 0$.

PROOF. Let $r(x) < n - k$. If $x \vee z = 1$, then by the semimodular inequality,

$$r(z) \geq r(1) - r(x) + r(x \wedge z) > n - (n - k) = k.$$

Since f is in $\mathcal{L}_k^*$, $f(z)$ is zero and $Cf(x)$ is a sum of terms $f(z)$ all of which are zero. Hence, $Cf(x) = 0$. □

Note. The full strength of the semimodular inequality is not used in (2.5.4). Only the weaker inequality

$$r(z) + r(x) \geq r(x \vee z)$$

is used. The results in this subsection hold under this weaker assumption, provided that the appropriate Möbius functions are still non-zero.

FIRST PROOF OF (2.5.3) (Dowling and Wilson 1975). Using Equation (2.2) in §2.3 and the previous lemma, we obtain

$$(2.3) \qquad f(x) = \sum_{\substack{y:y\geq x \text{ and} \\ y\in \mathcal{U}_k}} \left[\sum_{z:z\leq x\wedge y} \frac{\mu(z,x)\mu(z,y)}{\mu(z,1)} \right] Cf(y).$$

Hence, the function f can be reconstructed from the upper complement transform Cf restricted to $\mathcal{U}_k$. By (2.2.2), the matrix $J(\mathcal{U}_k|\mathcal{L}_k)$ with x,y-entry 1 if $x \vee y = 1$ and 0 otherwise has rank $|\mathcal{L}_k|$. We conclude that $|\mathcal{U}_k| \geq |\mathcal{L}_k|$. □

The first proof yields the following matching theorem (Dowling and Wilson 1975).

(2.5.5) PROPOSITION. There exists an injection $\gamma:\mathcal{L}_k \to \mathcal{U}_k$ such that for all x, $x \vee \gamma(x) = 1$.

SECOND PROOF (Dowling and Wilson 1975). Using Equation (2.1) in §2.1, we can reconstruct Cf(x) for all x in $\mathcal{U}_k$ from the Radon transform $Tf:\mathcal{U}_k \to \mathbb{Q}$ restricted to $\mathcal{U}_k$ by first reconstructing $Cf:\mathcal{U}_k \to \mathbb{Q}$ and then applying Equation (2.3) in the first proof. Hence, by (2.2.2), the incidence matrix $I(\mathcal{U}_k|\mathcal{L}_k)$ has rank $|\mathcal{L}_k|$ and we conclude that $|\mathcal{U}_k| \geq |\mathcal{L}_k|$. □

The second proof also yields a matching theorem (Dowling and Wilson 1975).

(2.5.6) PROPOSITION The incidence matrix $I(\mathcal{U}_k|\mathcal{L}_k)$ has rank $|\mathcal{L}_k|$ and there exists an injection $\sigma:\mathcal{L}_k \to \mathcal{U}_k$ such that for all x, $x \leq \sigma(x)$.

THIRD PROOF. We shall show that for a function f in $\mathcal{L}_k{}^*$, the Radon transform $Tf:L \to \mathbb{Q}$ can be reconstructed given its restriction $Tf:\mathcal{U}_k \to \mathbb{Q}$ to $\mathcal{U}_k$. This is done by going inductively down the lattice. Because Tf(x) is given if $r(x) \geq n - k$, we can assume that $r(x) < n - k$. Further, by induction, we can assume that the values Tf(y), y > x, are already computed. Using (2.1.2) and the fact that [by (2.5.4)] Cf(x) = 0, we obtain

$$\mu(x,1)Tf(x) = \sum_{y:x<y\leq 1} \mu(y,1)Tf(y).$$

As $\mu(x,1) \neq 0$ in a geometric lattice, Tf(x) can be computed from this equation. By (2.2.2), the incidence matrix $I(\mathcal{U}_k|\mathcal{L}_k)$ has rank $|\mathcal{L}_k|$ and $|\mathcal{U}_k| \geq |\mathcal{L}_k|$. □

The Dowling-Wilson inequalities are false for semimodular lattices in general. However, by analyzing the third proof, something can be salvaged. Let x be an element in a semimodular lattice L and let x* be the join of all the elements in L covering x. The *reach* of x is defined by

$$\text{reach}(x) = r(x^*) - r(x).$$

Thus the reach of x is the rank of the geometric lattice generated by the atoms in $[x,1]$. It is evident that the reach of x is at most the corank of x, and the reach and the corank are equal in a geometric lattice. Now, let

$$\mathcal{R}_k = \{x:\ \mathrm{reach}(x) \le k\}.$$

For example, $\mathcal{R}_1$ is the subset of meet-irreducibles in L. Here, an element x is a *meet-irreducible* if $x = a \wedge b$ implies $x = a$ or $x = b$. Using the notion of reach, the Dowling-Wilson inequalities can be generalized as follows (Kung pre).

(2.5.7) THEOREM. Let L be a semimodular lattice. Then the incidence matrix $I(\mathcal{R}_k|\mathcal{L}_k)$ has rank $|\mathcal{L}_k|$. In particular,

$$|\mathcal{R}_k| \ge W_0 + W_1 + \ldots + W_k.$$

PROOF. We shall show that the Radon transform $Tf:L \to \mathbf{Q}$ of a function in $\mathcal{L}_k^*$ can be reconstructed given its restriction $Tf:\mathcal{R}_k \to \mathbf{Q}$. As earlier, we proceed inductively down the lattice. If $\mathrm{reach}(x) \le k$, then $Tf(x)$ is given. Thus, we can assume that $\mathrm{reach}(x) > k$. By induction, we can also assume that $Tf(y)$ has already been computed for all y greater than x. Now by (2.1.2) applied to the interval $[x,x^*]$, we have

$$\mu(x,x^*)Tf(x) = -\sum_{y:x<y\le x^*} \mu(x,y)Tf(y) + \sum_{z:x\vee z=x^*} f(z).$$

Using the same argument as in (2.5.4),

$$\sum_{z:x\vee z=x^*} f(z) = 0.$$

Since $\mu(x,x^*) \ne 0$ by (2.4.3), $Tf(x)$ can be calculated using the equation above. The theorem now follows from (2.2.2). □

(2.5.8) COROLLARY. In a semimodular lattice, there exists an injection $\sigma:\mathcal{L}_k \to \mathcal{R}_k$ such that for all x, $x \le \sigma(x)$.

Another variation on the Dowling-Wilson inequalities is to look at modular elements. Let L be a semimodular lattice. The *modular Whitney numbers* M_k are defined by

M_k = number of modular elements of rank k.

The numbers M_k and W_k are related by inequalities (Kung pre; the case $k = 1$ was proved in Brylawski 1975) similar to the Dowling-Wilson inequalities.

(2.5.9) THEOREM. Let L be a geometric lattice of rank n. Then for $0 \leq k \leq n/2$,

$$W_0 + W_1 + \dots + W_k \geq M_{n-k+1} + \dots + M_{n-1} + M_n.$$

PROOF. Apply the argument in the third proof of (2.5.3) to the order dual of L, making use of the fact that if x is modular, then, for every y in L,

$$r(y) - r(x \wedge y) = r(x \vee y) - r(x) \leq \mathrm{corank}(x).$$

Details can be found in Kung pre. □

There is also a version of (2.5.9) for semimodular lattices analogous to (2.5.7); see Kung pre.

2.6. FAMILIES OF MAXIMAL CHAINS

One of the fundamental notions of extremal set theory is the notion of a chain decomposition (see, for example, Greene and Kleitman 1978). Ideally, one wants to partition a partially ordered set into a collection of disjoint chains satisfying certain "nice" conditions. Such chain decompositions do not necessarily exist in geometric lattices. One reason for this is the fact that geometric lattices do not necessarily satisfy the Sperner property (for definitions and examples of non-Sperner geometric lattices, see Dilworth and Greene 1971, Canfield 1978, or Kahn 1980). The best result in this direction for geometric lattices is the maximal chain theorem due to Mason (1973).

Let L be a geometric lattice of rank n, where $n \geq 2$. A *maximal chain* in L is a chain

$$x_1 < x_2 < x_3 < \dots < x_{n-1}$$

where x_i is an element of rank i. Let m(L) be the maximum size of a collection of pairwise disjoint maximal chains in L. As every maximal chain contains an atom, $m(L) \leq W_1$. This upper bound can always be attained.

(2.6.1) THEOREM. Let L be a geometric lattice of rank at least 2. Then there exists a collection of W_1 pairwise disjoint maximal chains.

The original proof in Mason 1973 uses Menger's theorem. We shall present a proof based on Radon transforms (Kung 1978).

Let L be a geometric lattice of rank n and let $\mathcal{W}_i$ be the subsets of L defined by

$$\mathcal{W}_i = \{x: r(x) = k\}$$

Thus, $\mathcal{W}_1$ is the set of atoms and $\mathcal{W}_{n-1}$ the set of coatoms. Let $f:L \to \mathbb{Q}$ be a function in L*. The *complementary Radon transform* (briefly, *c-transform*) of the function f is the function $Hf:L \to \mathbb{Q}$ defined by

$$Hf(x) = \sum_{y:y\not\leq x} f(y) = Tf(1) - Tf(x).$$

Since Hf(1) always equals zero and f(0) never appears in any of the sums for Hf(x), the c-transform is not reconstructible. Fortunately, this is the only way in which reconstructibility fails.

(2.6.2) COMPLEMENTARY MÖBIUS INVERSION FORMULA. Let $f:L \to \mathbb{Q}$ be a function defined on a lattice L. Then, if $x \neq 0$,

$$f(x) = -\sum_{y:y\leq x} \mu(y,x)Hf(y).$$

PROOF. Because Hf(y) = Tf(1) - Tf(y), the right hand side of the equation equals

$$-\sum_{y:y\leq x} \mu(y,x)Tf(1) + \sum_{y:y\leq x} \mu(y,x)Tf(y).$$

Since Tf(1) is constant and the sum

$$\sum_{y:y\leq x} \mu(y,x)$$

equals 0 unless x = 0, (2.6.2) follows immediately from the usual Möbius inversion formula (2.1.1). □

Now let $\mathcal{W}_1{}^*$ be the subspace in L* consisting of functions $f:L \to \mathbb{Q}$ such that $f(x) = 0$ unless $r(x) = 1$.

(2.6.3) LEMMA. A function f in $\mathcal{W}_1{}^*$ can be reconstructed from its c-transform $Hf:\mathcal{W}_{n-1} \to \mathbb{Q}$ restricted to the coatoms.

PROOF. It suffices to prove that the *non-incidence* matrix $I^c(\mathcal{W}_1|\mathcal{W}_{n-1})$ with x,y-entry 1 if $x \nleq y$ and 0 otherwise has rank $|\mathcal{W}_1|$. Consider the incidence matrix $I(\mathcal{U}_1|\mathcal{L}_1)$. By subtracting the row indexed by 0 from every row, we obtain the matrix

$$\begin{bmatrix} 1 & 1 & 1 & \cdots & 1 & 1 & 1 \\ & & & & & & 0 \\ & & -I^c(\mathcal{W}_1|\mathcal{W}_{n-1}) & & & & \vdots \\ & & & & & & 0 \end{bmatrix},$$

where the first row is indexed by 0 and the last column is indexed by 1. Since $I(\mathcal{U}_1|\mathcal{L}_1)$ has rank $|\mathcal{L}_1|$ by (2.5.6), the matrix $I^c(\mathcal{W}_1|\mathcal{W}_1)$ must have rank $|\mathcal{W}_1|$. □

There is another, more explicit, way to invert the c-transform of a function in $\mathcal{W}_1{}^*$. It is based on the following observation.

(2.6.4) LEMMA. Let x be an element in L and let $u_1, u_2, \ldots, u_m$ be all the elements in L covering x. Then,

$$Hf(x) = \frac{1}{m-1} \sum_{i=1}^{m} Hf(u_i).$$

PROOF. Consider the following sets of atoms:

$$x^c = \{p\colon \text{p is an atom and } p \nleq x\},$$
$$u_i{}^c = \{p\colon \text{p is an atom and } p \nleq u_i\},$$
$$u_i - x = \{p\colon \text{p is an atom, } p \leq u_i \text{ but } p \nleq x\}.$$

If p is in x^c, then $x \vee p$ covers x and $x \vee p = u_i$ for a unique element u_i covering x. Hence, the sets $u_i - x$ partition x^c and every atom in x^c appears in all but one of the sets $u_i{}^c$. From this, we conclude that

$$(m - 1)Hf(x) = \sum_{i=1}^{m} Hf(u_i). \qquad \square$$

Using (2.6.4), we can reconstruct the restriction $Hf:\mathcal{W}_1 \to \mathbb{Q}$ of the c-transform to the atoms from the restriction $Hf:\mathcal{W}_{n-1} \to \mathbb{Q}$ to the coatoms by proceeding inductively down the lattice. More explicitly, we have

(2.4) $$I^c(\mathcal{W}_1|\mathcal{W}_1) =$$

$$T(\mathcal{W}_1|\mathcal{W}_2)T(\mathcal{W}_2|\mathcal{W}_3) \ldots T(\mathcal{W}_{n-3}|\mathcal{W}_{n-2})T(\mathcal{W}_{n-2}|\mathcal{W}_{n-1})I^c(\mathcal{W}_{n-1}|\mathcal{W}_1),$$

where $T(\mathcal{W}_{i-1}|\mathcal{W}_i)$ is the matrix with rows indexed by $\mathcal{W}_{i-1}$ and columns indexed by $\mathcal{W}_i$ whose x,y-entry is 1/(m-1) [where m is the number of elements in L covering x] if y covers x, and 0 otherwise. Since $I^c(\mathcal{W}_1|\mathcal{W}_1)$ is the matrix with 0's on the diagonal and 1's elsewhere, it has rank $|\mathcal{W}_1|$ and we can find the inverse of $I^c(\mathcal{W}_{n-1}|\mathcal{W}_1)$ explicitly from Equation (2.4).

Using (2.4), we can now prove (2.6.1). We need one tool from linear algebra.

(2.6.5) THE BINET-CAUCHY FORMULA. Let X = MN, where X is an $n \times n$ matrix, M an $n \times p$ matrix, N a $p \times n$ matrix, and the columns of M and the rows of N are labelled by the same index set $\{1,2,\ldots,p\}$. Then,

$$\det X = \sum_{i_1,\ldots,i_n} \det M(i_1,\ldots,i_n) \det N(i_1,\ldots,i_n),$$

where the sum is over all n-subsets $\{i_1,\ldots,i_n\}$ of $\{1,2,\ldots,p\}$, $M(i_1,\ldots,i_n)$ equals M restricted to the columns indexed by $\{i_1,\ldots,i_n\}$, and $N(i_1,\ldots,i_n)$ equals N restricted to the rows indexed by $\{i_1,\ldots,i_n\}$.

PROOF. See, for example, Muir 1933, p. 213. $\square$

In Equation (2.4), $I^c(\mathcal{W}_1|\mathcal{W}_1)$ is non-singular and its determinant is non-zero. Applying the Binet-Cauchy formula iteratively to the product on the right hand side, we can find subsets $\mathcal{V}_i \subseteq \mathcal{W}_i$, $|\mathcal{V}_i| = |\mathcal{W}_1|$, such that

$$\det T(\mathcal{V}_{i-1}|\mathcal{V}_i) \neq 0$$

for $2 \leq i \leq n - 1$. By (2.2.1), there exist injections $\sigma_i : \mathcal{V}_{i-1} \to \mathcal{V}_i$ such that $x \leq \sigma_i(x)$. Putting together these injections, we obtain a collection of disjoint maximal chains

$$\{p < \sigma_2(p) < \sigma_3(\sigma_2(p)) < \ldots < \sigma_{n-1}(\sigma_{n-2}\ldots(\sigma_2(p))): p \in \mathcal{W}_1\}$$

of size $|\mathcal{W}_1|$. This completes the proof of (2.6.1).

3. COVERING INEQUALITIES IN LATTICES

3.1. DILWORTH'S COVERING THEOREM

In contrast to §2, where we were concerned with subsets defined using the rank function, our attention in this section will be focused on subsets defined by the covering relation.

Let x and y be elements in a lattice L. We say that x *covers* y if x > y and there is no element z such that x > z > y. The founding theorem in the theory of covering relations in lattices is Dilworth's covering theorem (Dilworth 1954).

(3.1.1) DILWORTH'S COVERING THEOREM. In a finite modular lattice, the number of elements covering exactly k elements equals the number of elements covered by exactly k elements.

Apart from Dilworth's original proof (which uses the Möbius function), there have been several more elementary proofs (Ganter and Rival 1973, Kurinnoi 1973, and Reuter 1984). All of these proofs proceed in two steps. First, prove the theorem for complemented or atomic modular lattices and then extend it to all modular lattices by an inductive or Möbius function argument.

When faced with two sets of the same cardinality, one's combinatorial instinct is to try to find a "natural" bijection between them. In the case of Dilworth's covering theorem, this problem was first posed, for the case $k = 1$, by Rival in 1972 (see Rival 1974). Recall that an element j in a lattice L is a *join-irreducible* if $j = a \vee b$ implies that $j = a$ or $j = b$; equivalently, $j = 0$ or j covers exactly one element. A *meet-irreducible* is a join-irreducible in the order dual. More generally, let k be a positive integer. An element j is said to be *k-covering* if it covers at most k elements. Thus, the 1-covering elements are precisely the join-irreducibles. Dually, an element m is said to be *k-covered* if it is covered by at most k elements. Let $\mathcal{J}(k,L)$ be the set of k-covering elements and $\mathcal{M}(k,L)$ be the set of k-covered elements.

(3.1.2) RIVAL'S CONJECTURE. In a finite modular lattice, there exists a bijection $\sigma:\mathcal{J}(1,L) \to \mathcal{M}(1,L)$ such that for all x in $\mathcal{J}(1,L)$, $x \le \sigma(x)$.

This conjecture was proved in Kung 1985, using the inductive technique for reconstructing Radon transforms presented in §2. Rather gratifyingly, this technique turns out to be powerful enough to prove the following theorem, which is a strengthening of both Dilworth's theorem and Rival's conjecture.

(3.1.3) THEOREM. Let L be a finite modular lattice. Then the incidence matrix $I(\mathcal{M}(k,L)|\mathcal{J}(k,L))$ is a non-singular square matrix.

Rather than proving (3.1.3) directly, we first consider the following generalization.

(3.1.4) PROPOSITION. Let L be a finite semimodular lattice and $\mathcal{K}$ the set of modular elements in L. Then the incidence matrix $I(\mathcal{M}(k,L)|\mathcal{J}(k,L) \cap \mathcal{K})$ has rank $|\mathcal{J}(k,L) \cap \mathcal{K}|$.

PROOF. By (2.2.2), it suffices to show that the Radon transform $Tf:L \to \mathbf{Q}$ of a function $f:L \to \mathbf{Q}$ such that $f(x) = 0$ unless x is in $\mathcal{J}(k,L) \cap \mathcal{K}$ is reconstructible given its restriction $Tf:\mathcal{M}(k,L) \to \mathbf{Q}$. As earlier, we go inductively down the lattice.

The value $Tf(1)$ is given because 1 is in $\mathcal{M}(k,L)$. Now let x be an element in L. If x is in $\mathcal{M}(k,L)$, then $Tf(x)$ is given. If not, x is covered by at least $k + 1$ elements. By induction, we can assume that $Tf(y)$ for $y > x$ has already been calculated. We shall calculate $Tf(x)$ using Equation (2.1) in §2.1:

$$(3.1) \qquad \sum_{y:x\le y\le x^*} \mu(y,x^*)Tf(y) = \sum_{z:z\vee x=x^*} f(z),$$

with x^* equal to the join of all the elements in L covering x.

The interval $[x,x^*]$ contains the geometric sublattice L^x generated by the elements covering x. The geometric lattice L^x contains at least $k + 1$ atoms and hence, by case $k = 1$ of (2.5.3), L^x contains at least $k + 1$ coatoms. By (2.4.4), these coatoms are covered by x^*: thus, x^* covers at least $k + 1$ elements.

Consider now an element j in $\mathcal{J}(k,L) \cap \mathcal{X}$ and suppose that $j \vee x = x^*$. Because j is modular, the intervals $[x \wedge j, j]$ and $[x,x^*]$ are isomorphic by (2.4.1). As x^* covers at least $k + 1$ elements, j covers at least $k + 1$ elements, contradicting the assumption that j is in $\mathcal{J}(k,L)$. We conclude that for all j in $\mathcal{J}(k,L) \cap \mathcal{X}$, $j \vee x \ne x^*$. Hence, the sum on the right hand side of Equation (3.1) is zero. Since $\mu(x,x^*) \ne 0$ by (2.4.4), we can calculate $Tf(x)$ using Equation (3.1). □

From (3.1.4), (3.1.3) follows immediately on observing that every element in a modular lattice is modular (!) and that the order dual of a modular lattice is also modular. We remark that this proof differs from the earlier proofs of Dilworth's covering theorem in that it bypasses the usual first step of proving the theorem for the special case of a complemented modular lattice.

The original proof in Kung 1985 of Rival's conjecture introduced the notion of a consistent join-irreducible. We shall end this subsection with a brief account of this notion.

A join-irreducible j in a lattice L is said to be *consistent* if for every element x in L, $x \vee j$ is a join-irreducible in the upper interval $[x,1]$. A lattice L is said to be *consistent* if every join-irreducible in L is consistent.

(3.1.5) THEOREM. Let $\mathcal{J}_c(1,L)$ be the set of consistent join-irreducibles in a finite lattice L. Then the incidence matrix $I(\mathcal{M}(1,L)|\mathcal{J}_c(1,L))$ has rank $|\mathcal{J}_c(1,L)|$.

The proof is similar to the proof of (3.1.4), with the exception that the element x* has to be chosen more carefully. An elementary account can be found in Kung 1985.

(3.1.6) COROLLARY. Let L be a finite consistent lattice. Then the incidence matrix $I(\mathcal{M}(1,L)|\mathcal{J}(1,L))$ has rank $|\mathcal{J}(1,L)|$.

Since modular lattices are consistent [use Dedekind's transposition principle (2.4.1)], (3.1.6) implies Rival's conjecture. However, semimodular lattices are not necessarily consistent. Thus, (3.1.6) and the case $k = 1$ of (3.1.4) are really quite different approaches to generalizing the case $k = 1$ of Dilworth's covering theorem.

We end this subsection with another application of (3.1.5).

(3.1.7) COROLLARY. Let L be a coatomic semimodular lattice of rank n. Then, $W_1 \le W_{n-1}$.

3.2 BREADTH AND REACH

Prior to the solution of Rival's conjecture, there were two partial solutions. One of them can be found in Duffus 1982 where the conjecture was verified for modular lattices whose skeleton has width 2 (see the paper for definitions). This paper also contains techniques for putting together matchings which are of independent interest.

The other partial solution is due to Rival himself. In Rival 1976, the conjecture was verified for modular lattices of breadth two using a matching theorem for semimodular lattices of breadth two.

Let L be a lattice. The *breadth* of L is the least positive integer b such that for every subset $\{a,b,\ldots,c\}$ of elements of L, there exists a subset $\{x,y,\ldots,z\}$ contained in $\{a,b,\ldots,c\}$ of size at most b such that

$$a \vee b \vee \ldots \vee c = x \vee y \vee \ldots \vee z.$$

(Note that this definition of breadth is order-dual to the definition in Birkhoff 1967.) Now suppose L is semimodular. The *reach* of L is defined by

$$\text{reach}(L) = \max \{\text{reach}(x) : x \in L\}.$$

Thus, reach(L) is the maximum rank of an interval of the form $[x,x^*]$ in L. For a semimodular lattice L, breadth and reach are related as follows:

$$\text{breadth}(L) \geq \text{reach}(L).$$

Equality is attained when L is geometric; indeed, for a geometric lattice L,

$$\text{breadth}(L) = \text{rank}(L) = \text{reach}(L).$$

We can now state Rival's matching theorem (Rival 1976).

(3.2.1) THEOREM. Let L be a semimodular lattice of breadth at most two. Then, there exists an injection $\tau:\mathcal{M}(1,L) \to \mathcal{J}(1,L)$ such that for all x in $\mathcal{M}(1,L)$, $x \geq \tau(x)$

Perhaps the most interesting feature of this theorem is that the inequality goes against the expected direction.

Rival's matching theorem can be generalized in the following way (Kung pre).

(3.2.2) THEOREM. Let L be a semimodular lattice of reach at most two. Then the incidence matrix $I(\mathcal{J}(1,L)|\mathcal{M}(1,L))$ has rank $|\mathcal{M}(1,L)|$.

PROOF. By (3.1.6), it suffices to show that the order dual of L is a consistent lattice. Let m be a meet-irreducible and x an element of L. Suppose $m \wedge x$ is not a meet-irreducible in $[0,x]$. Let a and b be two elements in $[0,x]$ covering $m \wedge x$. By semimodularity, $a \vee b$ covers a and b and $r(a \vee b) - r(m \wedge x) = 2$. Since $x \not\leq m$, there exists an element c in L but not in $[0,x]$ covering $m \wedge x$. Since $a \vee b \in [0,x]$ but $c \notin [0,x]$, $a \vee b \vee c$ is strictly greater than $a \vee b$. Thus, the element $(m \wedge x)^*$ obtained by taking the join of all the elements covering $m \wedge x$ satisfies

$$(m \wedge x)^* \geq a \vee b \vee c > a \vee b;$$

hence,

$$\text{reach}(m \wedge x) = r((m \wedge x)^*) - r(m \wedge x) \geq 3,$$

contradicting our assumption that L has reach at most 2. □

4. CONCORDANCE: AN AXIOMATIC APPROACH

Reading through §§2 and 3, one cannot help but notice that all the inductive proofs follow the same pattern. In this section, we shall describe a general theorem of which the preceding results are examples. Unlike composers of music, mathematicians committed to a historical approach must first present the variations and then the theme.

4.1. CONCORDANT SETS

In hindsight, the basic notion in the results on matchings in lattices presented in §§2 and 3 is the notion of a concordant pair of subsets.

(4.1.1) DEFINITION. Let $\mathcal{J}$ and $\mathcal{M}$ be subsets of a finite lattice L. The subset $\mathcal{J}$ is said to be *concordant with* $\mathcal{M}$ if for every element x in L, either x is in $\mathcal{M}$ or there exists an element x* such that

CS1. The Möbius function $\mu(x,x^*) \neq 0$, and

CS2. For every element j in $\mathcal{J}$, $x \vee j \neq x^*$.

When $\mathcal{J}$ is concordant with $\mathcal{M}$, and $\mathcal{M}$ is concordant with $\mathcal{J}$ in the order dual of L, we say that $\mathcal{J}$ and $\mathcal{M}$ are *concordant*.

Although x* equals the join of all the elements covering x in most of the examples, this is not required by the definition. However, for CS1 to hold, x* must be the join of elements covering x [use (2.4.5)].

By imitating any one of the inductive proofs in §§2 and 3, it is straightforward to prove the following theorem (Kung pre).

(4.1.2) THEOREM. Let $\mathcal{J}$ be concordant with $\mathcal{M}$. Then the incidence matrix $I(\mathcal{J}|\mathcal{M})$ has rank $|\mathcal{J}|$, and hence, there exists an injection $\sigma:\mathcal{J} \to \mathcal{M}$ such that $j \leq \sigma(j)$ for all j in $\mathcal{J}$.

We shall present two examples of concordant sets in §§4.2 and 4.3. A list of known concordant sets can be found in Kung pre.

4.2. MODULAR JOIN- AND MEET-IRREDUCIBLES

The *coreach* of an element x in a semimodular lattice is defined by:

$$\text{coreach}(x) = r(x) - r(x_*),$$

where x_* is the meet of all the elements covered by x. In this subsection, we shall be working with the following subsets in a semimodular lattice: $\mathcal{R}_k = \{x: \text{reach}(x) \leq k\}$, $\mathcal{S}_k = \{x: \text{coreach}(x) \leq k\}$, and $\mathcal{K} = \{x: x \text{ is modular}\}$.

(4.2.1) THEOREM. Let L be a semimodular lattice. Then $\mathcal{S}_k \cap \mathcal{K}$ is concordant with $\mathcal{R}_k$ in L.

PROOF. Let x be an element not in $\mathcal{R}_k$. Define x* to be the join of all the elements covering x. By (2.4.3), $\mu(x,x^*) \neq 0$ and so CS1 holds. Now let j be a modular element of coreach at most k. Suppose that $j \vee x = x^*$. Since the interval $[x,x^*]$ contains the geometric lattice L^x generated by the elements covering x and geometric lattices are coatomic, the meet of the elements covered by x* in L^x equals x and

$$\text{coreach}(x^*) \geq \text{reach}(x) > k.$$

Because j is modular, $[j \wedge x, j]$ and $[x,x^*]$ are isomorphic. Hence, coreach(j) > k, contradicting our assumption that j is in $\mathcal{S}_k$. We conclude that $j \vee x \neq x^*$. This proves CS2. □

This theorem has two special cases which are of particular interest. The first is a result about modular lattices which coincides with Dilworth's covering theorem when k = 1.

(4.2.2) COROLLARY. In a modular lattice, $|\mathcal{S}_k| = |\mathcal{R}_k|$.

The second is an analogue for semimodular lattices of the case k = 1 of (2.5.9).

(4.2.3) COROLLARY. In a semimodular lattice,

#modular join-irreducibles ≤ #meet-irreducibles.

Because a coatomic semimodular lattice is not necessarily atomic, the dual version of (4.2.1) cannot be proved by dualizing the argument. However, when k = 1, the dual argument does work and we have the following result. [This result can also be obtained by using (3.1.5).]

(4.2.4) PROPOSITION. In a semimodular lattice,

#modular meet-irreducibles $\leq$ #join-irreducibles,

Finally, recall that $\mathcal{L}_k = \{x: r(x) \leq k\}$. Since $\mathcal{L}_k \subseteq \mathcal{S}_k$ (but not every element in $\mathcal{L}_k$ is modular), there is a certain similarity between (4.2.1) and (2.5.7). In fact, they can be combined into the following result.

(4.2.5) THEOREM. Let L be a semimodular lattice. Then $\mathcal{L}_k \cup (\mathcal{S}_k \cap \mathcal{K})$ is concordant with $\mathcal{R}_k$.

PROOF. Combine the proofs of (4.2.1) and (2.5.7). □

4.3. TYPES AND DIAGRAMS

The reach of an element x is only one of many numerical invariants of the interval [x,x*]. One can therefore ask whether (4.2.2) carries over to other numerical invariants. That this - and more - can always be done follows from the next result, which is perhaps the ultimate sharpening of (4.2.2).

Let L and M be modular lattices. A function $\iota: M \to L$ is said to be an *embedding* if it is one-to-one and preserves rank and joins (i.e. for all $x, y \in M$, $r_M(x) = r_L(\iota(x))$ and $\iota(x \vee y) = \iota(x) \vee \iota(y)$). The lattice M is said to be a *lower subinterval* of L if there exists an embedding of M into a lower interval of L. Analogously, M is said to be an *upper subinterval* of L if there exists an embedding of M into a upper interval of L.

Let $\{M_\alpha\}$ be a collection (not necessarily finite) of isomorphism classes of finite atomic modular lattices. Let L be a finite modular lattice and let

$\mathcal{E}(M_\alpha) = \{x: [x,x^*]$ does not contain any of the lattices M_α as a lower subinterval$\}$,

$\mathcal{F}(M_\alpha)$ = $\{x : [x_*, x]$ does not contain any of the lattices M_α as an upper subinterval$\}$.

(4.3.1) THEOREM. Let L be a finite modular lattice and $\{M_\alpha\}$ a collection of isomorphism classes of finite atomic modular lattices. Then $\mathcal{F}(M_\alpha)$ and $\mathcal{E}(M_\alpha)$ are concordant.

PROOF. The proof is similar to, say, the proof of (4.2.1) and uses the following fact: if an atomic modular lattice L contains a lower interval isomorphic to a modular lattice M, then L also contains an upper interval isomorphic to M. □

Let M be an atomic modular lattice. An element x in a modular lattice L is said to be of *type* M if the atomic modular lattice L^x generated by all the elements covering x in L is isomorphic to M; x is said to be of *cotype* M if the coatomic (hence, atomic) modular lattice L_x generated by all the elements covered by x is isomorphic to M. We define

$$D^*(L;M) = \#\text{elements in } L \text{ of type } M$$

and

$$D_*(L;M) = \#\text{elements in } L \text{ of cotype } M.$$

We call these numbers the *upper* and *lower Dilworth numbers* of the modular lattice L.

(4.3.2) THEOREM. Let L be a finite modular lattice and M a finite atomic modular lattice. Then $D^*(L;M) = D_*(L;M)$.

PROOF. Let Z be the collection of isomorphism classes of atomic modular lattices contained as a lower subinterval in M, Y be the complement of Z in the collection of all isomorphism classes of atomic modular lattices, and X be the collection $Y \cup \{M\}$.

Consider the subsets $\mathcal{E}(X)$ and $\mathcal{E}(Y)$ in L. Since $Y \subseteq X$, $\mathcal{E}(X) \subseteq \mathcal{E}(Y)$. Let x be an element in $\mathcal{E}(Y) - \mathcal{E}(X)$. Because $x \notin \mathcal{E}(X)$, $[x,x^*]$ contains M as a lower interval. However, because $x \in \mathcal{E}(Y)$, $[x,x^*]$ cannot contain any atomic modular lattices strictly containing M. Hence, L^x is isomorphic to M. On the other hand, if L^x is isomorphic to M, then x is in $\mathcal{E}(Y) - \mathcal{E}(X)$.

We conclude that

$$\mathcal{E}(Y) - \mathcal{E}(X) = \{x\colon x \text{ has type } M\}.$$

Similarly,

$$\mathcal{F}(Y) - \mathcal{F}(X) = \{x\colon x \text{ has cotype } M\}.$$

Using these facts and (4.3.1), we obtain

$$\begin{aligned} D^*(L;M) &= |\mathcal{E}(Y)| - |\mathcal{E}(X)| \\ &= |\mathcal{F}(Y)| - |\mathcal{F}(X)| \\ &= D_*(L;M). \end{aligned}$$

□

We end this section with an application of (4.3.2) inspired by Reuter and Rival pre. Let P and Q be partially ordered sets. A P-*subdiagram* of Q is an injection $i:P \to Q$ which preserves the covering relation. If P has a (unique) minimum 0_P, the element $i(0_P)$ is called the *base* of the P-subdiagram. Dually, the element $i(1_P)$, where 1_P is the maximum of P (if it exists) is called the peak of the P-subdiagram. We denote by P^{op} the order dual of P. A partially ordered set P is said to be *reflexive (for modular lattices)* if

RE1. P has a unique minimum 0_P and every element in P is a join of atoms in P.

RE2. For every finite atomic modular lattice L, the number of P-subdiagrams in L with base 0_P equals the number of P^{op}-subdiagrams with peak 1_P.

For example, Boolean algebras, and more generally, self-dual atomic modular lattices are reflexive. Other examples can be found in Reuter and Rival pre.

(4.3.3) PROPOSITION. Let L be a finite modular lattice and P a reflexive partially ordered set. Then the number of P-subdiagrams in L equals the number of P^{op}-subdiagrams in L.

PROOF. Let $i:P \to L$ be a P-subdiagram with base x. By RE1, the range i(P) in L is contained in the atomic modular lattice L^x generated by x in L. Hence, the number of P-subdiagrams in L equals

$$(4.1) \qquad \sum_{x \in L} \#P\text{-subdiagrams with base } x \text{ in } L^x$$

$$= \sum_M (\#P\text{-subdiagrams in } M \text{ based at } 0_M) D^*(L;M),$$

where the second sum is over all isomorphism classes of finite atomic modular lattices. (This sum is, of course, finite.) Using the same argument on the order dual of L, we also have

$$\#P^{\underline{op}}\text{-subdiagrams in } L$$

$$= \sum_M (\#P^{\underline{op}}\text{-subdiagrams in } M \text{ with peak at } 1_M) D_*(L;M).$$

The proposition now follows from this equation, Equation (4.1), (4.3.2), and RE2. □

APPENDIX. SOME UNSOLVED PROBLEMS

A.1. COMPLEMENTS

In §2.1, it is shown that the upper complement transform is reconstructible if for all x, $\mu(x,1) \neq 0$ and the lower complement transform is reconstructible if for all x, $\mu(0,x) \neq 0$. Let us define the *complement transform* Kf of a function $f: L \to \mathbb{Q}$ on a lattice by

$$Kf(x) = \sum_{\substack{y:y \text{ is a complement} \\ \text{of } x}} f(y),$$

where a *complement* of x is an element which is both an upper and lower complement of x.

(A.1.1) PROBLEM. For which classes of lattices is the complement transform reconstructible?

The matching version of this problem was solved in Dowling 1977 using ideas from Crapo's complementation theorem for Möbius functions (Crapo 1966).

(A.1.2) THEOREM. Let L be a lattice in which $\mu(0,x) \neq 0$ and $\mu(x,1) \neq 0$ for all x in L. Then there exists a bijection $\gamma: L \to L$ such that $\gamma(x)$ is a complement of x for all x in L.

A.2. EXTREMAL PROBLEMS

Given a rank or covering inequality, a natural thing to do is to study the lattices for which equality is attained. The first result of this type is due to Greene (1970).

(A.2.1) THEOREM. Let L be a geometric lattice in which $W_1 = W_{n-1}$. Then L is modular.

This was later generalized in Dowling and Wilson 1975.

(A.2.2) THEOREM. Let L be a geometric lattice of rank n in which

$$W_0 + W_1 + \dots + W_k = W_{n-k} + W_{n-k+1} + \dots + W_{n-1} + W_n$$

for some k, $1 \le k \le n/2$. Then L is modular.

The many results in §§2, 3, and 4 yield inequalities, most of which are equalities for modular lattices.

(A.2.3) PROBLEM. For which inequalities is equality a necessary and sufficient condition for the lattice to be modular?

One inequality for which the answer to (A.2.3) is negative is the inequality implied by (3.1.6): for a consistent lattice L, $|\mathcal{J}(1,L)| \le |\mathcal{M}(1,L)|$. The 5-element non-modular lattice N_5 is a self-dual consistent lattice which is not modular. On the positive side, Race has recently proved the following result [see (3.1.5)]: Let L be a semimodular lattice in which the number of consistent join-irreducibles equals the number of meet-irreducibles. Then L is modular.

A.3. RANGE

The following problem, due to Ethan Bolker, is inspired by the importance of characterizing the range of Radon transforms in analysis (see Helgason 1982 for background).

(A.3.1) PROBLEM. Characterize the ranges of the Radon transforms defined in this paper, i.e. for a given Radon transform, find "natural" necessary and sufficient conditions for a function to be the Radon transform of another function.

Of course, for some transforms, such as the upper complement transform, the question is trivial.

A.4. CLUTTERS AND BLOCKERS

A *clutter* $\mathscr{C}$ on a finite set S is a collection of pairwise incomparable subsets of S. The *blocker* $\mathscr{B}$ of the clutter $\mathscr{C}$ is the clutter consisting of the minimal subets of S which have non-empty intersection with every member of $\mathscr{C}$. For the importance of clutters and blockers in combinatorial optimization and the duality between them, see Edmonds and Fulkerson 1970.

(A.4.1) PROBLEM. Is there a relation between the rank of the incidence matrix of a clutter and the rank of the incidence matrix of its blocker?

A.5. RADON TRANSFORMS OVER A FIELD OF POSITIVE CHARACTERISTIC

Throughout this paper, we have studied Radon transforms over fields of characteristic zero. What can be said about Radon transforms over other fields? One particularly interesting problem is the following.

(A.5.1) PROBLEM. What is the rank of the coatom-atom incidence matrix of a geometric lattice over a field of positive characteristics?

Consider, for example, a geometric lattice L of flats of a binary geometry G. Since the non-incidence matrix of coatoms versus atoms of L over GF(2) coordinatizes G, its rank is exactly rank(L). Thus, the rank of the coatom-atom incidence matrix of L equals rank(L) - 1, rank(L), or rank(L) + 1. All three possibilities can happen. Apart from this, no general results are known. The rank over a field of characteristic p of the coatom-atom incidence matrix of the lattice of subspaces of a vector space over GF(p) has been calculated. See Jamison 1977 or Smith 1969.

A.6. DIFFERENCES OF CONCORDANT SETS

To motivate this problem area, we begin with an example. A

lattice L is said to be *linearly indecomposable* if its Hasse or covering diagram cannot be disconnected by removing one edge. A natural extension of Rival's conjecture (3.1.2) is the following.

(A.6.1) CONJECTURE. Let L be a finite linearly indecomposable modular lattice. Then, there exists a matching from $\mathcal{J}(1,L) - \{0\}$ to $\mathcal{M}(1,L) - \{1\}$.

This conjecture was proved in Reuter pre, using (3.1.5) and the marriage theorem.

(A.6.1) is an example of the following very general problem.

(A.6.2) PROBLEM. Let $\mathcal{J}_1$, $\mathcal{J}_2$, $\mathcal{M}_1$, $\mathcal{M}_2$ be subsets of a finite lattice L such that $\mathcal{J}_1 \subseteq \mathcal{J}_2$, $\mathcal{M}_1 \subseteq \mathcal{M}_2$, $\mathcal{J}_1$ is concordant with $\mathcal{M}_1$, and $\mathcal{J}_2$ is concordant with $\mathcal{M}_2$. Does there exist a matching from $\mathcal{J}_2 - \mathcal{J}_1$ to $\mathcal{M}_2 - \mathcal{M}_1$?

An example is the "top-heaviness" conjecture (2.5.2) for geometric lattices. Another example is the following problem suggested by (A.6.1).

(A.6.3) PROBLEM. Let L be a finite modular lattice. What connectivity conditions on the Hasse diagram are sufficient to ensure that there exists a matching from $\mathcal{J}(k,L) - \mathcal{J}(k-1,L)$ to $\mathcal{M}(k,L) - \mathcal{M}(k-1,L)$?

ACKNOWLEDGEMENT. I would like to thank Curtis Greene for a very stimulating discussion in which he showed me the Möbius function identity in (2.1.2). I would also like to thank David Race and Tom Zaslavsky for their comments on the first draft of this paper.

REFERENCES

Items marked with an asterisk are papers on matchings or inequalities in lattices.

1. M. Aigner, 1979. *Combinatorial Theory*, Springer-Verlag, New York/Heidelberg/Berlin.

*2. J. G. Basterfield and L. M. Kelly, 1968. "A characterization of sets of *n* points which determine *n* hyperplanes," *Proc. Cambridge Philos. Soc.* 64, 585-588.

3. G. Birkhoff, 1967. *Lattice Theory* (3rd edition), Amer. Math. Soc. Colloq. Publ., Vol. 25, Amer. Math. Soc., Providence, Rhode Island.

4. E. Bolker, pre. "The finite Radon transform," *Proc. Conf. on Integral Geometry, Bowdoin College, 1984*, to appear.

*5. T. Brylawski, 1975. "Modular constructions for combinatorial geometries," *Trans. Amer. Math. Soc.* 203, 1-44.

6. E. R. Canfield, 1978. "On a problem of Rota," *Adv. Math.* 29, 1-10.

7. A. M. Cormack, 1982. "Computed tomography: Some history and recent developments," in Shepp 1982, pp. 35-42.

8. H. H. Crapo, 1966. "The Möbius function of a lattice," *J. Combin. Theory* 1, 126-131.

9. R. Dedekind, 1900. "Über die von drei Moduln erzeugte Dualgruppe," *Math. Ann.* 53, 371-403 (=*Gesammelte Werke*, Vol. 2, pp. 236-271).

10. R. A. DeMillo *et al.*, 1983. *Applied Cryptography, cryptographic protocols, and computer security models*, Proc. Symp. Appl. Math., Vol. 29, Amer. Math. Soc., Providence, Rhode Island.

11. P. Diaconis, 1983. "Projection pursuit for discrete data," Technical Reports No. 148, Department of Statistics, Stanford University.

12. P. Diaconis and R. Graham, 1985. "The Radon transform on Z_2^k", *Pacific J. Math.* 118, 323-345.

*13. R. P. Dilworth, 1954. "Proof of a conjecture on finite modular lattices," *Ann. Math.* (2) 60, 359-364.

14. R. P. Dilworth and C. Greene, 1971. "A counterexample to the generalization of Sperner's theorem," *J. Combin. Theory* 10, 18-21.

15. P. Doubilet, 1972. "On the foundations of combinatorial theory. VII: Symmetric functions through the theory of distribution and occupancy," *Stud. Appl. Math.* 51, 377-396.

*16. T. A. Dowling, 1977. "Complementing permutations in finite lattices," *J. Combin. Theory, Ser. B* 23, 223-226.

*17. T. A. Dowling and R. M. Wilson, 1975. "Whitney number inequalities for geometric lattices," *Proc. Amer. Math. Soc.* 47, 504-512.

*18. D. Duffus, 1982. "Matching in modular lattices," *J. Combin. Theory, Ser. A* 32, 303-314.

19. J. Edmonds and D. R. Fulkerson, 1970. "Bottleneck extrema," *J. Combin. Theory* 8, 299-306.

*20. B. Ganter and I. Rival, 1973. "Dilworth's covering theorem for modular lattices: A simple proof," *Algebra Universalis* 3, 348-350.

*21. C. Greene, 1970. "A rank inequality for finite geometric lattices," *J. Combin. Theory* 9, 357-364.

*22. C. Greene, 1975. "An inequality for the Möbius function of a geometric lattice", *Stud. Appl. Math.* 54, 71-74.

23. C. Greene and D. J. Kleitman, 1978. "Proof techniques in the theory of finite sets," in *Studies in Combinatorics* (G.-C. Rota, ed), Math. Assoc. Amer., Washington, D. C., pp. 22-79.

*24. L. H. Harper and G.-C. Rota, 1971. "Matching theory, an introduction", in *Advances in Probability, Vol. 1* (P. Ney, ed), Marcel Dekker, New York, pp. 169-215.

25. S. Helgason, 1982. "Ranges of Radon transforms," in Shepp 1982, pp. 63-70.

*26. A. P. Heron, 1973. "A property of the hyperplanes of a matroid and an extension of Dilworth's theorem," *J. Math. Anal. Appl.* 42, 119-131.

27. R. E. Jamison, 1977. "Covering finite fields with cosets of subspaces," *J. Combin. Theory, Ser. A* 22, 253-266.

28. J. Kahn, 1980. "Some non-Sperner paving matroids," *Bull. London Math. Soc.* 12, 268.

*29. J. P. S. Kung, 1979. "The Radon transforms of a combinatorial geometry, I.," *J. Combin. Theory, Ser. A* 26, 97-102.

*30. J. P. S. Kung, 1985. "Matchings and Radon transforms in lattices. I. Consistent lattices", *Order*, to appear.

*31. J. P. S. Kung, pre. "Matchings and Radon transforms in lattices. II. Concordant sets", in preparation.

*32. G. C. Kurinnoi, 1973. "A new proof of Dilworth's theorem," *Vestnik Char'kov Univ.* 93, Mat. Nr. 38, 11-15 (in Russian).

*33. J. H. Mason, 1972. "Matroids: Unimodal conjectures and Motzkin's theorem," in *Combinatorics (Proc. Conf. Combinatorial Math., Math. Inst., Oxford, 1972)*, Inst. Math. Appl., Southend-on-Sea, pp. 207-220.

*34. J. H. Mason, 1973. "Maximal families of pairwise disjoint maximal proper chains in a geometric lattice," *J. London Math. Soc.* (2) 6, 539-542.

*35. T. S. Motzkin, 1951. "The lines and planes connecting the points of a finite set," *Trans. Amer. Math. Soc.* 70, 451-464.

36. T. Muir, 1933. *A Treatise on the Theory of Determinants*, Longmans, London (reprinted, Dover, New York, 1960).

*37. K. Reuter, 1985. "Counting formulas for glued lattices," *Order* 1, 265-276.

*38. K. Reuter, pre. "A matching result for modular lattices," preprint.

*39. K. Reuter and I. Rival, pre. "Subdiagrams equal in number to their duals," preprint.

*40. I. Rival, 1974. Contributions to combinatorial lattice theory, Doctoral thesis, Univ. Manitoba, Winnipeg, Manitoba.

*41. I. Rival, 1976. " Combinatorial inequalities for semimodular lattices of breadth two," *Algebra Universalis* 6, 303-311.

42. G.-C. Rota, 1964, "On the foundations of combinatorial theory. I. Theory of Möbius functions," *Zeit. Wahrscheinlichkeittheorie und Verw. Gebeite* 2, 340-368.

43. L. Shepp (ed.), 1982. *Computed Tomography*, Proc. Symp. Appl. Math., Vol. 27, Amer. Math. Soc, Providence, Rhode Island.

44. K. J. C. Smith, 1969. "On the p-rank of the incidence matrix of points and hyperplanes in finite projective geometry," *J. Combin. Theory* 7, 122-129.

45. R. P. Stanley, 1985. "Reconstruction from vertex-switching," *J. Combin Theory, Ser. B* 38, 132-138.

*46. D. R. Woodall, 1976. The inequality $b \geq v$, in *Proc. Fifth British Combinatorial Conf. (Univ. Aberdeen, Aberdeen, 1975), Congressus Numerantium, No. 15*, Utilitas Math., Winnipeg, Manitoba, pp. 661-664.

DEPARTMENT OF MATHEMATICS
NORTH TEXAS STATE UNIVERSITY
DENTON, TEXAS 76203

Contemporary Mathematics
Volume 57, 1986

RECURSIVE ORDERED SETS

Henry A. Kierstead[1]

§ 1 Introduction

Many mathematicians working in finite combinatorics are highly suspicious of infinite sets and structures. This suspicion does not extend to all infinite structures. The set of natural numbers with the usual order is considered quite acceptable and, on the whole, so is the set of real numbers with its usual order. The problems arise when some unusual structure is placed on these sets as in well ordering the real numbers. Other examples arise from extending theorems about finite structures to theorems about infinite structures. Once Dilworth's Theorem has been proven for finite ordered sets it is routine to extend it to infinite ordered sets of finite width via the Compactness Theorem. However the resulting structure may be quite unusual. The proof only shows the existence of a chain cover; it does not produce the chain cover. This is the source of the uneasiness in both cases: The structures whose existence is asserted have not been shown to be available for inspection.

Rather than ignoring these existence results one should go one step further and ask whether there is a satisfactory structure available for inspection. This leads to further questions. What does it mean to be available for inspection? How would you show that there was no satisfactory structure available for inspection? For countable structures these concerns can be made precise and can often be resolved using the theory of recursive (i.e. computable) functions. The resulting combinatorial theory is the subject of this article, which in particular will concentrate on antichain covers, chain covers, and realizers of recursive ordered sets.

Let us agree that a structure is available for inspection if it is recursive, where, roughly speaking, a relational structure **P** is recursive if

----

1980 Mathematics Subject Classification. Primary 03D45.
[1] Supported by ONR grant N00014-85K-0494

there is a finite algorithm that will provide yes-no answers to questions of the form: "Is x in the domain of **P**?" and "Does $\bar{x}$ hold for the relation R?" We shall see that the uneasiness resulting from the use of the Compactness Theorem is well founded, for there exist recursive ordered sets (necessarily infinite) with finite width w that cannot be covered by w recursive chains. Two reasonable responses to this negative result are the traditional recursion theoretic response of analyzing the "degree of non-computability" of such a cover and the more recent combinatorial response, which we shall pursue, of searching for a recursive chain cover that uses some finite number of additional chains. Thus we are asking whether there exists a function b such that every ordered set with finite width w, which is available for inspection, has a chain cover consisting of b(n) chains, which is available for inspection. From a finite point of view this approach leads to a very satisfying theory. There can be no doubt that the structures under consideration really exist. They are essentially finite since they can be stored using only finitely much space by storing the finite algorithms that describe them. They are amenable to the type of explicit construction often used on finite structures. The results we prove are quite interrelated. Recursive antichain covers and recursive realizers are used to construct recursive chain covers while recursive chain covers are used to construct recursive realizers. The proofs rely most heavily on algorithmic techniques and the theory of finite ordered sets. Indeed, essentially all the recursion theory needed can be reduced to two lemmas. The goal of this article is to present a collection of results, proofs, and problems on recursive ordered sets in a unified setting, which is accessible to discrete mathematicians with little or no background in recursion theory.

For the most part our notation is standard; a few special conventions are mentioned here. The set of natural numbers is represented by **N**. Incomparability in ordered sets is denoted by $||$. The restriction of a structure $(A,\bar{R})$ to the substructure generated by $B \subset A$ is denoted by $(A,\bar{R})\restriction B$.

§ 2 **K** - **L** Expansion Theorems

In this section we use the language of general relational structures to define the notion of an expansion theorem. Next we review the basics of elementary recursion theory and present the notion of a recursive expansion theorem. Finally we introduce expansion games and prove two lemmas which reduce the proofs of recursive expansion theorems to finding winning strategies in expansion games. These games are interesting in their own right and are related to on line algorithms.

A (relational) structure is a system $\mathbf{A} = (A,\bar{R})$ where $\bar{R}$ is a sequence of relations on the domain A. To avoid certain technical difficulties we will only consider structures with finitely many relations. Structures will be denoted with bold face letters; the same letter in standard form will denote the domain of the structure. Two structures $(A,\bar{R})$ and $(B,\bar{S})$ are similar provided that $\bar{R}$ and $\bar{S}$ have the same length and corresponding entries in $\bar{R}$ and $\bar{S}$ have the same rank. The structure $(A,R_1,\ldots,R_m,\ldots,R_n)$ is an expansion of the structure $(A,R_1,\ldots,R_m)$. Let $\mathbf{K}$ be a class of similar structures and let $\mathbf{L}$ be another class of similar structures. A $\mathbf{K} - \mathbf{L}$ expansion theorem is an assertion of the form: Every structure in $\mathbf{K}$ has an expansion in $\mathbf{L}$. Dilworth's Theorem is an example of a $\mathbf{K} - \mathbf{L}$ expansion theorem. Matching theorems, coloring theorems, and dimension theorems for ordered sets also fall into this category.

There are many equivalent ways to define the concept of a partial recursive or computable function. For our purposes it will be enough to say that a partial recursive function ϕ is a function that can be computed by an algorithm of finite length - if x is in the domain of ϕ then the algorithm eventually produces the output $\phi(x)$ when started on the input x; if x is not in the domain of ϕ the algorithm produces no output when started on the input x. A recursive set is a set whose characteristic function is recursive. Notice that every finite set is recursive. A precise definition of these concepts can be found in Machtey and Young [1978]. The domain of a partial recursive function may not be recursive; if it is we will say that the function is recursive. This is a slight variance from the normal definition. A structure $\mathbf{A} = (A,\bar{R})$ is recursive if there is a recursive function ϕ such that both $\phi(0,x) = 1$ iff $x \in A$ and $\phi(R,\bar{x}) = 1$ iff $\bar{x} \in R$, whenever R is in $\bar{R}$. We say that ϕ defines $\mathbf{A}$. If $\mathbf{A}$ is a recursive structure the domain of $\mathbf{A}$ and each of its relations are recursive sets. To test whether x is in A, first check whether $(0,x)$ is in the recursive set $\mathrm{dom}(\phi)$ and if it is check to see whether $\phi(0,x) = 1$. It is also easy to see that if each of the relations of $\mathbf{A}$ is recursive and the domain of $\mathbf{A}$ is recursive then $\mathbf{A}$ is recursive. Thus the domain and each of the relations of a recursive structure are completely described by a single algorithm of finite length. A recursive $\mathbf{K}$-$\mathbf{L}$ expansion theorem is an assertion of the following form: Every recursive structure in $\mathbf{K}$ has can be expanded to a recursive structure in $\mathbf{L}$.

It is possible to enumerate the countably many possible algorithms. Let ϕ_e be the partial recursive function computed by the eth algorithm in such a list. The execution of an algorithm on a particular input occurs in a step by step manner. If the execution of the eth algorithm on x halts after at most n steps and outputs y we say that $\phi_e^n(x)$ is defined and equals y;

otherwise we say that $\phi_e^n(x)$ is undefined. With special care (Kleene [1936], Turing [1936]) we can arrange the list so that the predicate $T = \{(e,x,n): \phi_e^n(x)$ is defined$\}$ is recursive. Thus in writing algorithms we can use statements of the form: if $\phi_e^n(x) = y$ then ... . In general we can not use statements of the form: if $\phi_e(x) = y$ then..., since there is no algorithm that will decide for arbitrary e whether the eth algorithm eventually produces output when started on the input x. For further details the reader should consult Machtey and Young [1978].

The **K** – **L** <u>expansion game</u> is an infinite game played by two players, the **K**-player and the **L**-player, in the following fashion: The play alternates between the two players with the **K**-player moving first. At the start of the i+1st round of play the **K**-player will be confronted with a finite structure $\bar{\mathbf{B}} = (B,R_1,\ldots,R_m,\ldots,R_n)$ in **L**, which is the expansion of a **K**-structure $\mathbf{B} = (B,R_1,\ldots, R_m)$. The **K**-player must define new relations $R_1^+,\ldots,R_m^+$ on $B^+ = B \cup \{i\}$ such tha $\mathbf{B}^+ = (B^+,R_1^+,\ldots,R_m^+)$ is in **K** and **B** is a substructure of $\mathbf{B}^+$, i.e., $\bar{x} \in R_i$ iff $\bar{x} \in R_i^+$ for all $\bar{x}$ in B and $1 \leq i \leq m$. The **L**-player responds by defining new relations $R_{m+1}^+,\ldots,R_n^+$ on B^+ such that $\bar{\mathbf{B}}^+ = (B^+,R_1^+,\ldots,R_m^+,\ldots,R_n^+)$ is in **L** and $\bar{\mathbf{B}}$ is a substructure of $\bar{\mathbf{B}}^+$. The **K**-player <u>wins</u> a **K** – **L** expansion game if after finitely many rounds of play the **L**-player has no legal response; the **L**-player wins if the game continues for infinitely many rounds or in the unlikely event that the **K**-player cannot play legally.

A <u>strategy for the **K**-player</u> is a function S which assigns to each finite **L**-structure $\bar{\mathbf{B}}$ with $B = \{0,1,\ldots,i-1\}$ a structure $S(\bar{\mathbf{B}}) = \mathbf{B}^+$. The strategy S is a <u>winning strategy</u> if regardless of how the **L**-player responds the **K**-player can eventually win the **K** – **L** expansion game by continuing to play $S(\bar{\mathbf{B}})$ when confronted with $\bar{\mathbf{B}}$. Similarly a <u>strategy</u> for the <u>**L**-player</u> is a function S that assigns to each pair of structures $\bar{\mathbf{B}}$ in **L** and $\mathbf{B}^+$ in **K** such that $B^+ = B \cup \{i\}$ a structure $S(\bar{\mathbf{B}},\mathbf{B}^+) = \bar{\mathbf{B}}^+$. The strategy S is a <u>winning strategy</u> if regardless of how the **K**-player moves the **L**-player wins by always responding with $S(\bar{\mathbf{B}},\mathbf{B}^+)$ when confronted with $\bar{\mathbf{B}}$ and $\mathbf{B}^+$.

It is a consequence of König's Lemma that if the **K**-player has a winning strategy for the **K** – **L** expansion game then there is a fixed n such that regardless of how the **L**-player responds the **K**-player can win in at most n rounds: Consider the tree, whose nodes are the **L**-structures that can start some round of a **K** – **L** expansion game in which the **K**-player follows his winning strategy, and whose nodes are ordered by the substructure relation. This tree is finitely branching. If the **L**-player could force arbitrarily long games, then the tree would be infinite, and by König's Lemma would contain an infinite branch, which would correspond to a winning game for the

L-player playing against the **K**-player's winning strategy. Thus if the **K**-player has a winning strategy that strategy can be chosen to be finite and thus recursive. On the other hand, it is conceivable that the **L**-player has a winning strategy, which is so complex that it cannot be described by an algorithm.

The remainder of this section is devoted to reducing the recursive theorems to follow to game theoretic results. In order to state the crucial lemmas we need to define several operations on structures. The <u>union of a chain</u> of similar substructures $(A_0,\bar{R}_0) \subset (A_1,\bar{R}_1) \subset \ldots$, denoted by $(A,\bar{R}) = \bigcup_{i\varepsilon N}(A_i,\bar{R}_i)$, is defined by $A = \bigcup_{i\varepsilon N} A_i$ and $\bar{x} \varepsilon R$ iff $\bar{x} \varepsilon R_i$ for some $i \varepsilon N$, where R and R_i are corresponding relations in $\bar{R}$ and $\bar{R}_i$. Let $\mathbf{A} = (A,\bar{R})$ be a structure and f be a bijection from A to B. The <u>isomorphic image</u> of $\mathbf{A}$ under f is the structure $\mathbf{B} = (B,\bar{S})$ defined by $(x_1,\ldots x_r) \varepsilon R$ iff $(f(x_i),\ldots,f(x_r)) \varepsilon S$, where R and S are corresponding relations in $\bar{R}$ and $\bar{S}$. Let $\mathbf{A} = (A,\bar{R})$ be a structure and suppose that for each $a \varepsilon A$, $\mathbf{B}_a = (B_a,\bar{S}_a)$ is a structure similar to $\mathbf{A}$ such that $B_a \cap B_{a'} = \emptyset$ if $a \neq a'$. The <u>sum over</u> $\mathbf{A}$ of the $\mathbf{B}_a$, denoted by $(B,\bar{S}) = \sum_{a\varepsilon \mathbf{A}} \mathbf{B}_a$ is defined by $B = \bigcup_{a\varepsilon A} B_a$ and $\bar{x} = (x_1,\ldots, x_n)\varepsilon S$ iff either $\bar{x} \subset B_a$ and $\bar{x}\varepsilon S_a$, for some $a\varepsilon A$ or $\bar{x} \not\subset B_a$ for any $a\varepsilon A$, $x_i \varepsilon B_{a(i)}$, and $(a(1),\ldots,a(n))\varepsilon R$, where S, S_a, and R are corresponding relations in $\bar{S}$, $\bar{S}_a$, and $\bar{R}$. Let **K** be a class of similar structures and let $\mathbf{A}$ be a structure similar to the structures in **K**. We say that **K** is closed under the formation of, respectively, substructures, unions of chains, isomorphic images, and sums over $\mathbf{A}$, provided $\mathbf{B}$ is in **K** whenever $\mathbf{B}$ is, respectively, a substructure of a structure in **K**, a union of a chain of substructures in **K**, the isomorphic image of a structure in **K**, or a sum over $\mathbf{A}$ of structures in **K**.

<u>Lemma</u> 1. Suppose **K** is a class of similar structures closed under the formation of substructures and isomorphic images and **L** is a class of similar structures closed under the formation of unions of chains. If the **L**-player has a recursive winning strategy for the **K** - **L** expansion game, then every recursive structure in **K** can be expanded to a recursive structure in **L**.

<u>Proof</u>. Let $\mathbf{B} = (B,\bar{R})$ be a recursive structure in **K**. The recursive expansion $\bar{\mathbf{B}}$ of $\mathbf{B}$ is obtained from a run of the **K**-**L** expansion game. For notational simplicity, using that **K** is closed under the formation of isomorphic images, assume that $B = N$, and suppose that at each round s the **K**-player plays $\mathbf{B}^s = \mathbf{B}\lceil\{0,1,\ldots,s-1\}$ while the **L**-player uses his recursive winning strategy to respond with $\bar{\mathbf{B}}^s$. Since **K** is closed under the formation of substructures each play by the **K**-player is legal and the game continues until each point of B has been enumerated. The resulting structure $\bar{\mathbf{B}}$ is clearly

an expansion of **B**. Since $\bar{\mathbf{B}}$ is the union of the chain of structures $\bar{\mathbf{B}}^s$ played by the **L**-player and **L** is closed under unions of chains, $\bar{\mathbf{B}}$ is in **L**. The domain B is recursive since **B** is. Since **B** and **S** are recursive one can effectively calculate the position of the game after any finite number of rounds. Thus we can effectively determine whether or not any of the relations of $\bar{\mathbf{B}}$ hold for a sequence $\bar{x}$ by allowing the game to run until each point in $\bar{x}$ has appeared. Hence $\bar{\mathbf{B}}$ is recursive.▒

Lemma 2. Suppose that **A** is an infinite recursive structure, **K** is a class of structures similar to **A** such that **K** is closed under the formation of isomorphic images and sums over **A**, and **L** is a class of similar structures closed under the formation of isomorphic images and substructures. If the **K**-player has a winning strategy in the **K** – **L** expansion game then there is a recursive structure in **K** that cannot be expanded to a recursive structure in **L**.

Proof. Let **S** be a finite winning strategy for the **K**-player in the **K** – **L** expansion game and p be a recursive bijection from **N** to **N** × **N**. Without loss of generality assume that A = **N**. Like a chess master, we shall simultaneously play infinitely many **K**-**L** expansion games against every possible recursive strategy for the **L**-player, using the **K**-players winning strategy. The order in which we visit the various boards will be governed by the function p. The result will be a recursive structure **B** in **K** which cannot be expanded to a recursive structure in **L**.

Algorithm for **B**: The algorithm proceeds in stages. At stage s we construct $\mathbf{B}^s$. After ω stages we let $\mathbf{B} = \bigcup_{s \in N} \mathbf{B}^s$.
Stage 0. Set $f_e^0 = \emptyset = \mathbf{B}_e^0$ for all $e \in \mathbf{N}$.
Stage s+1. Suppose p(s) = (e,i). If $d \neq e$ set $f_d^{s+1} = f_d^s$ and $\mathbf{B}_d^{s+1} = \mathbf{B}_d^s$. If ϕ_e^s defines an expansion of $\mathbf{B}_e^s$ to a structure $\bar{\mathbf{B}}_e^s$ whose preimage $\bar{\mathbf{C}}_e^s$ under f_e^s is in **L** set $f_e^{s+1} = f_e^s \cup \{(|B_e^s|, s)\}$ and $\mathbf{B}_e^{s+1} = f_e^{s+1}[\mathbf{S}(\bar{\mathbf{C}}_e^s)]$; otherwise set $f_e^{s+1} = f_e^s$ and $\mathbf{B}_e^{s+1} = \mathbf{B}_e^s$. Let $\mathbf{B}_e^{s+1} = \sum_{e \in \mathbf{N}} \mathbf{B}^{s+1}$.

The predicate "ϕ_e^s defines an expansion of $\mathbf{B}_e^s$, whose preimage $\bar{C}_e^s$ under f_e^s is in **L**" is easily seen to be recursive since $\mathbf{B}_e^s$ is finite, "$\phi_e^s(x)$ is defined" is a recursive predicate, and if $\bar{C}_e^s$ is in **L** then $\bar{C}_e^s$ is one of the only finitely many legal positions which can be reached during a run of the **K** – **L** expansion game with the **K**-player using his finite winning strategy **S**. Using also that **A** is recursive we see that we can effectively construct each $\mathbf{B}^s$. Thus as in the proof of Lemma 1, $\mathbf{B} = \bigcup_{s \in N} \mathbf{B}^s$. is recursive.

To see that $\mathbf{B}$ is a K-structure note that $\mathbf{B} = \bigcup_{s\varepsilon N} \Sigma_{e\varepsilon \mathbf{A}} \mathbf{B}_e^s = \Sigma_{e\varepsilon \mathbf{A}} \bigcup_{s\varepsilon N} \mathbf{B}_e^s$ and that for each e there exists a stage t such that $\bigcup_{s\varepsilon N} \mathbf{B}_e^s = \mathbf{B}_e^t$. Set $\mathbf{B}_e = \mathbf{B}_e^t$. Since $\mathbf{K}$ is closed under the construction of images of isomorphisms and sums over $\mathbf{A}$, each $\mathbf{B}_e$ is in $\mathbf{K}$ and thus $\mathbf{B}$ is in $\mathbf{K}$.

Finally suppose that $\bar{\mathbf{B}}$ were a recursive expansion of B. Then $\bar{\mathbf{B}}$ would be defined by some partial recursive function ϕ_e, which would also define an expansion $\bar{\mathbf{B}}_e$ of $\mathbf{B}_e$. Since $\mathbf{L}$ is closed under the formation of substructures and images of isomorphisms, $\bar{\mathbf{B}}_e$ and its preimage $\bar{\mathbf{C}}_e$ would be in $\mathbf{L}$, which would contradict $\bigcup_{s\varepsilon N}\mathbf{B}_e^s = \mathbf{B}_e$. ▒

§ 3 Recursive Antichain Covers

We begin our study of recursive ordered sets with recursive antichain covers, since as one might expect from the finite case, the results are relatively easy to prove. Every ordered set of finite height (the number of points in a maximum chain) h can be covered by h antichains. Schmerl proved the following recursive version of this result.

Theorem 1.(Schmerl [1979]) Every recursive ordered set of finite height h can be covered by $\binom{h+1}{2}$ recursive antichains; moreover for all positive integers h there are recursive ordered sets of height h that cannot be covered by less than $\binom{h+1}{2}$ recursive antichains.

Proof. Let $\mathbf{K}$ be the class of ordered sets $\mathbf{P} = (P,R)$ of height at most h and let $\mathbf{L}$ be the class of all structures of the form $\bar{\mathbf{P}} = (P,R,A_1,\ldots,A_n)$ where (P,R) is in $\mathbf{K}$, $(A_1,\ldots,A_n)$ is an antichain cover of P, and $n = \binom{h+1}{2}$. Note that $\mathbf{K}$ and $\mathbf{L}$ satisfy the hypothesis of Lemma 1. To prove the first part of the theorem we provide the $\mathbf{L}$-player with a recursive winning strategy for the $\mathbf{K}$-$\mathbf{L}$ expansion game. Relabel the antichains $A_1,\ldots,A_n$ as $\ldots A_{d,u}\ldots$ where $0 \leq d+u \leq h-1$. When confronted with a new point i in the finite ordered set $\mathbf{P}^+$ the L-player should put i in the antichain $\mathbf{A}_{d,u}^+$, where d is the length of the longest chain in $\mathbf{P}^+$ strictly below i and u is the length of the longest chain in $\mathbf{P}^+$ strictly above i. This is a recursive winning strategy: We have (informally) provided the finite algorithm for computing the next move from any position, so the strategy is recursive. To see that it wins, it suffices to show that each $A_{d,u}^+$ is still an antichain. Suppose i is comparable to x and $x \varepsilon A_{d,u}$, say $i < x$. Then there is a chain of $u + 1$ elements strictly above x. Thus $i \not\varepsilon A_{d,u}^+$.

The second part of the theorem is contained in Theorem 2, a stronger (and later) result of Szemerédi, which will also be used in the proof of Theorem 4.▒

Before presenting Theorem 2, we remark that some care was necessary in designing the winning strategy for the L-player. Consider the greedy strategy: Put i in the antichain A_j^+ of least index such that $A_j \cup \{i\}$ is still an antichain. This is not a winning strategy for the L-player even if we allow him extra antichains. To see this, we show that for each positive n there is a sequence of plays S_n^+ (S_n^-) for the K-player ending with a structure $\mathbf{P}_n^+$ $(\mathbf{P}_n^-)$ of height 2 such that the last element x_n is maximal (minimal) and if the L-player uses the greedy strategy then x_n is put in A_n. The argument is by induction on n. Figure 1 illustrates the inductive step.

A collection of linear orders $\bar{L} = (L_1, \ldots, L_d)$ is a realizer of an ordered set $\mathbf{P} = (P,R)$ provided $x<y$ in R iff for all i, $x<y$ in L_i. $\bar{L}$ is a recursive realizer if the structure $(P,R,\bar{L})$ is recursive. The (recursive) dimension of $\mathbf{P}$ is the least number of linear extensions in a (recursive) realizer of $\mathbf{P}$.

Theorem 2. (Szemerédi [1982]) For every positive integer h there exists a recursive ordered set with height h and recursive dimension 2 which cannot be covered by fewer than $\binom{h+1}{2}$ recursive antichains.

Proof. Let $\mathbf{K}_h$ be the class of structures $\mathbf{P} = (P,R,L_1,L_2)$ where (P,R) is an ordered set of height at most h and (L_1,L_2) is a realizer of (P,R); let $\mathbf{L}_h$ be the class of structures $\bar{\mathbf{P}} = (P,R,L_1,L_2,A_1,\ldots,A_n)$ such that (P,R,L_1,L_2) is in $\mathbf{K}_h$ and $(A_1,\ldots,A_n)$ is an antichain cover of P, where $n = \binom{h+1}{2}-1$; and let $\mathbf{A} = (A,S,M_1,M_2)$ where $A = \mathbf{N}$, S is the empty order, M_1 is the natural order on $\mathbf{N}$ and M_2 is the dual of M_1. Then $\mathbf{K}_h$, $\mathbf{L}_h$ and $\mathbf{A}$ satisfy the hypothesis of Lemma 2. To prove the theorem we provide a winning strategy for the $\mathbf{K}_h$-player in the $\mathbf{K}_h$ - $\mathbf{L}_n$ expansion game by recursion on h. If h=1 the $\mathbf{L}_1$-player cannot respond to any move, so assume h = g+1 and let S_g be a winning strategy for the $\mathbf{K}_g$-player in the $\mathbf{K}_g$ - $\mathbf{L}_g$ expansion game.

Note that $\binom{h+1}{2} - \binom{g+1}{2} = h$. The $\mathbf{K}_h$-players strategy consists of two stages. The strategy creates a structure $\mathbf{P} = (P,R,L_1,L_2)$, with $P = P_1 \cup P_2$, where P_1 is the set of points played during the first stage and P_2 is the set of points played during the second stage. The goal of the first stage is to arrange that the top h points, T, of P_1 in L_1 are all incomparable in R, while tricking the L-player into putting the elements of T into h distinct antichains. The goal of the second stage is to arrange that: $(P_2,R\lceil P_2)$ has height g; every element of P_2 is under every element of T in R and incomparable in R to every element of $P_1 - T$; and the $\mathbf{L}_h$-player is forced to use $\binom{g+1}{2}$ distinct antichains to cover P_2. Then the **L** player will be forced to use $h + \binom{h}{2}$ antichains in all and loose.

The second stage is easy. The **K**-player simply follows the strategy $\mathbf{S}_g$ (modified for the points in P_2). At the same time he relates each new point i in P_2 to the points in P_1 by putting i over all the points of $P_1 - T$ and under all the points of T in L_1^+ and putting i under all the points of P_1 in L_2^+.

It remains to explain the $\mathbf{K}_h$-players strategy for the first stage. Suppose the points Q_r from P_1 have been played so that $\mathbf{P}\lceil Q_r$ has height at most r, the top r points T_r of Q_r in L_1 are all incomparable in R, and the $\mathbf{L}_h$-player has used exactly r antichains to cover Q_r, each of which contains an element of T_r. The **K**-player should add new points $c_1, \ldots, c_i, \ldots, c_j$ to P_1 one at a time, so that c_i is the r + 1st point in L_1^+ and the top point in L_2^+, until the $\mathbf{L}_h$-player puts $c_i = c_j$ into a new antichain. Note that this will occur at or before a point at which c_j is in a chain of length r + 1 and that such a point will be reached since the set $\{c_1, .., c_i\}$ is a chain. Continuing in this manner, after replacing r by r+1, Q_r by $Q_r \cup \{c_1, \ldots, c_j\}$, and T_r by $T_r \cup \{c_j\}$, the **K**-player will eventually obtain the goal of the first stage.▒

It is reasonable to ask whether the order relation or just its comparability graph is needed for the hypothesis of Theorem 1, i.e., can every recursive comparability graph with clique size h be recursively $\binom{h+1}{2}$ colored? An early result of Bean [1976] shows that this is not the case - there is a recursive comparability graph of clique size 2 that cannot be recursively k-colored for any finite k. However the full strength of the order relation is not needed to obtain a finite bound on the recursive chromatic number of the structure. Chvatal [*] proved that any digraph without odd cycles C_n, n>3, or induced subgraphs of the form o→o→o←o is perfect. Such digraphs extend the class of ordered sets. The next theorem is a recursive version of Chvatal's result. First we state a combinatorial lemma which is a slight variation of the key lemma in Chvatal [1981].

<u>Lemma</u> 3. Let $\mathbf{D} = (D, \rightarrow)$ be a digraph which induces neither a directed 3-cycle nor a the digraph o→o→o←o. Let K be a clique and I an independent set in **D** such that for each $k \varepsilon K$ there is $i \varepsilon I$ such that $i \rightarrow k$. Then there is some $i \varepsilon I$ such that $K \cup \{i\}$ is a clique in **D**.▒

<u>Theorem</u> 3. (Kierstead [1984]) If G is a recursive digraph that contains neither a directed 3-cycle nor o→o→o←o nor o←o→o→o as induced

subgraphs then G is recursively 2^n-1-colorable, where $n = \omega(G)$, the clique size of G.

Proof. Let $\mathbf{K}$ be the class of digraphs $\mathbf{D} = (D,\to)$ with clique size n that contain neither a directed 3-cycle nor $o\to o\to o\leftarrow o$ nor $o\leftarrow o\leftarrow o\to o$ and let $\mathbf{L}$ be the class of structures $\bar{\mathbf{D}} = (D,\to,I_1,\dots,I_t)$, where $t = 2^n - 1$, such that $(D,\to)$ is in $\mathbf{K}$ and $(I_1,\dots,I_t)$ is a covering of D by independent sets. It suffices to show that the $\mathbf{L}$-player has a recursive winning strategy in the $\mathbf{K}$-$\mathbf{L}$ expansion game. Relabel $I_1,\dots,I_t$ as $\dots I_\sigma \dots$, where σ is a (possibly empty) sequence of zeros and ones of length at most $n-1$. Suppose the $\mathbf{L}$-player is confronted with $\bar{\mathbf{D}}$ and $\mathbf{D}^+$, where $D^+ = D\cup\{i\}$. He should put i in I_σ^+, where σ is the lexicographically least sequence such that $I_\sigma \cup \{i\}$ is independent and for each proper initial subsequence τ of σ there exists $y \varepsilon I_\tau$ such that $i\to y$ if $\tau\hat{}0$ is an initial subsequence of σ and $i\leftarrow y$ if $\tau\hat{}1$ is an initial subsequence of σ. Notice that if we do not restrict the length of σ, then such a sequence will exist: If $I_\emptyset \cup \{i\}$ is not independent then either $i \to y$ or $i \leftarrow y$ for some $y \varepsilon I_\emptyset$, say $i \to y$. Then try $I_{\langle 0\rangle} \cup \{i\}$. Suppose this set is not independent, say $i \leftarrow y$ for some $y \varepsilon I_{\langle 0\rangle}$. Next try $I_{\langle 0,1\rangle} \cup \{i\}$. Continuing in this fashion, eventually an acceptable σ will be found of length at most i.

To show that this strategy wins it is enough to prove that if i is put into I_σ^+ where $\sigma = \langle s_1,\dots,s_m\rangle$, then there is an $m+1$-clique $K=\{x_1,\dots,x_{m+1}=i\}$ such that (*) $x_j \varepsilon I_{\tau^j}$, where τ^j is the initial subsequence of σ of length $j-1$, for $1\leq j\leq m+1$. We construct K by reverse recursion on j. Suppose that the clique $\{x_j,\dots,x_{m+1}\}$, $j>1$, has been constructed and satisfies (*). Then for each x_k, $j\leq k\leq m+1$, there exists $z_k \varepsilon I_{\tau^{j-1}}$ such that $x_k \to z_k$ if $s_{j-1}=0$ and $x_k \leftarrow z_k$ if $s_{j-1} = 1$. Thus by Lemma 3 or its dual version there exists z in the independent subset $\{z_k : j\leq k\leq m+1\}$ of $I_{\tau^{j-1}}$ such that $\{z,x_j,\dots,x_{m+1}\}$ is a clique satisfying (*).▓

It is not known whether the bound in Theorem 3 is best possible. By Theorem 2, it must be at least $\binom{h+1}{2}$.

§4 Recursive Chain Covers

Covering recursive ordered sets with recursive chains presents a considerably more difficult problem. The author showed the existence of an exponential function b, such that any recursive ordered set of width w can be covered by $b(w)$ recursive chains, and that no such function was linear. Later Szemerédi proved that any such function is at least quadratic.

Theorem 4. (Kierstead [1981a] and Szemerédi [1982]) Every recursive ordered set of finite width w can be covered by $(5^w - 1)/4$ recursive chains; moreover for all positive integers w there exist recursive ordered sets with width w that cannot be covered by fewer than $\binom{w+1}{2}$ recursive chains.

Proof. Theorem 2 was formulated specifically to prove the second part of Theorem 4. Let $\mathbf{P} = (P,R)$ be a recursive ordered set of height w which has a recursive realizer (L_1,L_2) and which cannot be covered by fewer than $\binom{w+1}{2}$ recursive antichains. Define $\hat{\mathbf{P}}$ to be $(P,L_1 \cap L_2^*)$, where L_2^* is the dual of L_2. Since L_1 and L_2 are recursive, L_2^* and $\hat{\mathbf{P}}$ are also recursive. Any subset of P is a chain (antichain) in $\hat{\mathbf{P}}$ iff it is an antichain (chain) in $\mathbf{P}$. Thus the width of $\hat{\mathbf{P}}$ is w and $\hat{\mathbf{P}}$ cannot be covered by fewer than $\binom{w+1}{2}$ recursive chains.

The proof of the first part of the theorem is more complicated. Here we give the main line of the argument, but leave many of the details for the reader to check. Let $\mathbf{K}_w$ be the class of ordered sets $\mathbf{P}$ of width at most w and let $\mathbf{L}_w$ be the class of structures $\bar{\mathbf{P}} = (P,R,C_1,\ldots,C_n)$, where $n = (5^w-1)/4$ and $(C_1,\ldots,C_n)$ is a chain cover of P. Both $\mathbf{K}_w$ and $\mathbf{L}_w$ satisfy the hypothesis of Lemma 1. Thus it suffices to provide a winning strategy for the $\mathbf{L}_w$-player. The strategy is defined by recursion on w. When $w=1$ we ca nnot go wrong, so consider the step $w = v+1$. Let $\mathbf{S}_v$ be a winning strategy for the $\mathbf{L}_v$-player. First the $\mathbf{L}_w$-player trys to put the new point i in a distinguished maximal chain B. If this is impossible i is put in a set A on which a width v order S extending R has been defined. This order is extended to i. Next the $\mathbf{L}_v$-player's strategy is used to put i in one, say D_j, of $m = (5^v-1)/4$ S-chains $D_1,\ldots,D_m$. Finally, using special properties of the order S, i is put in one of five R-chains $C_{j,1},\ldots,C_{j,5}$, which cover D_j. To give a precise description of this strategy we define a more involved expansion game. Note that $(5^w-1)/4 = 1+5(5^v-1)/4$ and relabel the chains $C_1,\ldots,C_n$ as $B,C_{1,1},\ldots,C_{1,5},\ldots,C_{m,1},\ldots,C_{m,5}$. Let $\mathbf{L}$ be the class of structures of the form $\mathbf{P} = (P,R,A,S,D_1,\ldots D_m,B,C_{1,1},\ldots, C_{m,5},\sim)$ such that:

(0) $(P,R) \in \mathbf{K}_w$;

(1) (A,B) is a partition of P such that

- (a) B is a maximal R-chain,
- (b) $R \restriction A$ is a subset of $S \restriction A$,
- (c) $(A,S,D_1,\ldots, D_m) \in \mathbf{L}_v$, and
- (d) if $x<z$ in S and $x||b||z$ in R, for some $b \in B$ which was played before the latter of x and z, then $x<z$ in R, and moreover, if $x<y<z$ in S, then $y||b$ in R.

(2) $(A,\sim)$ is an equivalence relation such that:

(a) if $x \sim y$ then x is comparable to y in R and

(b) the equivalence classes of $(A,\sim)\restriction D_j$ are convex in S, for $1 \le j \le m$, i.e. if $x,y,z \in D_j$, $x \sim z$, and $x<y<z$ in S, then $x \sim y$;

(3) for $1 \le j \le m$, $(C_{j,1},\ldots,C_{j,5})$ is a chain cover of D_j in R such that if $x \sim y$ then $x \in C_{j,k}$ iff $y \in C_{j,k}$, for $1 \le k \le 5$.

Any winning strategy S for the L-player in the K_w-L expansion game can be trivially modified to a winning strategy for the L_w-player in the K_w-L_w expansion game, since $(B,C_{1,1},\ldots,C_{m,5})$ is a chain cover of P in R. We present such a strategy below.

Suppose that after the ith round of play the K-player is confronted with $\hat{P}$ in L and plays P^+ in K_w, where $P^+ = P \cup \{i\}$. The L-player should respond with $\hat{P}^+$ as follows.

Step 1: Construction of $A^+,B^+,S^+,D_1^+,\ldots,D_m^+$. If $B \cup \{i\}$ is a chain in R^+ put i into B^+; otherwise put i into A^+ and let $x<i$ $(i<x)$ in S^+ iff at least one of the following holds:

(4) $x<i$ $(i<x)$ in R^+;

(5) for all $b, c \in B$, if $b||x$ and $c||i$ in R then $b<c$ $(c<b)$ in R;

(6) there exists $a \in A$ such that $x<a$ $(a<x)$ in S and (4) or (5) holds when x is replaced by a.

Once (A^+,S^+) has been played view it as a play by the K_v-player in the K_v-L_v expansion game and put i into D_j^+, where j is chosen by the winning strategy S_v.

Step 2: Construction of $\sim^+$, $C_{1,1}^+,\ldots,C_{m,5}^+$. Suppose that $i \in D_j^+$. Let i^- be the largest element of D_j less than i in S^+ and i^+ be the smallest element of D_j greater than i in S^+. Extend $\sim$ to $\sim^+$ by:

(7) i is in the equivalence class of i^- if there exists $b \in B$ such that $i||b||i^-$ in R^+;

(8) i is in the equivalence class of i^+ if (7) does not apply and there exists $b \in B$ such that $i||b||i^+$ in R^+;

(9) i is in a new equivalence class if neither (7) nor (8) apply.

Notice the bias in favor of (7) over (8), which will be important later. If $i \sim^+ x$ and $x \in C_{j,k}$ put i into $C_{j,k}^+$; otherwise put i into $C_{j,e}^+$, where e is chosen so that neither the two equivalence classes immediately over i in $(A^+,S^+)\restriction D_j^+$ nor the two equivalence classes immediately below i in $(A^+,S^+)\restriction D_j^+$ are contained in $C_{j,e}$.

To complete the proof it must be verified that $\hat{P}^+$ is indeed in L^+. Most of this verification is left to the reader, as it is tedious, but routine, when done in order. We will however show (1c) and that given (1) and (2), (3) holds. For (1c) the main point that must be checked is that (A^+,S^+) is an ordered set of width at most v. Reflexivity and antisymmetry are easy. For transitivity the only problem arises when $p<i<q$ in S^+. Choose p_0 and q_0 in A so that p_0 is as large as possible in S^+ and q_0 is as small as possible in S^+ subject to $p\leq p_0<i<q_0\leq q$ in S^+. By the transitivity of (A,S) it suffices to show that $p_0<q_0$ in S^+. By the choice of p_0 and q_0, $p_0<i<q_0$ in S^+ by conditions (4) or (5), but not (6). clearly there is no problem if $p_0<i$ and $i<q_0$ in S^+ by the same condition. So suppose $p_0<i$ in S^+ by (4) and $i<q_0$ in S^+ by (5). Then in R^+, i, and thus p_0, is under any element of B which is incomparable to q_0. Thus $p_0<q_0$ in S by (5).

To see that the width of (A^+,S^+) is at most v we show that for any antichain I in S^+ there exists $b\varepsilon B$ such that $I\cup\{b\}$ is an antichain in R^+. We argue by induction on $|I|$. The case $|I| = 1$ is trivial, so suppose $j,k \varepsilon I$. Since $j||k$ in S^+, there exists $b\varepsilon B$ such that $j||b||k$ in R^+. By the inductive hypothesis there exist $c,d \varepsilon B$ such that c is incomparable to every element of $I-\{j\}$ in R^+ and d is incomparable to every element of $I-\{k\}$ in R^+. If b, c, and d are not distinct we are clearly done; otherwise the middle element of b, c, and d in R^+ is incomparable to every element of I in R^+.

To show that each $C^+_{j,k}$ is a chain, suppose i and x are in $C^+_{j,k}$. If $i\sim^+x$ then by (2a) i is comparable to x in R^+. So suppose that $i\not\sim^+x$. Without loss of generality suppose that $i<x$ in S^+. Choose u and v in D^+_j such that $i<u<v<x$ in S^+, $i\not\sim u\not\sim v\not\sim x$, and u and v are each the first elements of their respective equivalence classes to be played. Since B^+ is maximal there are elements u′ and v′ in B^+ such that $u'||u$ and $v'||v$ in R^+. In fact u′ and v′ can be chosen so that u′ was played before u and v′ was played before v. We claim that $v'<x$ in R: Certainly not $x<v'$ in R since $v<x$ in S. If x were incomparable to v′ then x would be equivalent to v. This uses (1d), the bias for (7), and the fact that v was the first element of its equivalence class played. Similarly $u<v'$. Here, lacking the bias for (7), we need that both u and v are the first elements of their respective equivalence classes. Also $u'<v'$. Finally either $i<u'$ or $i<u$ in R^+: Certainly not $u'<i$ in R^+. If $u'||i$ in R^+ then using (1d) $i<u$ in R^+. In either case $i<x$ in R^+. This is illustrated in Figure 2.▒

Again we note that the greedy strategy **S** fails miserably. Let **K** be the class of ordered sets of width 2 and let $\mathbf{L}_n$ be the class of structures of the form $(P,R,C_1,\ldots,C_n)$ such that $(P,R) \varepsilon$ **K** and $(C_1,\ldots,C_n)$ is a chain cover of (P,R). Figure 3 illustrates a strategy for the **K**-player which will defeat any L_n-player who uses strategy **S**, in the **K** - L_n expansion game. Each point is labeled (i,j) where i is the name of the point and j is the index of the chain C_j into which the greedy strategy puts i.

Closing the enormous gap between the quadratic lower bound and exponential upper bound of Theorem 4 remains an important open problem. In certain special cases more can be said. The author's original argument gives the better lower bound of 5 in the case of width 2. Even here there is a nagging gap between the lower bound and the upper bound of 6. For recursive interval orders the author and Trotter were able to give an exact linear answer. These results are presented below.

Theorem 5. (Kierstead [1981a] There exists a recursive ordered set of width 2 which cannot be covered by fewer than 5 recursive chains.

Proof. Let **K** be the class of ordered sets of width 2; let **L** be the class of structures $(P,R,c_1,\ldots,c_4)$ such that (P,R) is in **K** and $(c_1,\ldots c_4)$ is a chain cover of (P,R); and let $\mathbf{A}=(\mathbf{N},S,\mathbf{N},\ldots,\mathbf{N})$, where S is the natural order on **N**. Then **K**, **L**, and **A** satisfy the hypothesis of Lemma 2. Thus it suffices to provide the **K**-player with a winning strategy in the **K** - **L** expansion game.

We begin by identifying two positions from which the **K**-player can win. These positions are shown in Figure 4(a) and 4(b). The loops represent chains of unspecified length. An upper case letter next to a point indicates, up to permutation, in which chain the **L**-player has placed that point. An uppercase letter next to a loop indicates that at least one point from the loop is in the chain represented by the letter. The point with the arrow represents the winning move. There may be other points entirely above or entirely below all the specified points. Of course the duals of these positions are also winning positions.

The **K**-player can obtain either the first or the second position. First he plays his first $53 = 4\cdot 13 + 1$ points, forming a linear order. The **L**-player is forced to put at least 14 of these, say $a_1,\ldots,a_{14}$, where $a_i < a_j$ if $i<j$, in the same chain A. We can ignore the remaining points by letting them behave in the same manner as the nearest a_i. Next the **K**-player plays b_7 and b_8 such that b_i is only incomparable to a_i, $7 \leq i \leq 8$. Then depending on which chains the **L**-player chooses for b_7 and b_8, the **k**-player plays c as shown in Figure 4(c) or 4(d). In the first case the **L**-player must put c into

a new chain C; in the second case we may assume by duality that he does. The next sequence of plays is shown in Figure 4(e). First the **K**-player plays b_1. The **L**-player dare not put b_1 in chain D. Without loss of generality suppose he puts b_1 in B. The **K**-player responds with b_3. If the **L**-player puts b_3 in chain C, the **K**-player obtains position 4(b) by playing d; so suppose the **K**-player puts b_3 in chain B and consider Figure 4(f). The **K**-player plays c′. To avoid position 4(a), the **L**-player must put c′ in chain C, to which the **K**-player responds with b_5. As before, putting b_5 into chain D or C leads to position 4(a) or 4(b), so the **L**-player puts b_5 into chain B. Finally the **K**-player obtains position 4(a) by playing d′. ∎

<u>Theorem</u> 6. (Kierstead and Trotter [1981]) Every recursive interval order of width w can be covered by 3w-2 recursive chains; moreover there are recursive interval orders with width w which cannot be covered by fewer than 3w-2 recursive chains.

<u>Proof</u>. For the first part of the theorem, let $\mathbf{K}_w$ be the class of interval orders $\mathbf{P} = (P,R)$ of width at most w and let $\mathbf{L}_w$ be the class of structures $\bar{\mathbf{P}} = (P,R,C_1,\ldots,C_{3w-2})$ such that (P,R) is in **K** and $(C_1,\ldots,C_{w-2})$ is a chain cover of P. Both $\mathbf{K}_w$ and $\mathbf{L}_w$ satisfy the hypothesis of Lemma 1 so it suffices to provide a recursive winning strategy **S** for the $\mathbf{L}_w$-player in the $\mathbf{K}_w$ - $\mathbf{L}_w$ expansion game.

The strategy is defined by recursion on w. There is no problem when w=1, so consider the case w=v+1. Let $\mathbf{S}_v$ be a winning strategy for the $\mathbf{L}_v$-player in the $\mathbf{K}_v$ - $\mathbf{L}_v$ expansion game. When the $\mathbf{L}_w$-player is confronted with $\mathbf{P}^+$ and $\bar{\mathbf{P}}$, where $P^+ = P \cup \{i\}$ he first tries to place i in a subset B^+ of width v. If he succeeds he then uses the strategy $\mathbf{S}_v$ to put i in one of the chains $C_1^+,\ldots,C_{3v-2}^+$. Otherwise he puts i in one of the remaining chains C_{3v-1}^+, C_{3v}^+, or C_{3v+1}^+.

To show that this is indeed a winning strategy, we prove that if i is in $A^+ = P^+ - B^+$ then there are at most two other elements of A^+ which are incomparable to i. We begin by making two observations about interval orders. First, interval orders do not contain suborders of the form $\underline{2} + \underline{2}$ = . Second, a linear order can be defined on the maximum antichains of an interval order by setting $I < J$ iff there exist $i \varepsilon I$ and $j \varepsilon J$ such that $i<j$ in the interval order.

Next we show that $P^+ \restriction A^+$ has width 2. Consider x,y, and z in A^+. There exist antichains X, Y, and Z of width v contained in B such that, respectively, x, y, and z, are incomparable to every element of, respectively, X, Y, and Z. Suppose $X \leq Y \leq Z$. If $x||y$ and $y||z$ then x is comparable to z: There exists x′ and z′ in Y such that $x<x'$ in R and $z'<z$ in

R. If $x||z$ then x, x', z', z is a suborder of type $\underline{2} + \underline{2}$. Thus x is comparable to z.

Finally, suppose i is in A^+ and i is incomparable to three other elements x, y, and z in A. Then $\{x,y,z\}$ is a chain, say $x<y<z$. Let Y be an antichain of width v contained in B such that y is incomparable to every element of Y. Then i is comparable to some element i' in Y. Without loss of generality $i<i'$ in R. But this is a contradiction, since i, i', y, z is a suborder of type $\underline{2} + \underline{2}$.

For the second part of the theorem, let $\mathbf{L}'_w$ be the class of structures $\bar{B} = (P,R,C_1,\ldots,C_{3w-3})$ such that (P,R) is in $\mathbf{K}_w$ and $(C_1,\ldots,C_{3w-3})$ is a chain cover of P and let $\mathbf{A} = (\mathbf{N},S)$, where S is the natural order on $\mathbf{N}$. Then $\mathbf{K}_w$, $\mathbf{L}'_w$ and $\mathbf{A}$ satisfy the hypothesis of Lemma 2. Thus it suffices to show that the $\mathbf{K}_w$-player has a winning strategy $\mathbf{S}$ in the $\mathbf{K}_w$ - $\mathbf{L}'_w$ expansion game.

We define $\mathbf{S}$ by recursion on w. If $w=1$ then the $\mathbf{L}'_w$-player cannot even respond to the first play, so suppose $w=v+1$ and S_v is a winning strategy for the $\mathbf{K}_v$ - player in the $\mathbf{K}_v$ - $\mathbf{L}'_v$ expansion game. The $\mathbf{K}_w$-player starts the game by using the strategy $\mathbf{S}_v$ to play a chain of $n=3\cdot\binom{3w-3}{3v-2}+1$ subinterval orders $\mathbf{P}_i$ in $\mathbf{K}_v$ so that the $\mathbf{L}_w$-player is forced to use $3v-2$ distinguished chains to cover each $\mathbf{P}_i$. Then there exist at least four $\mathbf{P}_i$, say $\mathbf{P}_a$, $\mathbf{P}_b$, $\mathbf{P}_c$, $\mathbf{P}_d$, on which the $\mathbf{L}_w$-player has used the same $3v-2$ distinguished chains. The rest of the $\mathbf{K}$-player strategy is illustrated in Figure 5, where loops represent possibly empty chains of suborders $\mathbf{P}_i$ and upper case letters represent chains not among the distinguished chains. The $\mathbf{K}$-player plays x and y, as shown in Figure 5(a), and then waits to see whether the $\mathbf{L}$-player puts them both in the same chain. Depending on the $\mathbf{L}$-player's choice the $\mathbf{K}$-player wins by playing as shown in Figure 5(b) or 5(c). ▒

It is not known whether there exists a function b such that every comparability graph with independence number i can be covered by $b(i)$ recursive cliques. However in the proof of Theorem 6 we only needed the comparability graph. Thus we have the following:

<u>Corollary</u>. Every recursive interval graph of clique size ω can be recursively $3\omega-2$ - colored. ▒

We end this section by proving a strong chain covering theorem for ordered sets with small recursive dimension.

Theorem 7. (Kierstead, McNulty, and Trotter [1984]) Every recursive ordered set with recursive dimension d and width w can be covered by $\binom{w+1}{2}^{d-1}$ recursive chains.

Proof. We argue by induction on d. Let $\mathbf{P} = (P,R)$ be an ordered set with recursive dimension d and width w. The case $d=1$ is trivial, so suppose $d = e+1$ and $(L_1,\ldots,L_{e+1})$ is a recursive realizer of $\mathbf{P}$. Let $S = L_1\cap\ldots\cap L_e\cap L^*_{e+1}$, where L^*_{e+1} is the dual of L_{e+1}. Then $\hat{\mathbf{P}} = (P,S)$ is recursive and the height of $\hat{\mathbf{P}}$ is bounded by w. Thus by Theorem 1, P can be covered by $\binom{w+1}{2}$ recursive antichains in $\hat{\mathbf{P}}$. If A is a recursive antichain in $\hat{\mathbf{P}}$ then $\mathbf{P}\lceil A$ is a recursive ordered set with width at most w and realizer $(L_1\lceil A,\ldots,L_e\lceil A)$. Thus by the inductive hypothesis, each of the $\binom{w+1}{2}$ recursive antichains covering P can be covered by $\binom{w+1}{2}^{e-1}$ recursive chains in $\mathbf{P}$.▒

§ 5 Recursive Realizers

The concept of recursive dimension, which was introduced in §3, is particularly rewarding because of its interaction with the theories of recursive chain and antichain covers. We have already seen that this concept can be combined with results on recursive antichain covers to prove results on recursive chain covers. In this section we will use recursive chain covers to construct recursive realizers of certain ordered sets. As is well known, the ordinary dimension of an ordered set is bounded by its width w. A realizer is produced by first covering the ordered set with w chains and then for each chain C constructing a linear extension which puts each point a in C over all points incomparable to a. Here we follow a similar approach, but the construction will require more linear extensions and be considerably more involved.

Before proceeding with the main result we must narrow the scope of ordered sets to be considered. A c-crown $(c>2)$ is an ordering on $2c$ elements of the form shown in Figure 6(a). An ordered set is crown-free if it does not contain any c-crowns as suborders. Our first result shows that we must restrict our attention to ordered sets without 3-crowns.

Theorem 8. (Kierstead, McNulty, and Trotter [1984]) For every positive integer d there s a recursive ordered set $\mathbf{P}$ with width 3, which can be covered by 4 recursive chains, but has recursive dimension greater than d.

Proof. Let $\mathbf{K}$ be the class of structures of the form $\mathbf{P} = (P,R,C_1,C_2,C_3,C_4)$ such that (P,R) is an ordered set with width at most 3 and (C_1,C_2,C_3,C_4) is a chain cover of (P,R); let $\mathbf{L}$ be the class of structures of the form $\hat{\mathbf{P}} = (P,R,C_1,C_2,C_3,C_4)$ such that (P,R) is in $\mathbf{K}$ and $(L_1,\ldots,L_d)$ is a realizer of (P,R); and let $\mathbf{A} = (\mathbf{N},S,\mathbf{N},\mathbf{N},\mathbf{N},\mathbf{N})$, where S is the natural order on $\mathbf{N}$. Then $\mathbf{K}$, $\mathbf{L}$, and $\mathbf{A}$ satisfy the hypothesis of Lemma 2, so it suffices to provide the $\mathbf{K}$-player with a winning strategy in the $\mathbf{K}$-$\mathbf{L}$ expansion game.

Let $\bar{\mathbf{P}} = (P,R,C_1,\ldots,C_4,L_1,\ldots,L_d)$ be in $\mathbf{L}$. A pair of R-chains $u<v$ and $x<y$ is said to be pointlike in $\bar{\mathbf{P}}$ provided that in each linear extension L_j either both u and v are below both x and y or vice versa. The $\mathbf{K}$-player's strategy consists of two stages. During the first stage he continues to add new points to the bottom of C_1 and to the top of C_2 so that there are no comparabilities between points in C_1 and points in C_2, until the $\mathbf{L}$-player is forced to make the bottom two points $c_1^{i+1} < c_1^i$ of C_1 and the top two points $c_2^i < c_2^{i+1}$ of C_2 be a pair of point like chains. After at most 2d rounds this will be accomplished: Let $M_i = \{k\colon\ c_1^i < c_2^i \text{ in } L_k\}$. Observe that $M_i \subset M_{i+1}$ and that if $M^i = M^{i+1}$ then $c_1^{i+1} < c_1^i$ and $c_2^i < c_2^{i+1}$ are a pair of point likechains. Since $(L_1,\ldots,L_d)$ is a realizer of (P,R), $1 \le |M_i| \le |M_{i+1}| \le d-1$. Thus after $i+1 \le d$ sequences of two rounds $M_i = M_{i+1}$ and we are done.

To complete his victory in the second stage the $\mathbf{K}$-player adds points $x \in C_3^+$ and $y \in C_4^+$ as shown in Figure 6(b). Since p is pointlike it is impossible for the $\mathbf{L}$-player to put x over y in any L_j, $1 \le j \le d$.▒

It is not hard to iterate the above construction to obtain a single recursive width 3 ordered set that has infinite recursive dimension. The main result of this section provides a bound, in terms of width, on the recursive dimension of crown-free ordered sets, when combined with Theorem 4.

Theorem 9. (Kierstead, McNulty, and Trotter [1984]) Every recursive crown-free ordered set, which can be covered by c recursive chains, has recursive dimension at most $c!$; moreover there exist recursive crown-free ordered sets, which can be covered by c recursive chains, but have recursive dimension at least $c\binom{c-1}{t}$, where $t = \lfloor (c-1)/2 \rfloor$.

Proof. For the first part, let $\mathbf{K}$ be the class of structures $\mathbf{P} = (P,R,C_1,\ldots,C_c)$ such that (P,R) is an ordered set and $(C_1,\ldots,C_c)$ is a chain cover of P. Let $\mathbf{L}$ be the class of structures $\bar{\mathbf{P}} = (P,R,C_1,\ldots,C_c,L_1,\ldots,L_{c!})$ such that $(P,R,C_1,\ldots,C_c)$ is in $\mathbf{K}$ and $(L_1,\ldots,L_{c!})$ is a realizer of (P,R). Since $\mathbf{K}$ and $\mathbf{L}$ satisfy the hypothesis of Lemma 1, it suffices to provide the $\mathbf{L}$-player with a recursive winning strategy in the $\mathbf{K}$-$\mathbf{L}$ expansion game.

Suppose that in the i+1st round of play the L-player is confronted by P^+ in $\mathbb{K}$ and $\bar{P}$ in $\mathbb{L}$, where $P^+ = P \cup \{i\}$. He must decide where to place i in each of the linear extensions L_j^+, $1 \leq j \leq c!$. In each L_j there is an interval into which i can be inserted: The upperbound of this interval (if any) is the L_j-least element of P that is larger than i in R^+, while the lower bound (if any) is the L_j-largest element of P that is less than i in R^+. Positioning i anywhere in this interval will produce an extension of both R^+ and L_j. The exact position chosen will be based on the chain cover $(C_1^+,\ldots,C_c^+)$. For each $x \varepsilon P^+$ let $\alpha(x)$ be the index of the chain into which the $\mathbb{K}$-player has placed x. Relabel the linear extensions $L_1,\ldots,L_{c!}$ as $\ldots L_\sigma \ldots$, where σ is a linear order (permutation) on the set $\{1,\ldots,c\}$. The L-player's strategy is: Let x be the L_σ-least element of P such that not $x<i$ in R^+ and either $i<x$ in R^+ or $\alpha(i) < \alpha(x)$ in σ. Insert i immediately below x in L_σ^+. If there is no such x, put i at the top of L_σ^+.

It remains to show that $(\ldots L_\sigma^+ \ldots)$ is a realizer of (P^+,R^+). It suffices to choose an arbitrary incomparable pair (x,y) and show that there is at least one σ for which $y<x$ in L_σ^+. Observe that if x is adjacent to y in L_σ^+ and $x||y$ then $x<y$ in L_σ^+ iff $\alpha(x) < \alpha(y)$ in σ. Of course we should choose σ so that $\alpha(x)$ is the greatest element in σ and $\alpha(y)$ is the least element in σ. But this may not be enough. For example, if $P = \{x,u,v,y\}$, $x<u||v<y$ in R, $\alpha(u)<\alpha(v)$ in σ, and u and v are inserted before x and y, then $x<u<v<y$ in L_σ. Thus we also need $\alpha(v)<\alpha(u)$ in σ. More generally the interval between x and y in L_σ could be: $x = w_0 < w_1 < \ldots < w_{n+1} < w_n = y$. Then for all i, either $w_i < w_{i+1}$ in R or $w_i || w_{i+1}$ and $\alpha(w_i) < \alpha(w_{i+1})$ in σ. Such an interval will be called an (x,y)-<u>blocking interval</u> for σ. It suffices to find a σ which does not have an (x,y)-blocking interval.

An ordered set on 2n points $(n \geq 1)$ of the form: 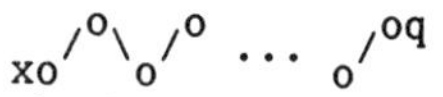 is called an <u>even up-fence</u> from x to q. Let Δ be the binary relation on $\{1,\ldots,n\} - \{\alpha(x),\alpha(y)\}$ defined by $\beta\Delta\gamma$ if there exist $p \varepsilon C_\beta$ and $q \varepsilon C_\gamma$ such that in R $p<y$ and there exists an even up-fence F from x to q, all of whose points are incomparable to both p and y. We say that the pair (F,p) witnesses $\beta\Delta\gamma$. Observe that Δ is defined in terms of representatives of the chains C_β and C_γ and that the conditions need not be satisfied by every pair of representatives.

<u>Claim</u> 1. The relation Δ is a strict interval order.

<u>Proof</u>. It suffices to show that Δ is irreflexive and does not induce 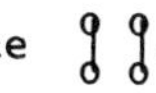. The former is obvious since each C_α is a chain. Suppose Δ contains 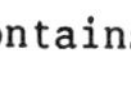. Let (F,p) and (H,r) witness $\beta\Delta\gamma$ and $\delta\Delta\varepsilon$. Since

neither $\beta\Delta\varepsilon$ nor $\delta\Delta\gamma$, p and r are both under elements of H and F. Let s' be the element of H closest to x that is over p and q' be the element of F closest to x that is over r. See Figure 7(a). Then $\{s',p,y,r,q'\}$ induces the suborder $^{s'\circ}\backslash_{\circ}/^{\circ}\backslash_{\circ}/^{\circ}$, which can be extended to a crown in $F \cup H \cup \{y,p,r\}$. With this contradiction the proof of the claim is complete.

Let $\bar{\Delta}$ be the extension of Δ formed by making $\alpha(y)$ a minimum element and $\alpha(x)$ a maximum element. Then $\bar{\Delta}$ is still an interval order. Chose an interval representation of $\bar{\Delta}$ and let σ be the linear extension of $\bar{\Delta}$ by left end points with respect to this representation.

<u>Claim</u> 2. (a) If $\alpha(v) < \alpha(w)$ in $\bar{\Delta}$ and $\alpha(w) < \alpha(z)$ in σ then $\alpha(v) < \alpha(z)$ in $\bar{\Delta}$. (b) If $\begin{smallmatrix} y\circ & \circ z \\ v\circ & \circ w \end{smallmatrix}$ in R^+ and $\alpha(v)<\alpha(w)$ in Δ, then $\alpha(v)$ $\bar{\Delta}$ $\alpha(z)$.

<u>Proof</u>. Property (a) is a distinctive property of extension by left end points. To see that (b) is valid even when $z \notin C_{\alpha(x)}$, choose $v' \in C_{\alpha(v)}$ and $w' \in C_{\alpha(w)}$ and an even up fence F from x to w' such that (F,v') witnesses $\alpha(v)<\alpha(w)$ in Δ. Let $v'' = \max\{v,v'\}$ in R^+ and let F' be an even up-fence from x to z with $F' \subset F \cup \{w,z\}$. Then (F',v'') witnesses $\alpha(v)<\alpha(z)$ in Δ.

Finally we are ready to show that there is no (x,y)-blocking interval. Suppose to the contrary that $x = w_o < w_1 <\ldots< w_n = y$ is a blocking interval in L^+_α. Then $x<w_1||w_{n-1}<y$ in R^+. Let r be the least index such that $w_r<y$ in R^+. Clearly $2\le r\le n-1$, $w_i||w_r$ for $1\le i\le_r -1$, $\alpha(w_{r-1})<\alpha(w_r)$ in σ, and $w_r<w_1$ in Δ. Let s be the greatest index such that s<r and $\alpha(w_r)<\alpha(w_s)$ in $\bar{\Delta}$. Clearly $1\le s\le r-2$. For a contradiction we show that $\alpha(w_r)<\alpha(w_{s+1})$ in $\bar{\Delta}$. First suppose $w_s<w_{s+1}$ in R^+. Either $\alpha(w_r)<\alpha(w_s)$ in Δ or $\alpha(w_r) = \alpha(y)$ or $\alpha(w_s) = \alpha(x)$. In the first case the conclusion follows from (b), the second is trivial, and in the third it follows from the fact that $\begin{smallmatrix} y\circ & \circ z \\ v\circ & \circ x \end{smallmatrix}$ is a subordered set of R^+, where $v=w_r$ and $z = w_{s+1}$. On the other hand if $\alpha(w_s)<\alpha(w_{s+1})$ in σ then the conclusion follows from (a).

For the second part of the theorem let **K** be as in the first part; let **L** be the class of structures of the form $\bar{P} = (P,R,C_1,\ldots,C_c,L_1,\ldots,L_{d-1})$, where $d = c\binom{c-1}{t}$ such that $(P,R,C_1,\ldots,C_c)$ is in **K** and $(L_1,\ldots,L_{d-1})$ is a realizer of (P,R); and let $\mathbf{A} = (\mathbf{N},S,\mathbf{N},\ldots,\mathbf{N},S,\ldots,S)$ where S is the natural order on **N**. By Lemma 2 it suffices to show that the **K**-player has a winning strategy in the **K**-**L** expansion game.

The first step is to see that the K-player can force a position in which points from different chains C_i and C_j are incomparable for all $i \neq j$ and each chain C_i contains a pair p_i such that for each $i \neq j$ p_i and p_j are a pair of pointlike chains. First the K-player uses the technique developed in the proof of Theorem 8 to assure that there exist $x_i, y_i \varepsilon C_i$ for $1 \leq i \leq c$ such that the L-player has made $x_1 < y_1$ and $x_j < y_j$ a pointike pair in $\bar{P}$ for $2 \leq j \leq c$. Next we observe that if $x<y$ and $u<v$ are a pointlike pair in $\bar{P}$ and $x<x'<y'<y$ and $u<u'<v'<v$ in R^+ then $x'<y'$ and $u'<v'$ are also a pair of pointlike chains in $\bar{P}$. Thus we can repeat the technique inside the chains $x_i<y_i$ to assure the existence of $x_j', y_j' \varepsilon C_j$, $2 \leq j \leq c$, such that $x_j<x_j'<y_j'<y_j$, $x_1<y_1$ and $x_j'<y_j'$ are a pointlike pair of chains in $\bar{P}$ for $2 \leq j \leq c$, and $x_2'<y_2'$ and $x_k'<y_k'$ are a pointlike pair of chains in $\bar{P}$ for $3 \leq k \leq c$. Continuing in this manner the K-player can obtain the goal of the first stage.

In the second stage the K-player plays two more elements to win. He first notes that each linear extension L_i induces a linear order on $Q = \{p_1, \ldots, p_c\}$ since the p_i behave pointlike. For $1 \leq i \leq d-1$ let q_i be the t+1st largest element of Q in L_i and let $X_i = \{C_j : q_i < p_j \text{ in } L_i\}$. Choose an element $p_k \varepsilon Q$ and a t element subset X of $\{C_1, \ldots, C_c\} - \{C_k\}$ such that $(p_k, X) \neq (q_i, X_i)$ for any $1 \leq i \leq d-1$. This is possible since there are d ways to pick (p_k, X). Let C_m be any element of X and C_n be any element of $Y = \{C_1, \ldots, C_c\} - (X \cup \{C_k\})$. The K-player completes his win by playing u in C_m so that u is under every element of every chain in X and under the top element of p_k and v in C_n so that v is over every element of every chain in Y and v is over the bottom element of p_k. This is depicted in Figure 6(b). The L-player cannot respond: He must put $v<u$ in some L_i, but this requires that $(q,X) = (q_i, X_i)$. ▩

It is not known whether it is necessary to forbid all crowns or just 3-crowns in order to obtain a bound on the recursive dimension of an ordered set in terms of its width. Interval orders are an important class of crown-free ordered sets. Kierstead, McNulty, and Trotter [1984] proved that every recursive interval order that can be covered by c recursive chains has recursive dimension at most 2c. When combined with Theorem 4 this shows that every recursive interval order of width w has recursive dimension at most 6w-4. Hopkins improved the second bound with the next theorem.

Theorem 10. (Hopkins [1981]) Every recursive interval order of width w has recursive dimension at most 4w-4; moreover there are recursive interval orders of width w that have recursive dimension $\left\lceil \frac{4}{3}w \right\rceil$.

Proof. Let $\mathbf{K}$ be the class of interval orders $\mathbf{P} = (P,R)$ with width at most w. For the first part let $\mathbf{L}$ be the class of structures of the form $\bar{\mathbf{P}} = (P,R,L_1,\ldots,L_{4w-4})$ such that (P,R) is in $\mathbf{K}$ and $(L_1,\ldots,L_{4w-4})$ is a realizer of (P,R). It suffices to show that the L-player has a recursive winning strategy in the $\mathbf{K}$-$\mathbf{L}$ expansion game. Before this strategy can be described it is necessary to introduce some notation. The up, down, and incomparable sets of an element x in $\mathbf{P}$ are denoted by $U(x) = \{y\varepsilon P : x<y \text{ in } R\}$, $D(x) = \{y\varepsilon P : y<x \text{ in } R\}$ and $I(x) = \{y\varepsilon P: y||x\}$. It is a property of interval orders that for all x and y in P, either $D(x) \subset D(y)$ or $D(y) \subset D(x)$ and either $U(x) \subset U(y)$ or $U(y) \subset U(x)$. Let $D^*(x) = \{y: D(x) \subset D(y) \text{ and } x\neq y\}$ and $U_*(x) = \{y: U(x) \subset U(y) \text{ and } x\neq y\}$. The element x is said to be down (up) in a linear extension L if x is less (greater) than every element of $D^*(x)$ $(U_*(x))$ in L. It is an easy exercise to see that if x is not down (up) in any L_j, $1\leq j\leq 4w-4$, then the $\mathbf{K}$-player can win the $\mathbf{K}$-$\mathbf{L}$ expansion game on the next move. With this in mind, let $\mathbf{L}'$ be the class of structures $\bar{\mathbf{P}} = (P,R,L_1,\ldots,L_{4w-4})$ such that $\bar{\mathbf{P}}$ is in $\mathbf{L}$ and every element of P is down in some linear extension $\sigma\varepsilon M_1$ and up in some linear extension $\tau\varepsilon M_2$, where $M_1 = \{L_1,\ldots,L_{2w-2}\}$ and $M_2 = \{L_{2w-1},\ldots,L_{4w-4}\}$. We will actually provide the $\mathbf{L}'$-player with a winning strategy in the $\mathbf{K}$-$\mathbf{L}'$ expansion game.

Suppose the $\mathbf{L}'$-player is confronted with $\mathbf{P}^+$ in $\mathbf{K}$ and $\bar{\mathbf{P}}$ in $\mathbf{L}'$, where $P^+ = P\cup\{i\}$. To assure that $M_1 \cup M_2$ is a realizer of (P,R) it suffices to arrange that:

(1) every element of $\max I(i)$ is under i in some $L \varepsilon M_1 \cup M_2$ and

(2) every element of $\min I(i)$ is over i in some $L \varepsilon M_1 \cup M_2$

where the minima and maxima are with respect to R. To assure that every element of P^+ is down (up) in some $L\varepsilon M_1$, $(L\varepsilon M_2)$ it is enough to have that:

(3) every element of $\{i\} \cup \min I(i)$ is down in some $L\varepsilon M_1$ and

(4) every element of $\{i\} \cup \max I(i)$ is up in some $L\varepsilon M_2$,

since the downess and upness of other elements will not be affected by the insertion of i. Finally, let $A = U_*(i) \cap \max I(i)$, $A' = (\max I(i)) - A$, $B = D^*(i) \cap \min I(i)$, and $B' = (\min I(i)) - B$. Then (*) $a\varepsilon A$ iff $U(i) = U(a)$ for every $a \varepsilon \max I(i)$ and $b\varepsilon B$ iff $D(i) = D(b)$ for every $b \varepsilon \min I(i)$. It takes special care to put i over elements of A' or under elements of B'.

The L-player should play as follows: He first selects a subset $M_1' \subset M_1$ of minimal cardinality such that every element of $A' \cup B$ is down in some $L \varepsilon M_1'$. For each $L \varepsilon M_1'$ he forms L^+ by inserting i as high as possible. Then by (*) i is over every element of A in L. Also if $x \varepsilon I(i)$ is down in L then i is over x: If not there exists $y\varepsilon P$ such that $i<y$ in R and $y<x$ in L. Since $i||x$, not $D(y) \subset D(x)$ in R^+, and thus $D(x) \subset D(y)$ in R; but then x is not down in L. Thus (1) is

satisfied and every element of B is down in some L^+, with $L \varepsilon M_1$. Assume for now that $M_1 - M_1'$ is non-empty. For each $L \varepsilon M_1 - M_1'$ insert i as low as possible in L^+ subject to $b'<i$ in L^+ if $b' \varepsilon B'$ and b' is down in L. Then both i and any $b' \varepsilon B'$, which is down in L are down in L^+. Thus (3) is satisfied. By playing dully on M_2, the L-player can also satisfy (2) and (4). Thus it only remains to show that $M_1 - M_1'$ and dually $M_2 - M_2'$ are non-empty.

It suffices to prove that $|M_1'| \leq 2w-3$. The width of $I(i) \leq w-1$. If $|A' \cup B| \leq 2w-3$ we are done; otherwise $|A'| = |B| = w-1$ and $A' \cap B = \emptyset$. In the latter case we show that if $a' \varepsilon A'$ is down in L then some $b \varepsilon B$ is also down in L. Suppose not. Let b be the least element of B in L. Since b is not down in L there exists $p \varepsilon P$ such that $p<b$ in L and $p||b$ (and $D(b) \subset D(p)$). Thus p is not over any element of B in R. Also p is not over i in R: Since $|B| = w-1$, a' is over some element of B in R and thus $p<a'$ in L. Thus $D(p) \subset D(a')$ and the conclusion follows. By (*) every element of $B \cup \{i\}$ has the same down set. Thus p is not under any element of $B \cup \{i\}$ in R. But then $B \cup \{i,p\}$ contradicts the width of **P** being w.

For the second part of the theorem let **K** be as before; let **L** be the class of structures of the form $\bar{P} = (P,R,L_1,\ldots,L_t)$, where $t = \lceil \frac{4}{3}w \rceil - 1$ such that (P,R) is in **K** and $(L_1,\ldots,L_t)$ is a realizer of (P,R); and let $\mathbf{A} = (\mathbf{N},S,S,\ldots,S)$ where S is the natural ordering on **N**. It suffices to show that the **K**-player has a winning strategy in the **K-L** expansion game. We shall need the following combinatorial result.

<u>Lemma</u> 4. (Trotter and Monroe [1982]) If f and g are functions from $\{1,\ldots,t\}$ to $\{1,\ldots,w\}$, where $t = \lceil \frac{4}{3}w \rceil - 1$ then there exist $i, j \varepsilon \{1,\ldots,w\}$ such that if $f(m) = i$ and $g(n) = j$ then m=n.▒

The **K**-player begins by playing the ordered set **P** = (P,R) consisting of n element antichains A and B such that every element of A is over every element of B in R. After the **L**-player has responded with $\bar{\mathbf{P}}$, the **K**-player lets f(m) be the unique element of A that is down in L_m and g(m) be the unique element of B that is up in L_m. By the lemma there exist $a \varepsilon A$ and $b \varepsilon B$ and m such that a is only down in L_m and b is only up in L_m. The **K**-player now wins by adding a new point c which is under every element of A except a and over every element of B except b. Then c can only be over a in L_j and can only be under b in L_j, but cannot be both over a and under b in the same extension. The case w=3 is shown in Figure 8. ▒

§ 6 Open Problems and Concluding Remarks

As mentioned in the introduction there are two reasonable responses to the negative result that a certain class of structures has no recursive member. The recursion theoretic response has not been discussed here. The interested reader can consult Jockusch [1972], Jockusch and Soare [1972], and Rosenstein [1982], among others. We took the combinatorial response and centered our discussion on recursive ordered sets. A similar set of results exists for (highly) recursive graphs. These results can be found in Bean [1976], Kierstead [1981b], [1983], Manaster and Rosenstein [1973] Schmerl [1980], [1982], and Tverberg [*]. Other types of combinatorial structures should also yield interesting recursive results.

We close by presenting a short list of extremal problems for recursive ordered sets.

(1) Let $b(w)$ be the least number such that every recursive ordered set with width w can be covered by w recursive chains. Is $b(w)$ polynomial? What is the value of $b(2)$?

(2) Does there exist a function $c(i)$ such that every recursive comparability graph with independence number i can be covered by $c(i)$ recursive cliques?

(3) Is there a finite bound in terms of width on the recursive dimension of 3-crown-free ordered sets?

(4) Give better bounds on the recursive dimension of crown-free ordered sets in terms of either width or the number of recursive chains required to cover the ordered set. The special case of interval orders is also of interest.

Bibliography

D. Bean
[1976] Effective coloration, J. Symbolic Logic **41**, 469-480.

V. Chvátal
[1981] Perfectly ordered graphs, Technical Report SOCS-81.28, McGill University.

R. Dilworth
[1950] A Decomposition theorem for partially ordered sets, Annals of Math. **51**, 161-166.

L. Hopkins
[1981] Some problems involving combinatorial structures determined by intersections of intervals and arcs, Ph.D. Thesis, University of South Carolina, 1-91.

C. Jockusch
[1972] Ramsey's Theorem and recursion theory, J. Symbolic Logic **37**, 268-279.

C. Jockusch and R. Soare
[1972] Π_1^0 classes and degrees of theories, Trans. Amer. Math. Soc. **173**, 33-55.

H. Kierstead
[1981a] An effective version of Dilworth's Theorem, Trans. Amer. Math. Soc. **268**, 63-77.

[1981b] Recursive colorings of highly recursive graphs, Can. J. of Math **33**, 1279-1290.

[1983] An effective version of Hall's theorem, Proc. Amer. Math. Soc. **88**, 124-128.

[1984] Unpuplished.

H. Kierstead and W. Trotter
[1981] An extremal problem in recursive combinatorics, Congressus Numerantium **33**, 143-153.

H. Kierstead G. McNulty and W. Trotter
[1984] Recursive dimension for partially ordered sets, Order **1**, 67-82.

A. Manaster and J. Rosenstein
[1973] Effective matchmaking and k-chromatic graphs, Proc. Amer. Math. Soc. **39** No 2, 371-378.

J. Rosenstein
[1982] Linear Orderings, (Academic Press).

J. Schmerl
[1979] Private communication.

[1980] Recursive colorings of graphs, Can. J. Math. **32** No. 4, 821-83.

[1982] The effective version of Brooks' Theorem, Can. J. Math. **34** No. 5, 1036-1046.

E. Szeméredi
[1982] Private communication.

W. Trotter and T. Monroe
[1982] A combinatorial problem involving graphs and matrices, Discrete Math. **39** 87-101.

H. Tverberg
[*] On Schmerl's effective version of Brooks' theorem, preprint.

DEPARTMENT OF MATHEMATICS
UNIVERSITY OF SOUTH CAROLINA
COLUMBIA, SOUTH CAROLINA, 29208

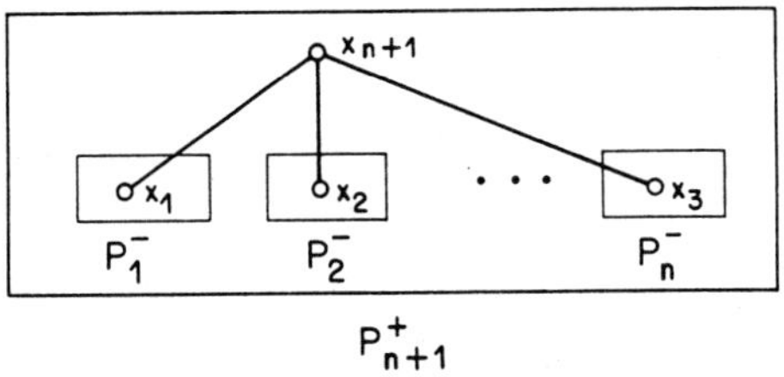

Figure 1

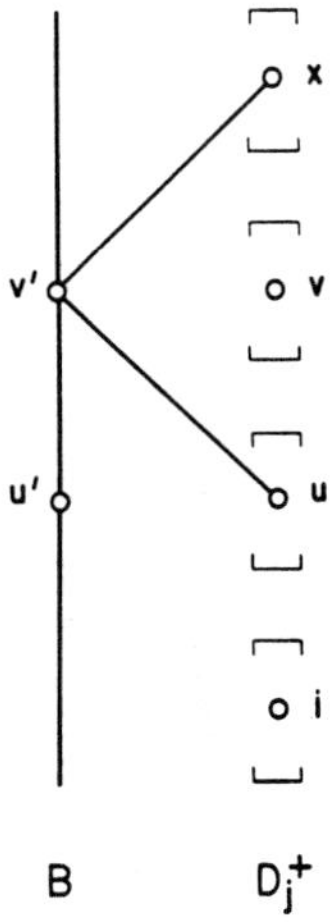

Figure 2

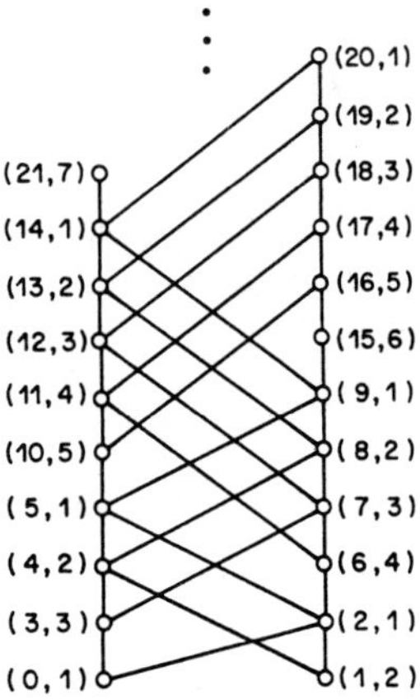

Figure 3

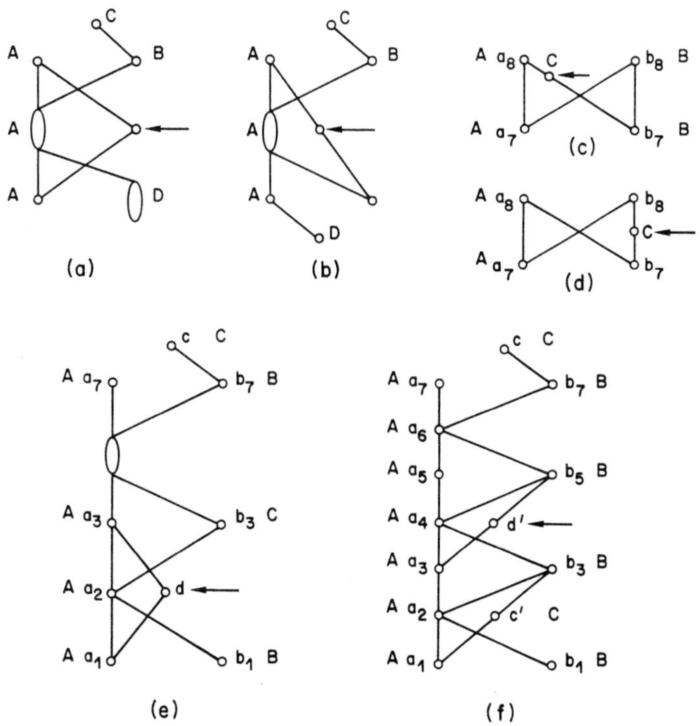

Figure 4

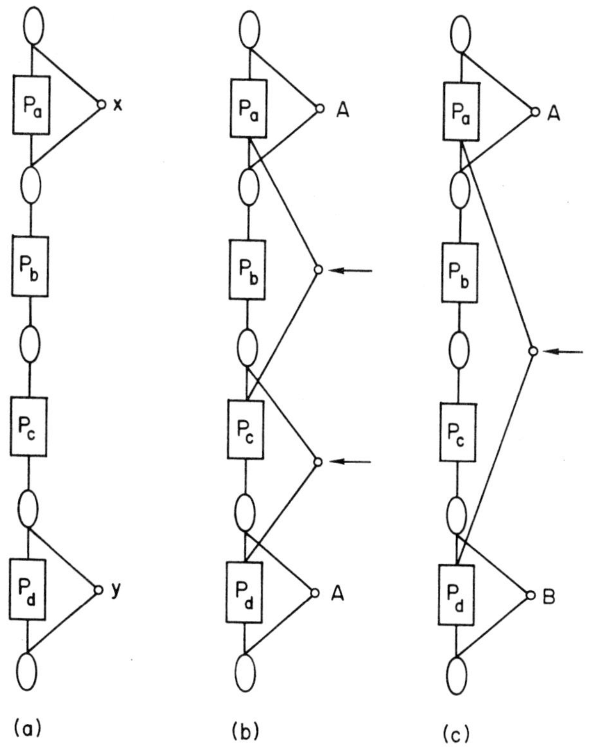

Figure 5

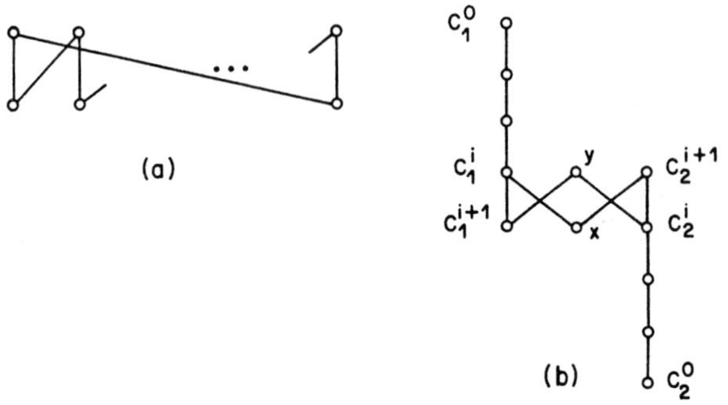

Figure 6

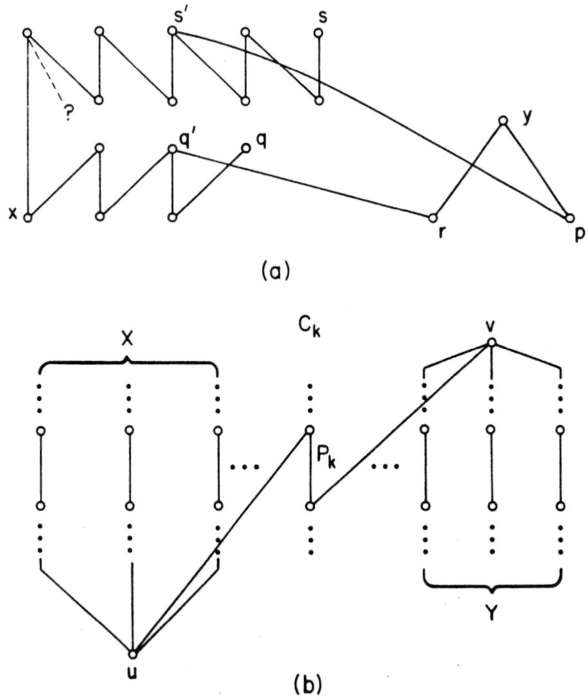

Figure 7

Contemporary Mathematics
Volume 57, 1986

ORIENTATIONS AND REORIENTATIONS OF GRAPHS

Oliver Pretzel

ABSTRACT. The orientations of principal interest in this survey are those orientations of finite graphs that make them into diagrams of ordered sets. We survey the questions of existence, recognition and construction of graphs that cannot be so oriented. There follows a discussion of theorems on graphs that can be oriented as diagrams of ordered sets or lattices, perhaps satisfying further conditions. Finally we discuss operations that change the orientations of a graph G while preserving the property that these orientations make G into the diagram of an ordered set.

The diagram of a finite ordered set S is an oriented graph with the points of S as its vertices and an edge $x \rightarrow y$ if y covers x in S (that is $x < y$ and for no $z \in S : x < z < y$). Usually the arrows on the edges are dispensed with and the graph is arranged so that all edges point upwards on the page. The diagram is as old as the theory of ordered sets and is the most common (but not the only) tool for representing them graphically (see Rival [1985a]). The unoriented graph C(S) underlying the diagram of S is called the covering graph of S . Questions about C(S) were asked almost as soon as the task of constructing a theory of ordered sets was undertaken. Two examples from masters of the theory may serve as an indication.

First G. Birkhoff in the second edition of his famous book on Lattice Theory [1948], asks as Problem 8 what the necessary and sufficient conditions on a lattice L are in order that every lattice M with the same covering graph as L is lattice isomorphic to L . This question was answered J. Jakubík [1954] for modular lattices but remains open for general lattices.

Instead of asking when isomorphism of covering graphs implies isomorphism of lattices one can ask which lattice invariants depend on the covering graph. Having gone this far, why not extend the question even further to general ordered sets? Some answers are known for lattices which we shall mention later but very little is known about the answer for general ordered sets.

1980 Mathematics Subject Classifications. 06A10, 05C15, 05C20, 05C38

The second example is in Oystein Ore's almost equally famous book on Graph Theory [1962]. On page 155 he asks when it is possible to orient a graph making it into a basis graph of a directed graph. Basis graphs generalize diagrams. They have the property that removing any edge changes the accessibility along directed walks. From this one sees immediately that acyclic basis graphs are diagrams. In the next chapter Ore specializes his questions to the acyclic case. The problem of characterizing covering graphs is still one of the major open problems of order theory, and a short article on it has recently appeared in Order (see Rival [1985b]).

Oystein Ore's book was the inspiration for the papers of Mosesian [1972 and 1973], in which he used an operation we shall call pushing down maximal vertices to modify diagram orientations and thus simplify his arguments. This operation turns out to be extremely useful and it is for this reason that this survey has the word reorientation in its title.

The survey is divided into five short sections as follows.

Section 1, Orientations, ordered sets and graph colouring, deals with elementary translations between these theories that are frequently needed when working with orientations.

Section 2, Constructions of non-covering graphs, introduces the known constructions and some results of Nešetřil and Rödl on the existence of such graphs and the complexity of the decision problem for covering graphs. This section is intended to give an indication of the phenomena a theory of orientations should explain and clarify.

Section 3, Theorems on covering graphs, first presents the sparse general results on covering graphs and then special classes of ordered sets for which better results are available, principally the case of modular lattices and their generalizations.

Section 4, introduces pushing down, explains its uses and gives necessary and sufficient conditions for acyclic orientations to be linked by pushing down.

The last section, Section 5, presents some new results on the relation between pushing down and other reorientations of graphs.

A COMMENT ON TERMINOLOGY. As we shall be dealing with varying orientations of a graph it is important to have a terminology that distinguishes oriented and unoriented concepts. So let graph, path and cycle have their usual unoriented meanings and let walk and circuit denote paths and cycles run through in a particular direction. We shall assume that paths and cycles are simple, i.e. that they have no repeated vertices, and that a path does not return to its starting point. For walks and circuits we make no such assumption. Given an orientation R of a graph G we can divide the edges of a walk (or circuit) W into forward and backward edges. If W has only forward edges. we call it a forward walk. Paths and cycles that correspond to forward walks or circuits are called monotone. R is acyclic if there are no monotone cycles.

Otherwise any non-standard terminology will be introduced as we need it. All sets will be assumed to be finite.

1. ORIENTATIONS, ORDERED SETS AND GRAPH COLOURING. Orientations of graphs form a link between two important areas of combinatorial theory, ordered sets and graph-colouring. If the edges of a graph $G = (V,E)$ are given an orientation R, this defines a preorder X on the vertices v by putting $v \leqslant w$ in X if there is a forward walk from v to w. If the orientation is acyclic then X will be an order and these are the orientations we shall be primarily concerned with. It is possible to read off all the properties of X from R and we shall therefore freely use order theoretic terms, such as "chain" when dealing with acyclic orientations. Conversely X only determines R if we know which vertices are adjacent in G. Thus there is a range of oriented graphs that can be used to represent the ordered set (V,X) and it is natural to look at the extremal graphs in this range.

At one end there is the comparability graph of X, so called because any pair of comparable vertices are adjacent. The transitivity of X places strong restrictions on the structure of a comparability graph and there is a well-developed theory of these graphs, starting with the characterization given by Ghouila-Houri and Gilmore and Hoffman.

THEOREM (GHOUILA-HOURI [1962], AND GILMORE AND HOFFMAN [1964]). A graph G is a comparability graph if and only if every odd cycle of length $\geqslant 5$ has a chord.

Perhaps the deepest results of this theory are to be found in a beautiful paper of Gallai [1967] in which he gives a complete forbidden subgraph characterization of comparability graphs. An excellent survey of the whole theory is given by Kelly [1985]. Ordered sets and their comparability graphs are very closely related, so it is not surprising that some important properties of an ordered set depend only on its (unoriented) comparability graph (see Kelly [1985] and Möhring [1985]).

At the other extreme is the covering graph of X in which v and w are only adjacent if one covers the other. The theory of these is still in its infancy and appears to be difficult. For instance, from results that appear later in this survey it seems to be hopeless to expect a concise forbidden subgraph characterization of covering graphs. The difficulty of the theory is in part due to the fact that ordered sets and their covering graphs are only loosely related. Indeed Ivan Rival has asked for interesting order-theoretical invariants that depend only on the covering graph, but so far such invariants have only been found in special cases.

The connection between orientations and colouring is less deep but it is very useful. It arises from the observation that if we label the vertices of a graph with numbers, say $0,\ldots,n$, so that adjacent vertices have distinct labels, then directing each edge towards the end with the higher label produces an acyclic orientation. Conversely, given an acyclic orientation of G , we can label each vertex v by the maximum of the lengths of the forward walks ending in v . If we refer to an orientation obtained from a colouring or vice-versa we shall always understand that it was obtained in the above way.

It is now a triviality to prove the following theorem if we restrict ourselves to acyclic orientations.

THEOREM (ROY[1967], GALLAI [1968]). Let G be a graph with chromatic number χ . For any orientation R of G define $h(R)$ to be the maximum length of an R-monotone path. Then $\chi = \min h(R) + 1$, where the minimum is taken over all orientations of G .

This result is not particularly difficult to prove, even for orientations with monotone cycles, but it is frequently useful in applications. There are two more difficult related results which link $\chi(G)$ with cycles. The first was conjectured by Las Vergnas [1975] and proved by J.A. Bondy [1976]. (This result is wrongly quoted in Bollabás [1978] p226). To state it we define an orientation R of G to be connected if for any pair of vertices (v,w) there is a forward walk from v to w . Then in the preorder defined by R all the vertices of G are equivalent, so such orientations are at the opposite extreme from acyclic orientations. For a graph to have such an orientation it is necessary and sufficient for it to be 2-edge connected (Robbins [1939]).

THEOREM (BONDY [1976]). Let G be a graph with connected orientations. For such an orientation R let $C(R)$ be the maximum of the lengths of R-monotone cycles. Then $\chi(G) \leqslant \min C(R)$, where the minimum is taken over all connected orientations of G .

Bondy also gives the following example, to show that $\min C(R) - \chi(G)$ can be arbitrarily large.

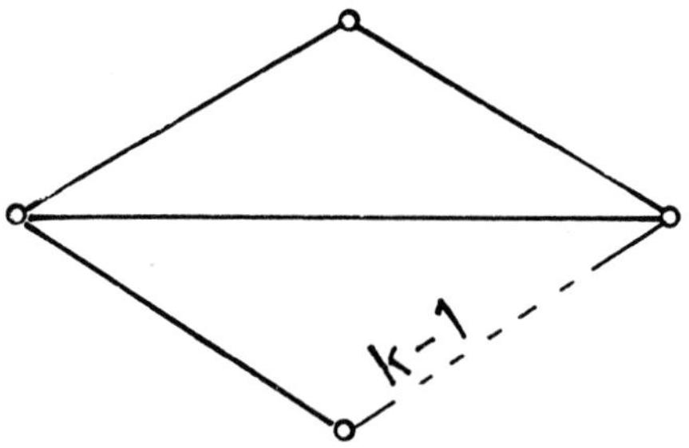

Figure 1

Here, if R is a connected orientation, $C(R) = k+1$ or $C(R) = k+2$ but $\chi(G) = 3$.

This result obviously tells us nothing about acyclic orientations, which are our main concern. But a subtle result of Minty [1962] does, and has been used effectively by Aigner and Prins [1980]. Again we need to make a definition. Let R be an acyclic orientation of G and let C be a circuit. The flow-ratio $f_R(C)$ is the number of forward edges of C divided by the number of backward edges of C, and $F(R)$ is the maximum of the flow ratios of all the circuits in G (this is easily seen to be attained by a simple circuit).

THEOREM (MINTY [1962]). Let G be a graph with cycles, then

$$\chi(G) = \min \lceil F(R) \rceil + 1$$

where the minimum is taken over all acyclic orientations of G.

As this result has not received the attention it deserves, and because we shall be discussing its relation to other work, we give a quick sketch of Minty's proof. The interesting direction is to produce, given a connected graph G with an acyclic orientation R, a colouring of G with $\lceil F(R) \rceil + 1$ colours. Choose an arbitrary starting point p and an arbitrary integer $n \geq F(R) + 1$. Assign a traveller the problem of finding for each point q a walk from p to q which maximizes his gain subject to the following rules: he is paid \$1 for each forward edge of the walk and is fined \$n-1 for each backward edge. As $n-1 \geq f(C)$ for any cycle C, traversing a cycle will never produce a positive benefit, so for each point q our traveler can find a maximum gain $g(q)$. Also, if $a \to b$ is an edge of G directed towards b by R, then $g(a)+1 \leq g(b) \leq g(a) + n-1$. So if we label the vertices q of the graph by $g(q) \bmod n$, we will have a graph colouring with at most n colours.

The most important result on graph colouring for our survey will be the following theorem of Erdös.

THEOREM (ERDÖS) [1959] AND [1961]). There exist graphs of arbitrarily large girth γ and chromatic number χ .

It would lead us too far astray to discuss the history of this problem, but as several constructions rely on it, we mention that Nešetřil and Rödl [1979] give a specific construction for such graphs.

2. CONSTRUCTIONS OF NON-COVERING GRAPHS. How does one characterize diagram orientations of a graph among the acyclic orientations? In a diagram there can be no edge connecting the top and bottom vertices of a chain of length ≥ 2 .

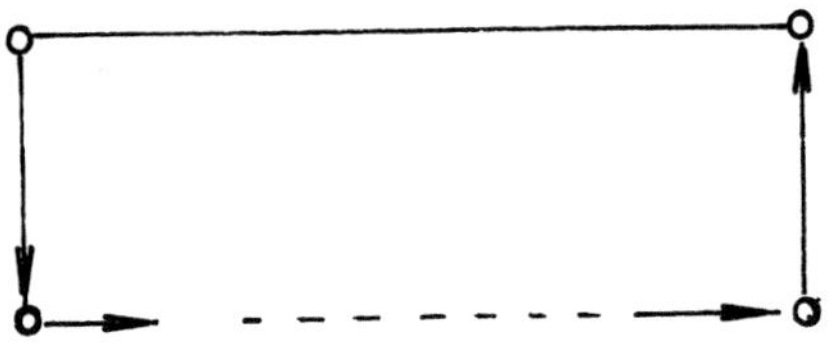

Figure 2

It follows immediately that every non-trivial circuit has at least two forward edges and two backward edges, and this is clearly also sufficient. So let us call an orientation R of a graph G <u>k-good</u> if every non-trivial circuit has at least k forward and k backward edges, and let us define the <u>orientation number</u> φ of G as the maximum k for which G has a k-good orientation. Then covering graphs are precisely those graphs with $\varphi \geq 2$. Now if G is to have a k-good orientation any cycle must have at least $2k$ edges. So the girth γ of G must be at least 2φ . Hence covering graphs have girth 4.

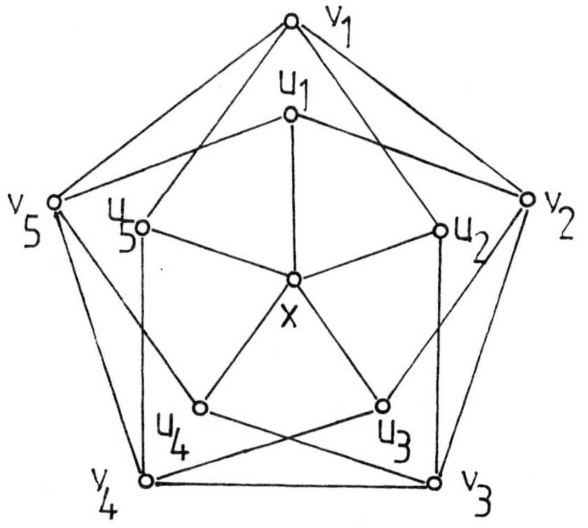

Figure 3: M_5

However, girth 4 is not in general sufficient for a graph to be a diagram. The smallest counter example (illustrated in Figure 3) is M_5 (the Grötsch or Mycielski Graph).

There are clearly very many orientations of M_5. How does one show that none of them are 2-good? Let us assume that the maximum vertex in some 2-good orientation R is x. Then from every vertex there is a forward walk to x. In particular all edges with x as one endpoint are directed towards x, but also all edges of the type v_iu_j are directed towards u_j, because otherwise if P is the forward monotone walk from v_i to x, the circuit $u_jv_iPxu_j$ has only one backward edge. Now the cycle $v_1 \dots v_5$ is odd, so it must have two adjacent edges oriented the same way, say $v_1 \to v_2 \to v_3$. Then the circuit $v_1v_2v_3v_2v_1$ has only one backward edge. So R cannot, after all, be 2-good. Later we shall show how we can use pushing down to reduce the general case to this one. Clearly the same idea can be applied to any odd cycle $v_1 \dots v_{2n+1}$.

We note that for M_5, $\chi = \gamma = 4$, and indeed if $\chi(G) < \gamma(G)$ then G is always a covering graph. For given a colouring of G with n colours in any cycle C, there will be two identically coloured vertices. Neither of the two disjoint paths along C connecting these vertices can be monotone in the orientation R obtained from the colouring. So if we take a circuit K with C as the underlying cycle it must have forward and backward edges in both the paths. Hence R is 2-good. In particular it follows that planar graphs with $\gamma \geqslant 4$, which have $\chi \geqslant 3$, are all covering graphs. This result has been noted many times, perhaps the earliest reference being Haff, Murty and Wilton (1970). Aigner and Prins (1980) note that the same argument can easily be extended to show that if $(k-1)\chi < \gamma$ for a graph G then $\varphi \geqslant k$. Pretzel [1985] contains a related condition which is sufficient for a colouring to produce a k-good orientation and which is also necessary in the case $k = 2$.

Having found one class of triangle-free non-covering graphs, it is natural to ask whether there are any others. The answer is yes, and a very simple idea leads to a large variety of non-covering graphs. Consider a chain in the diagram of an ordered set and a vertex a adjacent to both end points, x and y.

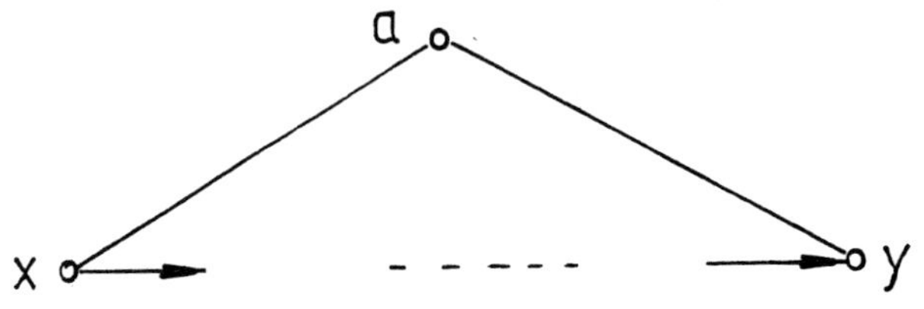

Figure 4

Then the edges xa and ay must be oriented $x \to a \to y$. So no diagram can contain a picture like Figure 5, in which we have a vertex a adjacent to three vertices x,y,z that lie on a chain.

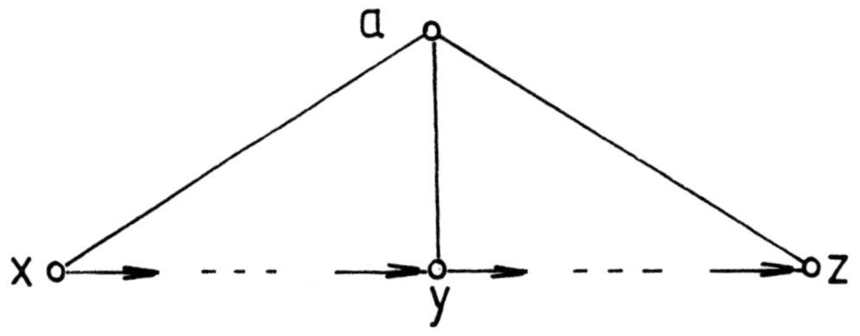

Figure 5

Now if xy or xz are adjacent adding a vertex a will introduce a triangle. But if the chain has length at least 4 we can avoid that. This gives the following construction.

CONSTRUCTION (PRETZEL [1985a]). If G is a graph with chromatic number $\chi \geqslant 5$ and H is obtained from G by adjoining vertices one adjacent to each triple of non-adjacent vertices of G, then $\varphi(H) = 1$, that is H is not a covering graph.

Thus every graph G of girth $\geqslant 4$ and $\chi \geqslant 5$ (of which there are infinitely many) can be embedded, in a non-covering graph H such that the complement $H - G$ is a stable set whose vertices have degree 3 in H. All the graphs produced by the construction have girth 4. This makes it unlikely that there will be an easy description of the minimal non-covering graphs, even if we only consider those of girth 4.

We can extend our observations about parts of diagrams of ordered sets a little further. Suppose a diagram contains the following picture, which we shall call a ladder.

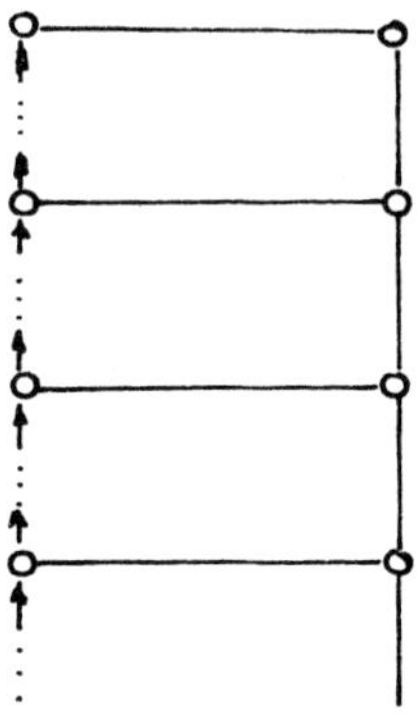

Figure 6:

Let us call the horizontal edges the rungs of the ladder. The two paths which remain when the rungs are removed will be called the left and right-hand uprights. Given that the left-hand upright is directed upwards as in the figure, at most one edge of the right-hand upright can be directed downwards. To see this, call the cycles containing successive rungs the holes of the ladder. Each hole can have only one clockwise directed edge that is not on the left upright. If this is <u>not</u> the upper rung then in the next hole up that rung (now the lower rung) is the unique such clockwise directed edge.

Recently Nešetřil has exploited ladders to show that the decision problem whether a graph is a covering graph is NP-complete. His method is as follows.

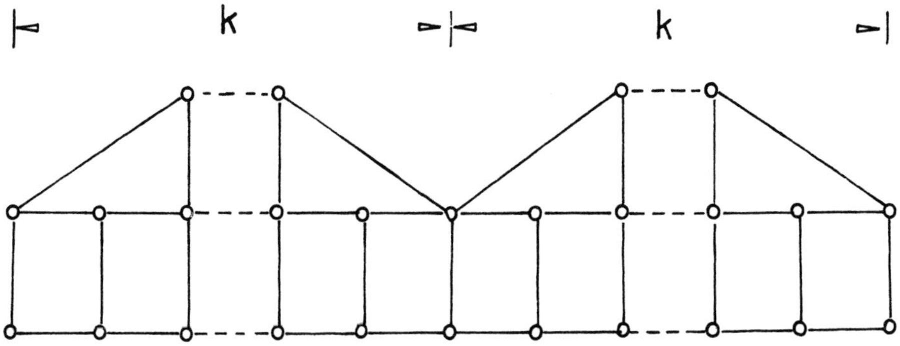

Figure 7: B_k

THEOREM (NEŠETŘIL [1985]). Let G be a graph with $\gamma \geq 2k \geq 4$ and on each path of length $2k$ erect a copy of B_k. Call the resulting graph G_k.

1. If $\chi(G) \leq k$ then $\varphi(G_k) \geq 2$ (G_k is a covering graph).
2. If $\chi(G) \geq 2k+1$ $\varphi(G_k)) = 1$ (G_k is not a covering graph).

As the problem of determining the chromatic number of a graph is NP-complete we have the following corollary.

COROLLARY [IBID]. The problem of determining $\varphi(G)$ is NP-complete. In particular it is a NP-complete problem to decide whether a graph is a covering graph.

So far we have only produced non-covering graphs of girth 4, but there is no limit on the girth of a non-covering graph:

THEOREM (NEŠETŘIL AND RÖDL [1978]). There exist non-covering graphs of arbitrarily large girth.

It would seem likely that a corresponding result holds for k-good orientations, but this has not been proved yet. So let us state it as a conejcture.

CONJECTURE. There exist graphs G of arbitrarily large girth with $\varphi(G) = n$ for every natural number n.

The proof of Nešetřil and Rödl uses an elegant probabilistic technique but yields no constructive examples. The current record for the girth of a constructive example is 6.

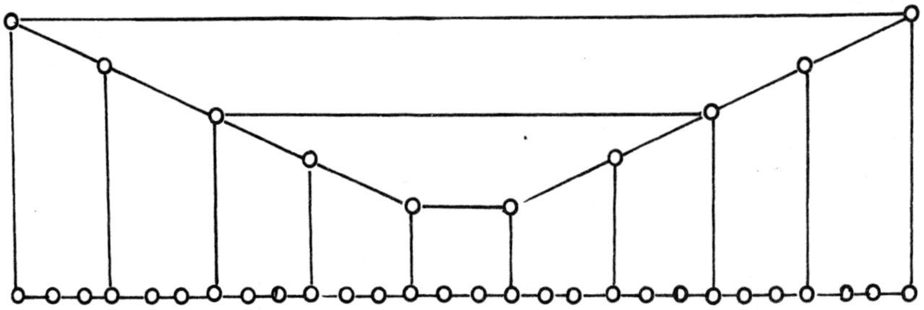

Figure 8: H

THEOREM (PRETZEL [1985b]). Let G be a graph with $\chi(G) \geq 28$ and $\gamma(G) \geq 30$, on each path of length 27 in G erect a copy of H. The resulting graph K has girth $\gamma(K) = 6$ and $\varphi(K) = 1$.

This result is constructive in so far as there is a method of Nešetřil and Rödl [1979] for constructing the underlying graph. However, that method has to be iterated many times and results in a huge graph. This graph is a covering graph (see Pretzel [1985a]) so the additional construction which produces an even more gigantic graph is necessary. The reason the girth drops to 6 is that it is impossible (in this construction) to prevent different copies of H being erected on paths which overlap and one then cannot avoid cycles lying in two copies of H, and having three edges in each.

3. THEOREMS ON COVERING GRAPHS.

GENERAL THEOREMS

Very little is known about covering graphs of arbitrary ordered sets and there are only two papers which go beyond the most elementary statements about them. They both relate the circumference Θ and the maximum length of a chord-free cycle β to the chromatic number χ. The first paper is by M. Aigner and G. Prins and has never been published. It is based on Minty's flow-ratio result, and uses it to prove stronger statements for k-good orientations. Their principal result is:

THEOREM (AIGNER AND PRINS [1980]). Let G be a graph and R an acyclic orientation of G. Then R is k-good if and only if for every simple circuit C

$$f(C) \leqslant \begin{cases} \dfrac{|C|}{k} - 1 & \text{if } |C| \leqslant \beta\ , \\[2ex] (\beta - 3) - (\beta - 2)\dfrac{(k-2)|C| + \beta - 2k}{(k-1)|C| + \beta - 2k} & \text{if } |C| \geqslant \beta\ , \end{cases}$$

Noting that the second inequality implies the first when $|C| \geqslant \beta$, they deduce the following corollaries,

COROLLARIES

1. If $\varphi(G) \geqslant 2$, then $\Theta(G) \geqslant 2\chi(G) - 1$;
2. If $\varphi(G) \geqslant 2$, then $\beta(G) \geqslant \chi(G) + 2$;
3. If $\varphi(G) \geqslant k$, then $\chi(G) \leqslant \left\lceil (\beta-2) - (\beta-2)\dfrac{(k-2)\Theta + \beta - 2k}{(k-1)\Theta + \beta - 2k} \right\rceil$.

They also introduce "higher" chromatic numbers χ_k where χ_k is the smallest number of colours in a graph colouring obtained from a k-good orientation. It is clear that if we put $\chi_k = \infty$ for $k > \varphi(G)$ the sequence $\chi_1, \chi_2 \cdots$ is non-decreasing. They consider the corresponding chromatic polynomials and ask whether the sequence χ_k can be strictly increasing.

The second paper is Pretzel [1985a]. It was inspired by Aigner and Prins and produces results analogous to theirs for χ_k, the main one being:

THEOREM (PRETZEL [1985a]). If $\varphi(G) \geqslant k$ and G has invariants β, Θ, χ_k then:

1. $\Theta > k(\chi_k - 1)$.
2. $\beta > (k-1)(\chi_k - 1) + 2$.
3. If $\beta \leqslant k(\chi_k - 1)$ then $\Theta > (2k-1)(\chi_k - 1)$.

Notice that for $k = 2$ and $\chi_2 = \chi_1$, (1) and (2) are equivalent to Corollary 1 and 2 of Aigner and Prins.

This theorem is proved by reorienting the graph by pushing down to show that in any k-orientation there is a cycle which has a flow-ratio $> \chi_k - 2$. That cycle is then used to derive all the statements.

As a test for these Theorems, let us consider whether they suffice to show that M_5 has $\varphi = 1$. The invariants of M_5 are $\chi = \gamma = 4$, $\beta = 6$ and $\Theta = 11$. These satisfy all these inequalities for $k = 2$ so the results don't score very highly. It is possible to elaborate the inequalities by taking account of the number of chords in various cycles, and then the techniques do produce inequalities violated by M_5 , but even these fail for M_9 , the analogous graph based on a cycle of length $\geqslant 9$ (see Pretzel [1985a]). One has to conclude these theorems do not represent very powerful tools for estimating $\varphi(G)$.

Pretzel [1985a] also constructs an example to show that χ_2 can increase arbitrarily while χ_1 remains constant, provided $\chi_1 \geqslant 4$ (for $\chi_1 = 3$ or $\chi_2 = \chi_1$ or $\chi_2 = \infty$ depending on the girth of G). Although his examples have $\gamma = 4$ and hence $\varphi = 2$, it seems likely that this technique can be used to answer Aigner and Prins' question in the affirmative.

CONJECTURE. For all k there exist graphs G such that

$$\chi_1(G) < \ldots\ldots < \chi_k(G) < \infty .$$

COVERING GRAPHS OF SPECIAL CLASSES OF ORDERED SETS. In contrast to the sparseness of results for arbitrary ordered sets, a fair amount is known about covering graphs of modular lattices. Perhaps the earliest result is

THEOREM (M. WARD [1939]). A modular lattice is distributive if and only if its covering graph does not contain $K_{2,3}$.

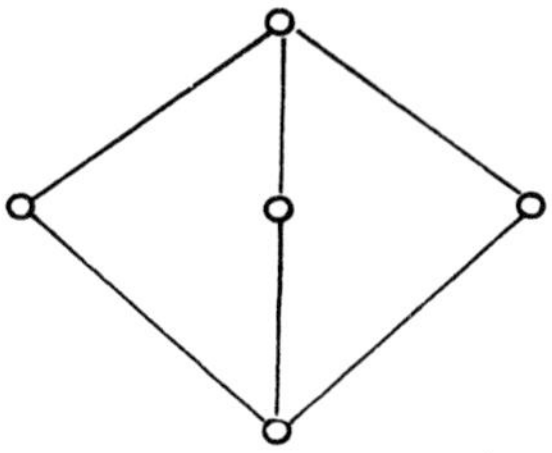

Figure 9: $K_{2,3}$

Obviously this result can be used to extend results on covering graphs of modular lattices to distributive lattices. Another early result is

PROPOSITION. The covering graph of a modular lattice is bipartite.

Alvarez [1965] derives a complete characterization of those bipartite graphs that are covering graphs of modular lattices and (using Ward's result) also characterizes covering graphs of distributive lattices. His conditions are somewhat technical. The exact statement can be found in Alvarez [1965] and in Rival [1985].

Alvarez does not discuss the question of deciding what the possible lattice orientations of the covering graph of a modular lattice are. This problem has been solved by J. Jakubík.

THEOREM (JAKUBÍK [1954]). Two modular lattices L and L' have isomorphic covering graphs if and only if there exist lattices A and B such that $L \simeq A \times B$ and $L' \simeq A^{opp} \times B$. This statement is not true in general for semi-modular lattices.

THEOREM (JAKUBÍK [1975]). If two lattices L and L' have isomorphic covering graphs and L is modular (distributive) then L' is modular (distributive).

One of Alvarez' technical conditions deals with the relation between certain subgraphs of the covering graph of a modular lattice and the covering graph $C(\underset{\sim}{2}^3)$ of the 3-cube.

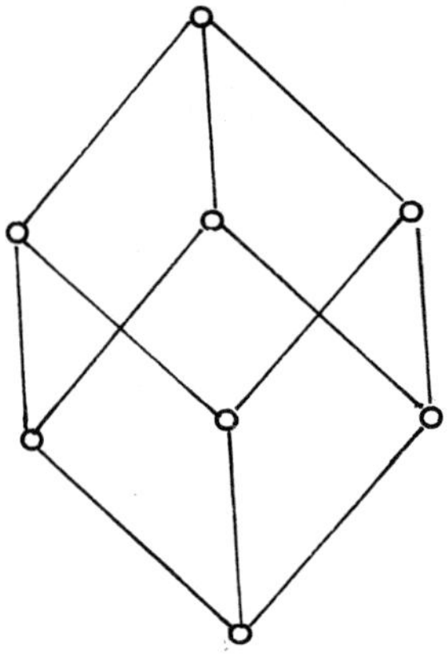

Figure 10 : $C(\underset{\sim}{2}^3)$

This theme is taken up from a different point of view in the following pretty result of D. Duffus and I. Rival [1983].

THEOREM (DUFFUS AND RIVAL [1983]). A graph is the covering graph of a distributive lattice of height n, if and only if it is a retract of $C(\underset{\sim}{2}^n)$ and has diameter n.

Duffus and Rival have also developed the path length techniques of Alvarez and prove results on graded lattices (which are essentially lattices satisfying the Jordan-Hölder condition). They extend the validity of Jakubík's Theorem [1954] to graded lattices in which every element is the join of atoms and the meet of coatoms and also prove the following two theorems.

THEOREM (DUFFUS AND RIVAL [1977]). Let L be a finite semi-modular lattice. Then L is dismantlable if and only if its covering graph does not contain a subgraph isomorphic to $C(\underset{\sim}{2}^3)$.

THEOREM (DUFFUS AND RIVAL [1977]). Let L and L' be graded with isomorphic covering graphs. If L is semi-modular and dismantlable then L' is dismantlable.

A number of authors, in particular H-J. Bandelt, J. Hedliková, O. Klaucová and M. Tomková have taken these investigations somewhat further. They have generalized Jakubík's results to multilattices and also examined lattices with properties intermediate between modularity and distributivity. The most notable such class are the median lattices, characterized by a betweenness axiom. For this class H-J. Bandelt generalizes the theorem of Duffus and Rival [1983].

THEOREM (BANDELT [1984b]). A graph is the covering graph of a median lattice if and only if it is a retract of $C(\underset{\sim}{2}^n)$.

The contrast between the range of results in these areas and those for general ordered sets is most marked. It is clear that the class of ordered sets with which these papers deal is quite restricted, but it might be worth stressing that the graphs and associated orientations have very special properties also. In particular they have a cycle basis of cycles oriented in the following manner.

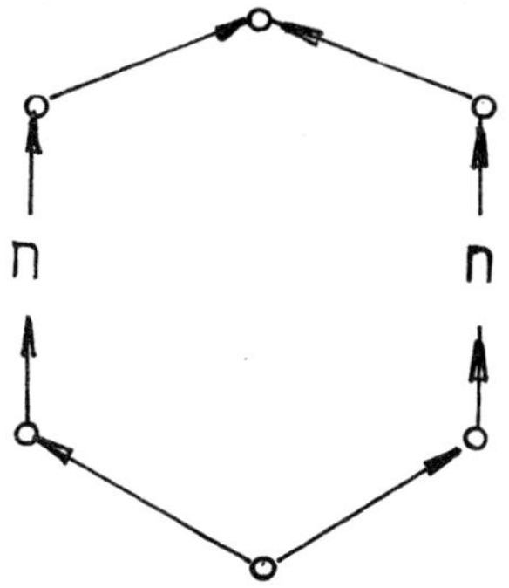

Figure 11

Such graphs are necessarily bipartite, and the available orientations are very heavily restricted. This allows the authors to develop and exploit far more powerful "forcing" arguments in which one can conclude from partial information about an orientation how certain edges must be directed. For general ordered sets, the only tool of this kind is the fact that every circuit must have at least two edges directed each way. For long circuits this is an extremely weak condition to satisfy, so "local" arguments tend to end inconclusively.

In a recent paper R. Jégou, R. Nowakowski and I. Rival have investigated the diagrams of a quite different special class of lattices, those that can be embedded in a plane. Their paper discusses what properties of such lattices are invariants of their covering graphs. They prove two theorems:

THEOREM (JEGOU, NOWAKOWSKI AND RIVAL [1985]). If P and P' are finite lattices with isomorphic covering graphs and P is planar, then P' contains doubly irreducible elements.

and

THEOREM [IBID]. If P and P' are finite lattices with isomorphic covering graphs and P has a planar embedding $e(P)$ in which all its doubly irreductible elements lie on the boundary, then for any planar e(P') embedding of P' the set of faces of e(P') is the same as the set of faces of e(P) .

(An element is doubly irreducible if it is not representable as a non-trivial join or meet in a lattice, and the "faces" of the second theorem, are to be taken in the usual geometric sense).

They also show that the converse of the second theorem is false and end their paper with an interesting conjecture on the range of planar lattice re-orientations of a planar lattice. They conjecture that every such reorientation can be obtained by a sequence of operations of three types. The description of these types is more technical so the reader is referred to their paper for the details. It is clear that a proof of this conjecture would provide a powerful tool for the investigation of planar lattices.

4. PUSHING DOWN MAXIMAL VERTICES. In 1972 K.M. Mosesian published a series of papers in the proceedings of the Armenian academy of sciences. These have never been translated and seem to have remained without echo. They deal with diagram orientations using a terminology based on O. Ore's book [1962]. In them he introduced an operation on orientations, pushing down maximal vertices. This consists of reversing the direction of all the edges incident with a maximal vertex, thus pushing it down and making it minimal.

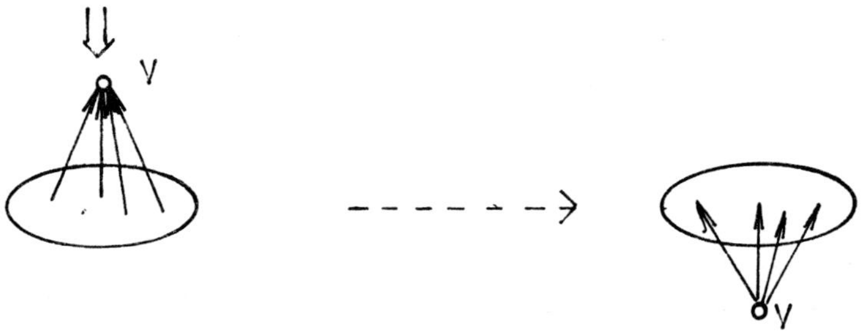

Figure 12

Since v is maximal the two edges incident with v in any circuit C containing v have opposite directions so reversing them both will not change the numbers of forward and backward edges of C . Hence pushing down preserves k-goodness. The reader will have no difficulty in verifying the elementary properties of pushing down listed in the following proposition.

PROPOSITION. Let R be an acyclic orientation of a connected graph G and S obtained from R by pushing down the vertices $v_1,\dots,v_n$ in that order (this includes the assumption that after $v_1,\dots,v_{i-1}$ have been pushed down v_i is maximal, but a vertex can occur several times in the sequence).

(a) If R is k-good, then so is S.

(b) If H is a subgraph of G, then $S|_H$ can be obtained from $R|_H$ by pushing down (in the same order) those vertices of $v_1,\dots,v_{i-1}$ that lie in H.

(c) If x and y are adjacent in G and $x = v_i = v_j$ for $i \neq j$, then $y = v_k$ for some $i < k < j$. So the number of occurences of x and y differ by at most 1 and the edge xy has the same direction in R and S if ond only if x and y occur equally often.

(d) $R = S$ if and only if all vertices of G occur equally often in the sequence $v_1,\dots,v_n$. The sequences in which they all occur once are precisely those obtained by taking the vertices of G in descending order according to some linear extension of R.

(e) If x and z are arbitrary vertices of G the number of occurences of x and z differ by at most $d(x,z)$.

It is possible to generalize (b) to subsets of ordered sets, but care is needed in the formulation of the hypotheses if the subsets are not convex. From (e) one can derive the following Theorem of Mosesian.

THEOREM (MOSESIAN [1972c]). Given a k-good orientation R of a connected graph G it is possible to make any vertex v of G into the (unique) maximum by pushing down.

This allows us to complete the proof given earlier that $\varphi(M_5) = 1$. For if M_5 has any 2-good orientation we can use pushing down to make x into the maximum without destroying 2-goodness. But we have already seen that M_5 has no 2-good orientation with x as its maximum.

We recall that Pretzel [1985a] uses pushing down to prove bounds relating χ_k into the circumference Θ and the chord free circumference β of G. Indeed, pushing down is such a powerful operation that it is worthwhile to find out which orientations can be obtained from a given acyclic orientation R by pushing down, and to ask what properties can be specified in such orientations. We know that if R is k-good then any orientation S obtained from R by pushing down is k-good. But that is not sufficient, for in any circuit C, R and S must produce the same numbers of forward edges. There are many numbers which allow one to calculate the number of forward edges in a circuit C

given the length of C. For instance we could choose Minty's flow-ratio $f(C)$, but it appears that the nicest number to choose is the flow-difference $d_R(C)$ = # (forward edges in C) − # (backward edges in C). The reason for this choice is that it is additive under concatenation and equal on circuits that differ only by retraced paths. So it can easily be calculated for all circuits from a knowledge of its values on a cycle basis. We denote by d_R the function assigning to each circuit its flow-difference. The following theorem shows that d_R characterizes the orientations that can be obtained from R by pushing down.

THEOREM (PRETZEL [1984]). Two acyclic orientations S and R of the same graph G can be obtained from each other by pushing down if and only if they have the same flow-differences: $d_R = d_S$.

The Theorem is proved via two lemmas that are of interest in themselves.

LEMMA. Pushing down is reversible: if S can be obtained from R by pushing down, then R can be obtained from S by pushing down.

This is an easy consequence of (d) in the Proposition above.

LEMMA. If R and S are orientations of G with the same maximum v and $d_R = d_S$, then $R = S$.

This lemma is a generalization of the observation that if in a tree a vertex v is specified as the maximum then the orientation of the tree is fixed. That is no longer true if two vertices are allowed to be maximal (see Figure 13). So the second lemma does not hold if we relax the condition that there should be a maximum.

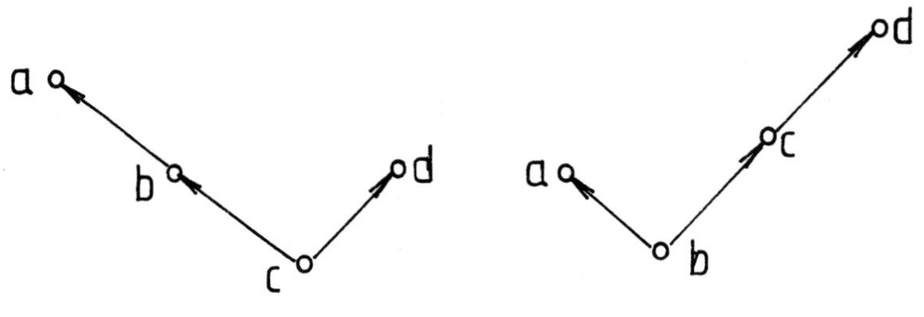

Figure 13

We can also generalize Mosesian's result on maximum vertices as follows.

THEOREM (PRETZEL [1984)]. Given an acyclic orientation R of a connected graph G and M a stable set of vertices such that for any circuit C, $M \cap C$ is at most the number of forward edges of C, then M can be made into the set of maximal vertices by pushing down.

Note that as C passes through a maximal vertex v one of the two edges of C incident with v must be a forward edge. So since pushing down does not change the number of forward edges in any cycle the condition on M is also necessary. If R is k-good the hypothesis of the theorem will be satisfied by any stable set M with $|M| \leqslant k$.

The major question these results leave open is:

QUESTION. What are the possible flow differences d_R of the orientations R of a given graph G?

From Nešetřil [1985] it follows immediately that finding a complete answer to this question is an NP-complete problem. But partial information would still be very valuable.

6. PUSHING DOWN AND OTHER REORIENTATIONS OF GRAPHS. A graph G with a pair (R,S) of orientations of G is a <u>reorientation</u>. We shall denote it by $G : R \to S$. If H is a subgraph of G there is an induced reorientation $H : R|_H \to S|_H$ of H. We shall call a class of reorientations <u>hereditary</u> if with $G : R \to S$ it contains all induced reorientations $H : R|_H \to S|_H$ of subgraphs of G. We shall say the class <u>preserves k-goodness</u>, if whenever R is k-good, so the same is true for S and that it <u>preserves goodness</u> if it preserves k-goodness for all k. For example, it is clear that the class of all reorientations obtained by pushing down is hereditary and preserves goodness. Conversely, the following theorem is easily seen to follow from the results of the previous section.

THEOREM (UNIVERSALITY OF PUSHING DOWN). If a class C of reorientations of finite graphs is hereditary and preserves goodness, then any reorientation $G : R \to S$ in C can be obtained by a sequence of pushdown operations.

As it is not always possible to obtain every k-good orientation of G from a given one by pushing down, it follows that an operation taht acts transitively on the k-good orientations of every graph G is either not hereditary or fails to preserve goodness. It is difficult to construct such operations that remain tame enough for one to predict their behaviour.

To end on a more positive note, we use our knowledge of pushing down to reexamine Minty's Flow-ratio Theorem. This implicitly uses a reorientation technique which can be made explicit as follows. Let $g : G \to \mathbb{Z}$ be a colouring such that for adjacent vertices a,b, $1 < |g(a) - g(b)| < n$ for some n, and let R be the associated orientation. We reorient R, by reducing the values of $g(a)$ to $g'(a) \in \{0,\dots,n-1\}$ so that $g(a) \equiv g'(a) \bmod (n)$. Let us consider the orientation R' associated with g', and the reorientation $G : R \to R'$. In this, the direction of an edge $a \to b$ is reversed if for some k we have $g(a) < kn \leqslant g(b)$ and if that holds, then $(k-1)n < g(a)$ and $(k+1)n > g(b)$. So for any circuit C the numbers of forward and backward edges that get reversed are identical. Hence this reorientation preserves goodness. Let us call these reorientations <u>modulo-n orientations</u>. As modulo-n reorientation is obviously hereditary we have

THEOREM. Modulo-n reorientations can be obtained by pushing down.

In particular, we can extend Minty's theorem as follows:

THEOREM. Let G be a graph with cycles, then

$$\chi_k(G) = \min \lceil F(R) \rceil + 1 \quad ,$$

where the minimum is taken over all k-good orientations of G.

This allows us to substitute χ_k for χ in the results of Aigner and Prins [1980], and also provides an alternative and shorter proof of the related Theorems in Pretzel [1985a]).

We close with four unsolved problems.

1. (I. Rival) What order-theoretic invariants depend only on the covering graph of an ordered set?
2. What is the smallest order and size of a graph of given orientation number φ and girth γ? In particular what is the answer to this question for $\varphi = 1$ and γ large?
3. What are the possible flow-differences of the orientations of a given graph?
4. Find a reversible operation, more general than pushing down, that preserves 2-goodness. Find a system of orientation invariants that characterize those reorientations obtainable by that operation.

BIBLIOGRAPHY

1. M. AIGNER AND G. PRINZ (1980). k-orientable graphs, unpublished preprint, Free University Berlin.

2. L.R. ALVAREZ (1965). Undirected graphs as graphs of modular lattices. Can. J. Math. 17, 923 - 932.

3. H-J. BANDELT (1984a). Discrete Ordered Sets whose covering graphs are median, Proc, A.M.S. 91, 6 - 8.

4. H-J. BANDELT (1984). Retracts of Hypercubes, J. Graph Theory, 8, 501 - 510.

5. H-J. BANDELT AND J. HEDLIKOVA (1983). Median algebras, Discr. Math 45, 1 - 30.

6. G. BIRKHOFF (1948). Lattice Theory, (Second Edition), Amer. Math. Soc. Col., 25, Providence R.I.

7. B. BOLLOBÁS (1978). Extremal Graph Theory, Lond, Math. Soc., Mon 11, Academic Press.

8. J.A. BONDY (1976). Disconnected orientations and a conjecture of Las Vergnas, J. Lon. Math. Soc. (2) 14, 277 - 282.

9. D. DUFFUS AND I. RIVAL(1977). Path length in the covering graph of a modular lattice, Discr. Math. 19, 139 - 158.

10. D. DUFFUS AND I. RIVAL (1983). Graphs orientable as distributive lattices. Proc. Amer. Math. Soc. 88, 197 - 200.

11. P ERDÖS (1959). Graph theory and probability, Can. J. Math 11, 34 - 38.

12. P ERDÖS (1961). Graph theory and probability II, Can. J. Math 13, 346 - 352.

13. T. GALLAI (1967). Transitiv orientierbare Graphen, Acta Math. Acad. Sci. Hungar. 18, 25 - 66.

14. T. GALLAI (1968). On directed paths and circuits, in: Theory of Graphs (eds. P. Erdös and G. Katona), Academic Press, 115 - 118.

15. A. GHOUILA-HOURI (1962). Characterisation des graphes non orientées dont on peut orienter les arètes de manière a obtenir le graphe d'une relation d'ordre, C.R. Acad. Sci. Paris 254, 1370 - 1371.

16. P.C. GILMORE AND A.J. HOFFMAN (1964). A characterization of comparability graphs and interval graphs, Can. J. Math., 16, 539 - 548.

17. C.E. HAFF, U.S.R. MURTY AND R.C. WILTON (1970). A note on undirected graphs realisable as p.o. sets, Can. Math. Bull. 13, 171 - 374.

18. J. HEDLIKOVÁ (1981). Betweenness isomorphisms of modular lattices, Arch. Math. 37, 154 - 162.

19. J. JAKUBÍK (1954). On lattices whose graphs are isomorphic [Russian]. Czech. Math., J. 4, 131 - 141.

20. J. JAKUBÍK (1975). Unoriented graphs of modular lattices, Czech. Math. J., 25, 240 - 246.

21. J. JAKUBÍK AND M. KOLIBIAR (1954). On certain properties of lattices [Russian], Czech Math. J. 4, 1 - 27.

22. R. JÉGOU, R. NOWAKOWSKI AND I. RIVAL (1985). The diagram invariant property for planar lattices, preprint, Ecole des Mines, St Etienne, France.

23. D. KELLY (1985). Comparability graphs, in: Graphs and Order (ed. I. Rival), Nato ASI Series 146, K. Reidel, 3 - 40.

24. O. KLAUCOVÁ (1976a). b-Equivalent multilattices, Math. Slovaca 26, 63 - 72.

25. O. KLAUCOVÁ (1976b). Characterization of distributive multilattices by a betweenness relation, Math. Slovaca 26, 119 - 129.

26. O. KLAUCOVÁ (1977). Lines in directed distributive multilattices Math. Slovaca 27, 55 - 64.

27. O. KLAUCOVÁ (1980). Pairs of multilattices defined on the same set, Math. Slovaca 30, 181 - 186.

28. LAS VERGNAS (1976). Problem, in: Proc. Fifth British Combinatorial Conference (eds. C. St-J, A. Nash-Williams and J. Sheehan), Utilitas Math. Winnipeg, p. 689.

29. G.S. MINTY (1962). A theorem on n-colouring the points of a linear graph, Amer. Math. Monthly 69, 623 - 624.

30. R.H. MÖHRING (1985). Algorithmic aspects of comparability graphs and interval graphs, in: Graphs and Order (ed. I. Rival), Nato ASI series 146, D. Reidel 41 - 102.

31. K.M. MOSESIAN (1972a). A minimal graph that is not strongly basable [Russian], Akad. Nauk, Armian. SSR. Dokl. 54, 8 - 12.

32. K.M. MOSESIAN (1972b). Strongly basable graphs [Russian], Akad. Nauk. Armian. SSR. Dokl. 54, 134 - 138.

33. K.M. MOSESIAN (1972c). Certain theorems on strongly basable graphs [Russian], Akad. Nauk. Armian. SSR. Dokl, 54, 241 - 245.

34. K.M. MOSESIAN (1972d). Basable and strongly basable graphs [Russian], Akad Nauk. Armian. SSR. Dokl. 55, 83 - 86.

35. K.M. MOSESIAN (1973a). Saturated graphs [RUssian], Akad. Nauk. Armian SSR. Dokl, 56, 257 - 262.

36. K.M. MOSESIAN (1973b). Basis graphs of certain orderings [Russian], Akad. Nauk. Armian. SSR. Dokl, 57, 264 - 270.

37. J. NEŠETŘIL (1985). Complexity of Diagrams, preprint, Charles University, Prague.

38. J. NEŠETŘIL AND V. RÖDL (1978). On a probabilistic graph-theoretical method, Proc. Amer. Math. Soc. 72, 417 - 421.

39. J. NEŠETŘIL AND V. RÖDL (1979). A short proof of the existence of highly chromatic graphs without short cycles, J. Comb. th. B27, 225 - 227.

40. O. ORE (1962). Theory of graphs, Amer. Math. Soc. Coll. 34, Providence, R.I.

41. O. PRETZEL (1984). On reorienting graphs by pushing down maximal vertices, preprint, Imperial College, London.

42. O. PRETZEL (1985a). On graphs that can be oriented as diagrams of ordered sets, Order 2, 25 - 40.

43. O. PRETZEL (1985b). Construction of a non-covering graph of girth six, preprint, Imperial College, London.

44. I. RIVAL (1985b). The diagram, in: Graphs and Order (ed. I. Rival) NATO ASI series 147, D. Reidel, 103 - 133.

45. I. RIVAL (1985b). The diagram, "Unsolved Problems", Order 2, 101-104.

46. H.E. ROBBINS (1939). A theorem on graphs iwth an application to a problem of traffic control, Amer. Math. Monthly 46, 281 - 283.

47. B. ROY (1967). Nombre chromatique et plus longs chemins d'un graphe, Rev. AFIRO, 1, 127 - 132.

48. M. TOMKOVÁ (1977). On the b-equivalence of multilattices, Math. Slovaca 27, 331 - 336.

49. M. TOMKOVÁ (1980a). Graph isomorphisms of modular multilattices, Math. Slovaca 30, 95 - 100.

50. M. TOMKOVÁ (1980b). On a local property of the unoriented graph of a modular multilattice, Math. Slovaca 30, 289 - 298.

51. M. TOMKOVÁ (1982). On multilattices with isomorphic graphs, Math. Slovaca 32, 63 - 73.

52. M. WARD (1939). The algebra of lattice functions, Duke J. Math. 5, 357 - 371.

DEPARTMENT OF MATHEMATICS
IMPERIAL COLLEGE
UNIVERSITY OF LONDON
LONDON SW7 2BZ

Contemporary Mathematics
Volume 57, 1986

Abstract convexity and meet-distributive lattices

Paul H. Edelman*

ABSTRACT: In this paper we discuss the recent development of structures called convex geometries which combinatorially abstract the notion of the convex hull of a set of points in Euclidean space in the same way that matroids abstract the idea of linear independence. The role of geometric lattices for matroids is played by meet-distributive lattices. We detail the history of these ideas and outline some of the interactions between the lattice theory and the combinatorics.

0. Introduction.

One of the most fruitful aspects of matroid theory is the equivalence of matroids with geometric lattices. This equivalence allows considerable interaction between combinatorics and lattice theory with both sides benefitting from the techniques of the other. Recent years have seen the development of structures which combinatorially abstract the notion of the convex hull of a set of points in Euclidean space in the same way that matroids abstract the idea of linear independence. In this study of abstract convexity the role of geometric lattices is played by meet-distributive lattices.

Meet-distributive lattices, although dating back to 1941, have not received the amount of attention that they deserve. We

---

1980 Mathematics Subject Classification. 06A15, 06B15, 52A01
*Supported in part by a grant from the NSF MCS-8301089

feel that for many problems they are the right level of generality in which to work. Moreover, we now have a combinatorial object which is equivalent to these lattices, so the stage is set for a similar interplay between combinatorics and lattice theory as that of matroids. It is this interaction on which we will focus in this paper.

The paper is organized in the following way: Section I is a discussion of the latticial characterizations of meet-distributive lattices, and some of their properties. In Section II we introduce the combinatorial objects which give rise to meet-distributive lattices. Section III presents three examples. The first of these examples is well-known. The second, to our knowledge, has not appeared in print in this context previously. The third example is not well-known and is particularly interesting. Finally in Section IV we give two examples of how the combinatorial and latticial points-of-view interact. One of these examples has appeared elsewhere. The last is a generalization of Dilworth's theorem on the dimension of distributive lattices, and has not appeared previously. Theorems are for the most part presented without proof, as most have appeared or will appear shortly in the literature.

We will restrict ourselves to finite sets. This is because the combinatorial development is primarily finite. The lattices have been studied in the infinite case. See [Di2], [CD], [Av4].

Most of our notation is standard. If L is a lattice, its maximum and minimum elements will be denoted $\hat{1}$ and $\hat{0}$, respectively. If all maximal chains in L have the same length, we will call L <u>graded</u>. For $x \in L$, let $rk(x)$, the <u>rank of x</u>, be the length of the longest chain from $\hat{0}$ to x.

Thus $rk(x)$ is defined even if L is not graded. The <u>rank of L</u> is equal to $rk(\hat{1})$.

By L^* we will mean the order dual of L . If $x,y \in L$ and $x \leq y$ then the interval $[x,y]$ is the set $[x,y] = \{z \in L | x \leq z \leq y\}$ with the induced ordering. By $[n]$ we mean the set $[n] = \{1,2,...,n\}$. If X is a finite set by $|X|$ we mean its cardinality.

I. Meet-distributive lattices.

The class of lattices which are central to this paper has undergone a number of name changes. They were first considered by Dilworth [Dil] who left them unchristened. Avann [Avl] subsequently named them <u>lower semi-distributive lattices</u>. This was altered to <u>lower locally distributive lattices</u> by Greene and Markowsky [GM], then to <u>locally meet-distributive lattices</u> by Stanley [St] and finally to <u>meet-distributive lattices</u> by ourselves [Edl]. We will retain this last name and define them as follows: A lattice L is called <u>meet-distributive</u> (m-d) if for every $y \in L$, if $x \in L$ is the meet of elements covered by y then $[x,y]$ is a boolean algebra.

If L is a lattice let $J(L)$ be the poset of join-irreducibles of L . For $x \in L$ we will let $T(x) = \{y \in J(L) | y \leq x\}$ and $\tau(x) = |T(x)|$. Finally, $\nu(x)$, the <u>join-excess function</u>, is defined by $\nu(x) = \tau(x) - rk(x)$. The function $\nu(x)$ was introduced by Avann in [Avl] where he proves the following theorem which is remarkable in its breadth of characterizations:

Theorem 1.1. The following are equivalent for an arbitrary finite lattice L

A) L is graded and $\nu(\hat{1}) = 0$.

B) L is graded and $\nu(x) = 0$ for all $x \in L$.

C) L is lower semi-modular and $\nu(\hat{1}) = 0$.

D) L is lower semi-modular and has no non-distributive modular sublattice of order 5 .

E) Every $x \in L$ can be uniquely expressed as the join of a minimal set of join-irreducibles in L .

F) L is lower semi-modular and every upper semi-modular sublattice is distributive.

G) L is lower semi-modular and every modular sublattice of L is distributive.

H) L is meet-distributive.

#

As Avann remarks, Dilworth had previously proven the equivalence of E and G (in their dual form) and that these two conditions imply the dual statements of B and H [Dil]. In subsequent papers [Av2-3] Avann develops more of the theory of the function ν .

For us it will be useful to rephrase some of these equivalences. Let B_J be the set of all subsets of J(L) ordered by containment. As is well-known, the map $T\colon L \to B_J$, defined above, embeds L into B_J so as to preserve the meet operation. The following theorem first appears explicitly in [GM] but follows from some of the equivalences above.

<u>Theorem 1.2</u>. L is meet-distributive if and only if L is graded and the map T embeds L into B_J preserving both meet and rank.

#

A consequence of Theorem 1.2 is that we can identify a m-d lattice L with a particular collection of subsets of $J(L)$ which is closed under intersection and contains both the empty set and $J(L)$. It is this collection which we study in Section II.

Subsequent to the work of Avann m-d lattices were rediscovered a number of times. Pfaltz [Pf1-2] in his study of convexity of directed graphs introduced the notion of a <u>G-lattice</u> which are exactly atomic m-d lattices. In [Pf1] he studies the way these lattices behave under lower semihomomorphisms, i.e. those functions $\sigma: L \to L'$ such that for all $x,y \in L$, $\sigma(x \wedge y) = \sigma(x) \wedge \sigma(y)$ and if $x \vee y$ covers both x and y then $\sigma(x \vee y) = \sigma(x) \vee \sigma(y)$.

Boulaye [Bo2] calls m-d lattices "treillis α-affaibli " and gives an inductive procedure for constructing them. (This paper has several errors in the statements of the theorems and contains no proofs.)

At this point we will close this section about lattices with some discussion of their combinatorial properties. Let μ be the Möbius function of the lattice L.

<u>Theorem 1.3</u> [EJ, Corollary 4.4]. If L is a meet-distributive lattice and $x,y \in L$ then

$$\mu(x,y) = \begin{cases} (-1)^{rk(y)-rk(x)} & \text{if } [x,y] \text{ is a boolean algebra.} \\ 0 & \text{otherwise.} \end{cases}$$

Proof. This follows immediately from the cross-cut theorem [Ro].

#

For L a lattice let $\overline{L} = L - \{\hat{0},\hat{1}\}$ and $\Delta(\overline{L})$ be the abstract simplicial complex of chains in $\overline{L}$. We will identify $\Delta(\overline{L})$ with its geometric realization. For other definitions see [BGS].

Theorem 1.4 [EJ, Corollary 4.9 and Theorem 4.10]. If L is a meet-distributive lattice then L is lexicographically shellable and hence Cohen-Macaulay. Moreover, if L is not a boolean algebra then $\Delta(\overline{L})$ is a ball of dimension $rk(L) - 2$.

#

II. Abstract Convexity.

So far we have only discussed latticial characterizations of m-d lattices. There is a more combinatorial approach to m-d lattices through some notions of abstract convexity. We will present two points-of-view. The first approach was developed jointly with R. Jamison after we had separately discovered many of the key ideas. The second is due to Korte and Lovasz and arises from their work on greedoids.

Let X be a finite set and $\mathcal{L}$ be a collection of subsets of X satisfying the conditions

A1) $\phi \in L$ and $X \in L$.

A2) $A \in L$ and $B \in L$ implies that $A \cap B \in L$.

L is called an <u>alignment</u> of X. We can alternatively think of L as being a closure operator. That is, for any subset $A \subseteq X$, we define the closure of A, $L(A)$, to be

$$L(A) = \bigcap_{\{C \in L \mid C \supseteq A\}} C .$$

It is easy to check that L is a closure operator on X, i.e. L is a function from 2^X to itself such that

(1) $A \subseteq L(A)$

(2) $A \subseteq B$ implies $L(A) \subseteq L(B)$

(3) $L(L(A)) = L(A)$

with the additional condition that $L(\phi) = \phi$. The subsets in L or, equivalently, those subsets of X of the form $L(A)$ for some $A \subseteq X$, will be called <u>convex sets</u>. Strictly speaking, it is not necessary to include the assumption that $\phi \in L$, but it simplifies matters to make it so. We will move between these two interpretations of L as is convenient.

We now make some definitions which will make sense for any alignments. Let A be a subset of X. A <u>basis</u> for A is a minimal subset $S \subseteq A$ such that $L(S) = L(A)$. A priori there may be many bases for a particular set. A point $p \in A$ is called an <u>extreme point</u> of A if $p \notin L(A-p)$. The set of extreme points of A is denoted $ex(A)$. Extreme points may or may not exist. Notice that ex(A) is contained in every basis of A .

Suppose $p \in X$. A <u>copoint C attached at p</u> is a maximal convex set in $X-p$. There may be more than one copoint attached at a point.

Finally we define the anti-exchange condition. We say that L is <u>anti-exchange</u> if, given any convex set K , and two unequal points p and q in X , neither in K , then $q \in L(K \cup p)$ implies that $p \notin L(K \cup q)$.

<u>Theorem 2.1</u> ([EJ, Theorem 2.1]). Let L be an alignment of a set X . Then the following are equivalent:

a) L is anti-exchange

b) For every convex set K , there exists a point $p \in X$ such that $K \cup p$ is convex.

c) For every point p and C a copoint attached at p , $C \cup p$ is convex.

d) Every subset $A \subseteq X$ has a unique basis.

e) For every convex set K , $K = L(\text{ex}(K))$.

f) For every convex set K and $p \notin K$, $p \in \text{ex}(L(K \cup p))$.

#

We will call the pair (X, L) satisfying the conditions in Theorem 2.1 a <u>convex geometry</u> (c.g.). If X is unambiguous then we will just refer to L as being a c.g. For a complete survey of properties of c.g.'s see [EJ].

Given any alignment L , we can partially order the convex sets in L by containment. This gives us a lattice $L(L)$, where for every $C, K \in \mathcal{L}$, $C \wedge K = C \cap K$ and $C \vee K = \mathcal{L}(C \cup K)$. Our main theorem relating this section with Section I of this paper is

<u>Theorem 2.2</u> [Edl, Theorem 3.3]. A lattice L is meet-distributive if and only if $L = L(\mathcal{L})$ for some convex geometry $\mathcal{L}$.

<u>Proof</u>: One direction of the theorem is proved by using the embedding in Theorem 1.2 to define an alignment $(J(L), \mathcal{L} = \{T(x) | x \in L\})$ for every m-d lattice. It then follows from Theorem 1.1E and Theorem 2.1d that $(J(L), \mathcal{L})$ is a c.g.

The other direction is obtained by realizing that in $L = L(\mathcal{L})$ the join-irreducibles are $J(L) = \{\mathcal{L}(p) | p \in X\}$. Applying Theorem 2.1d and Theorem 1.1E finishes the proof.

#

We can now play-off Theorems 1.1 and 2.1 against each other to gain some new insights. For instance, we can deduce from Theorem 2.1b and Theorem 1.1A that if $\mathcal{L}$ is an alignment and for each $K \in \mathcal{L}$ there <u>exists</u> a point $p \in X$ such that $K \cup p \in \mathcal{L}$, then in $L(\mathcal{L})$ K is covered <u>only</u> by convex sets of the form $K \cup q$ for some $q \in X$. It is not clear how all these equivalences interact. How conditions F and G in Theorem 1.1 are mirrored in the related c.g. is not obvious.

There is another combinatorial development that also leads to m-d lattices. This is through the theory of <u>greedoids</u> developed in a series of papers by Korte and Lovasz [KL 1-5]. We will briefly outline this now.

Let X be a finite set and $\mathcal{F}$ be a collection of subsets of X satisfying the two conditions

G1) $\phi \in \mathcal{F}$

G2) If $F, G \in \mathcal{F}$ and $|F| > |G|$ then there exists an $x \in F-G$ such that $G \cup x \in \mathcal{F}$.

The set system $(X, \mathcal{F})$ is called a <u>greedoid</u>. We should remark that this is just one of many descriptions of a greedoid.

A greedoid is said to have the <u>interval property without upper bounds</u> if for all $G,H \in \mathcal{F}$ with $G \subseteq H$ and $x \in X-H$ such that $G \cup x \in \mathcal{F}$ then it follows that $H \cup x \in \mathcal{F}$. A nonempty set system $(X, \mathcal{F})$ is called <u>accessible</u> if for all $F \in \mathcal{F}$ there exists $x \in F$ such that $F-x \in \mathcal{F}$.

<u>Theorem 2.3</u> [KL3, Lemmas 3.2 and 3.4]. For $(X,\mathcal{F})$ a set-system the following are equivalent:

i) $(X, \mathcal{F})$ is a greedoid with the interval property without upperbounds.

ii) $(X, \mathcal{F})$ is accessible and the collection $\mathcal{F}$ is closed under union.

iii) If $\mathcal{L} = \{X-F | F \in \mathcal{F}\}$ then $(X, \mathcal{L})$ is a convex geometry.

#

The greedoids satisfying any of the above conditions have been called <u>APS-greedoids</u>, <u>shelling structures</u>, <u>locally-free</u>

selectors [Cr], and most recently anti-matroids. This connection between shelling structures and convex geometries was first recognized by Björner.

III. Examples.

There are several places where one can find detailed examples of m-d lattices, convex geometries, and shelling structures. See for instance, [KL2, section 3] and [EJ, section III]. Here we will mention one standard example, one example which has not previously been discussed in this context, and one more that we feel deserves more attention.

Example I. Points in $\mathbf{R}^d$ and oriented matroids.

Let $X = \{x_1, x_2, \ldots, x_n\}$ be a set of points in $\mathbf{R}^d$. For $A \subseteq X$, $A = \{a_1, a_2, \ldots, a_k\}$ define

$$L(A) = \{x \in X \mid x = \sum_{i=1}^{k} \lambda_i a_i \text{ for some } \lambda_i \geq 0 ,\ \sum_{i=1}^{k} \lambda_i = 1\} .$$

In other words, $L(A)$ is the intersection of the convex hull of A with the set X. The Krein-Milman Theorem and Theorem 2.1E imply that (X, L) is a convex geometry. This has a natural generalization to oriented matroids. See [Ed3] for details.

Example II. Separating sets in graphs.

The next example comes from the work of Pym and Perfect [PP]. Their ideas were developed subsequently by Polat [Po]. Let G be a graph with vertex set V. Fix Y a subset of the vertices of G. For $E,F \subseteq V$, define $E \prec F$ if every path from a vertex in E to a vertex in Y contains a vertex in F, i.e. F separates E from Y. This ordering is easily seen to be reflexive and transitive. Unfortunately it is not anti-symmetric.

For a subset $E \subseteq V$ write

$$[E] = \{x \varepsilon E \mid \text{there is a path from some node } v \text{ in } E \text{ to } Y \text{ whose last node in } E \text{ is } x\}.$$

Then $[[E]] = [E]$ for all $E \subseteq V$.

Theorem 3.1 [PP, Lemma 7.2]. The relation $\prec$ is a partial ordering on $\{E \subseteq V \mid E = [E]\}$ and under this ordering this set is a lattice.

Theorem 3.2 [Po, page 743]. The lattice defined above is dually meet-distributive.

#

Example III. Ordered sets.

There are two c.g.'s related to k-families of a poset which we will discuss in this section. They are both little-known and poorly understood.

A subset A of a poset P is called a k-family if A

contains no chain of cardinality $k+1$. Thus an anti-chain is a 1-family. Let $\mathcal{A}_k(P)$ be the collection of k-families of P. We follow the development in [GK] where these ideas were introduced.

A k-family A can be uniquely partitioned into disjoint, possibly empty antichains $A_1, A_2, \ldots, A_k$ such that $A_k \leq A_{k-1} \leq \ldots \leq A_1$ where $A_i \leq A_j$ means that for every $a_i \in A_i$ there exists an $a_j \in A_j$ such that $a_i \leq a_j$ in P [GK, Lemma 2.3]. This is called the <u>canonical partition</u> of A. Given two k-families $A, B \in \mathcal{A}_k(P)$ define $A \leq B$ if and only if $A_i \leq B_i$ for all $i = 1, \ldots, k$, where $A_1, A_2, \ldots, A_k$ and $B_1, B_2, \ldots, B_k$ are the respective canonical partitions of A and B. This partially orders $\mathcal{A}_k(P)$. Let $\mathcal{A}_k^*(P)$ be the order dual of $\mathcal{A}_k(P)$ under this ordering.

By [GK, Theorem 2.32], $\mathcal{A}_k^*(P)$ is a m-d lattice. Curiously we know of no good description of the related anti-exchange closure. It is even a mystery as to the appropriate description of the underlying set X.

Our second example is due to M.E. Saks [Sa]. For fixed integer $k \geq 1$ define a closure $\mathcal{D}_P^k$ on the points of a poset P by

$$\mathcal{D}_P^k(A) = A \cup \{y \in P \mid y < x_1 < x_2 < \ldots < x_k \text{ for some chain } \{x_1, x_2, \ldots, x_k\} \subseteq A\}$$

for all subsets A of P. Note that $\mathcal{D}_P^1$ is just the set of order ideals of P. It is easy to check that $\mathcal{D}_P^k$ is a c.g.

The two c.g.'s $\mathcal{D}_P^k$ and $\mathcal{A}_k^*(P)$ are related in some ways. For instance

Theorem 3.3 [EJ, Corollary 3.4]. $|\mathcal{D}_P^k| = |A_k^*(P)|$.

#

On the other hand, these two c.g.'s have very different properties. The lattice $L(\mathcal{D}_P^k)$ is atomic and has rank $|P|$ where as the lattice $A_k^*(P)$ is not atomic and has rank

$$\sum_{\{x \varepsilon P\}} \min\{\delta_P(x), k\}$$

where $\delta_P(x)$ is the cardinality of the longest chain in P with minimum element x . The exact relationship between $\mathcal{D}_P^k$ and $A_k^*(P)$ remains a mystery which we think merits investigation.

IV. Applications.

In this section we will show how the interpretation of m-d lattices as lattices of convex sets of a convex geometry can be used to further understand the combinatorics of these lattices. We will discuss two such instances, a combinatorial interpretation of the zeta polynomial of an m-d lattice which generalizes a result of Stanley, and a new notion of dimension for m-d lattices that will lead to a generalization of a theorem of Dilworth. Some of the results in the latter section have not appeared previously.

For a lattice L and positive integer n define the zeta polynomial of L , $Z(L,n)$, to be the number of multichains in L of the form $\hat{0} \leq x_1 \leq x_2 \leq \ldots \leq x_n = \hat{1}$. It can be shown by computations in the incidence algebra of L that

$Z(L,n)$ is a polynomial in the variable n. (See [St, section 3] and [Ed2].) Since it is a polynomial $Z(L,n)$ can be evaluated at negative integers and for $n > 0$

$$Z(L,-n) = \Sigma \prod_{i=1}^{n} \mu(x_{i-1},x_i)$$

where $\mu(x_{i-1},x_i)$ is the Möbius function in L and the sum is over all the multichains in L of the form
$\hat{0} \leqslant x_1 \leqslant x_2 \leqslant \dots \leqslant x_n = \hat{1}$.

Let $(X, \mathcal{L})$ be a c.g., and f a function from X to the set $[n]$. If $K \in \mathcal{L}$ define $f_K = \max_{x \in K} f(x)$. We will call f <u>extremal</u> if for every convex set K, $\{x \in K | f(x) = f_K\} \cap ex(K) \neq \emptyset$, i.e. for every convex set K, f achieves its maximum on K at some extreme point. The function will be called <u>strictly extremal</u> if for every convex set K

$$\{x \in K | f(x) = f_K\} \subseteq ex(K)$$

i.e. for every convex set K, f achieves its maximum only on extreme points of K. The following theorem generalizes a duality theorem of Stanley [St, Proposition 2.1].

<u>Theorem 4.1</u> [EJ, Theorem 4.7]. For $(X, \mathcal{L})$ a convex geometry, L the lattice of convex sets of $\mathcal{L}$, and $n > 0$

$$Z(L,n) = \#\text{ of extremal functions } f\colon X \to [n]$$

and

$$(-1)^{|X|} Z(L,-n) = \#\text{ of strictly extremal functions } g: X \to [n] \;.$$

#

Finally, we want to discuss how the convex geometry approach leads us to a new notion of dimension for m-d lattices. We begin by defining a join operation on c.g.'s . Given two alignments on the same ground set $(X, \mathcal{L})$ and $(X, \mathcal{M})$ we define the join of $\mathcal{L}$ and $\mathcal{M}$, denoted $\mathcal{L} \vee \mathcal{M}$ to be the alignment

$$\mathcal{L} \vee \mathcal{M} = \{C \subseteq X \mid C = L \cap M \text{ for some } L \in \mathcal{L} \text{ and } M \in \mathcal{M}\} \;.$$

<u>Theorem 4.1</u> [Edl, Theorem 2.2]. If $(X, \mathcal{L})$ and $(X, \mathcal{M})$ are convex geometries then $(X, \mathcal{L} \vee \mathcal{M})$ is also a convex geometry.

#

Let CG(X) be the set of all c.g.'s on the set X . Partially order CG(X) by $(X, \mathcal{L}) \leq (X, \mathcal{M})$ if and only if $\mathcal{L} \subseteq \mathcal{M}$. Then Theorem 4.2 shows that this partial order is a join semi-lattice. Note that this is equivalent to a partial order on all m-d lattices which have their join-irreducibles labelled by the set X .

<u>Theorem 4.2</u> [EJ, Theorem 5.4]. The length of all maximal chains in an interval $[\mathcal{L}, \mathcal{M}]$ in CG(X) is $|\mathcal{M}| - |\mathcal{L}|$.

#

Let $E = x_1, x_2, \ldots, x_n$ be a total ordering of the set X. By $\mathcal{D}_E$ we mean the alignment $\mathcal{D}_E = \{C_i \mid C_i = \{x_1, x_2, \ldots, x_i\}\ 1 \leq i \leq n\}$. This we call the <u>monotone alignment</u> on E. It is easy to see that $(X, \mathcal{D}_E)$ is a minimal element in $CG(X)$ since $\mathcal{D}_E$ contains exactly one subset of X of each cardinality.

For $(X, \mathcal{L})$ a c.g. we define a <u>compatible ordering</u> of $\mathcal{L}$ to be a total ordering of the elements of X, $x_1, x_2, \ldots, x_n$ such that the set $C_i = \{x_1, x_2, \ldots, x_i\} \in \mathcal{L}$ for all i, $1 \leq i \leq n$. Note that a compatible ordering of $\mathcal{L}$ corresponds to a maximal chain in $L(\mathcal{L})$, i.e. $\emptyset \subseteq C_1 \subseteq \ldots \subseteq C_n = X$ is a maximal chain in $L(\mathcal{L})$ and every maximal chain in $L(\mathcal{L})$ gives rise to a compatible ordering. Denote by $\mathrm{Comp}(\mathcal{L})$ the set of compatible orderings of $\mathcal{L}$.

<u>Theorem 4.4</u> [EJ, Theorem 5.2]. For every convex geometry $(X, \mathcal{L})$ we have $\mathcal{L} = \bigvee_{E \in \mathrm{Comp}(\mathcal{L})} \mathcal{D}_E$. Thus the monotone alignments are exactly the minimal elements of $CG(X)$ and $CG(X)$ is an atomic join-semilattice.

#

As a consequence we are lead to the definition of convex dimension. A set of total orderings of X, $E_1, E_2, \ldots, E_k$ is said to <u>realize</u> $\mathcal{L}$ if $\mathcal{L} = \bigvee_{i=1}^{k} D_{E_i}$. Let $\mathrm{cdim}(\mathcal{L})$, the <u>convex dimension</u> of $\mathcal{L}$, be the minimum number of total orderings needed to realize $\mathcal{L}$, i.e. $\mathrm{cdim}(\mathcal{L})$ is the minimum number of atoms of $CG(X)$ whose join is $\mathcal{L}$. This can be extended to a definition of $\mathrm{cdim}(L)$ for a m-d lattice L by $\mathrm{cdim}(L) = \mathrm{cdim}(\mathcal{L})$ where $\mathcal{L}$ is a c.g. such that $L = L(\mathcal{L})$.

<u>Theorem 4.5 [ES]</u>. For L a meet-distributive lattice, $\mathrm{cdim}(L) = w(M(L))$ where $w(M(L))$ is the width of the poset of meet-irreducibles of L.

#

This theorem is a natural generalization of Dilworth's theorem that says that the dimension in the sense of Dushnik-Miller [DM] of a distributive lattice L is $w(M(L))$ (see [Ai,Theorem 8.17]). It is interesting that Theorem 4.5 comes not from analyzing the Dushnik-Miller dimension of a m-d lattice, but rather by keeping the answer as for distributive lattices and altering the notion of dimension.

<u>Corollary 4.6</u>. For any meet-distributive lattice L, $\mathrm{cdim}(L) \geq \dim(L)$, where $\dim(L)$ is the Dushnik-Miller dimension of L.

#

There are numerous examples where these two quantities are not equal, one of which is shown in figure 1.

Theorem 4.5 indicates that the poset $M(L)$ carries considerable information about a m-d lattice L. We can make this notion more precise as follows: Let P be a poset and $f: P \to [k]$. For $A \subseteq P$ define $f(A) = \{f(a) | a \in A\}$. For $x \in P$, let U_x, the filter generated by x, be $U_x = \{y \in P | y \geq x\}$.

<u>Theorem 4.7 [ES]</u>. A poset P is the poset of meet-irreducibles of a meet-distributive lattice of rank k if and only if there exists an onto function $f: P \to [k]$ such that for each

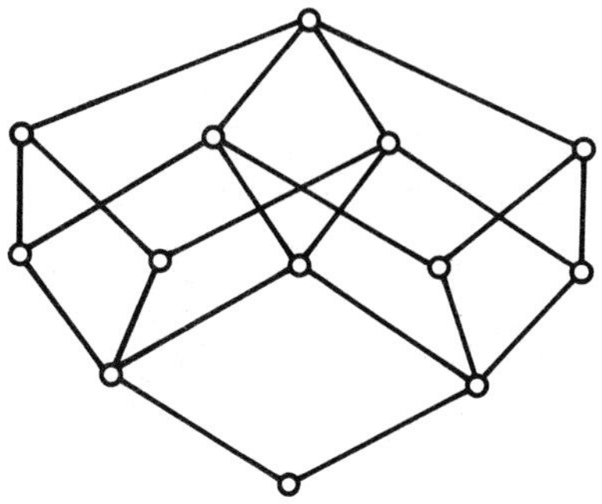

cdim = 4 dim = 3

Figure 1

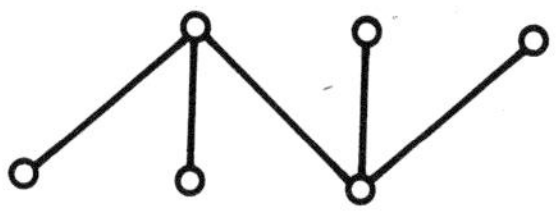

Figure 2

$x, y \in P$, $f(U_x) \neq f(U)$ for any filter U in P , $U \neq U_x$, and $f(x) \neq f(y)$ if $x \lessdot y$.

#

As an example consider the poset P in figure 2. The only value of k for which a function described in Theorem 4.7 exists is $k = 6$. This in fact implies that if $P = M(L)$ then L is distributive and $J(L) = P$ as well.

References

We have attempted to compile as complete a bibliography concerning the ideas in this paper as possible. Many of these references first appeared in a note of B. Monjardet [Mo] who uncovered several papers previously unknown to us. We would appreciate receiving information about papers we have overlooked.

[Ai] M. Aigner, Combinatorial Theory, Springer-Verlag, New York, 1979.

[Av1] S. P. Avann, Application of the join-irreducible excess function to semimodular lattices, Math. Annalen, 142 (1961), 345-354.

[Av2] ___________, Distributive properties in semimodular lattices, Math. Zeitschr. 76 (1961), 283-287.

[Av3] ___________, Increases in the join-excess function in a lattice, Math. Annalen, 154 (1964), 420-426.

[Av4] ___________, Locally atomic upper locally distributive lattices, Math. Annalen, 175 (1968), 320-336.

[Av5] ___________, The lattice of natural partial orders, Aequ. Math. 8 (1972), 95-102.

[BGS] A. Björner, A. Garsia, and R. Stanley, An introduction to Cohen-Macaulay partially ordered sets, in Ordered Sets (I. Rival, ed.), D. Reidel, Dordrecht, 1982.

[Bo1] G. Boulaye, Sous-arbres et homomorphismes a classes connexes dans un arbre, in Theory of graphs: International Symposium, Gordon and Breach, New York, 1967.

[Bo2] G. Boulaye, Notion d'extension dans les treillis et methodes booleennes, Rev. Roum. Maths. Pures et Appl., 13 (1968), 1225-1231.

[Bo3] __________, Sur l'ensemble ordonne des parties connexes d'un graphe connexe, Rev. Franc. Inf. Rech. Op., 2 (1968), 13-25.

[Cr] H. H. Crapo, Selectors, Adv. in Math., 54 (1984), 233-277.

[CD] P. Crawley and R. P. Dilworth, Decomposition theory for lattices without chain conditions, Trans. Amer. Math. Soc., 96 (1960), 1-22.

[DK] R. A. Dean and G. Keller, Natural partial orders, Canad. J. Math., 20 (1968), 535-554.

[Di1] R. P. Dilworth, Lattices with unique irreducible decompositions, Ann. of Math., 41 (1940), 771-777.

[Di2] __________, Ideals in Birkhoff lattices, Trans. Amer. Math. Soc., 49 (1941), 325-353.

[DM] B. Dushnik and E. Miller, Partially ordered sets, Amer. J. Math., 63 (1941), 600-610.

[Ed1] P. H. Edelman, Meet-distributive lattices and the anti-exchange closure, Alg. Univ., 10 (1980), 290-299.

[Ed2] __________, Zeta polynomials and the Möbius function, Europ. J. Comb., 1 (1980), 335-340.

[Ed3] __________, The lattice of convex sets of an oriented matroid, J. Combin. Th. (B), 33 (1982), 239-244.

[EJ] P. H. Edelman and R. Jamison, The theory of convex

geometries, Geom. Dedic., to appear.

[EK] P. H. Edelman and P. Klingsberg, The subposet lattice and the order polynomial, Europ. J. Comb., 3 (1982), 341-346.

[ES] P. H. Edelman and M. E. Saks, The convex dimension of meet-distributive lattices, to appear.

[FJ] M. Farber and R. Jamison, Convexity in graphs and hypergraphs, Report CORR 83-46, University of Waterloo, 1983.

[GK] C. Greene and D. J. Kleitman, The structure of Sperner k-families, J. Combin. Th. (A), 20 (1976), 41-68.

[GM] C. Greene and G. Markowsky, A combinatorial test for local distributivity, IBM Technical Report No. RC5129, November, 1974.

[J1] R. Jamison-Waldner, A convexity characterization of ordered sets, Proc. of the Tenth Southeastern Conf. on Combinatorics, Graph Theory, and Computing, (Boca Raton), Cong. Num., 24 (1979), 529-540.

[J2] ________, Copoints in antimatroids, Proc. of the Eleventh Southeastern Conf. on Combinatorics, Graph Theory and Computing (Baton Rouge), Cong. Num. 29 (1980), 535-544.

[J3] ________, Convexity and block graphs, Proc. of the Twelfth Southeastern Conf. on Combinatorics, Graph Theory and Computing, (Boca Raton), Cong. Num. 33 (1981), 129-142.

[J4] ________, Partition numbers for trees and ordered sets, Pac. J. Math., 96 (1981), 115-140.

[J5] ________, A perspective on abstract convexity, classifying alignments by varieties, in Convexity and Related Combinatorial Geometry, D. C. Kay and M. Breen, ed., Marcel Dekker, Inc., New York, 1982.

[J6] ____________________, Tietze's convexity theorem for semilattices and lattices, Semigroup Forum, 15 (1978), 357-373.

[KL1] B. Korte and L. Lovasz, Structural properties of greedoids, Combinatorica, 3 (1983), 359-374.

[KL2] ______________________, Greedoids, a structural framework for the greedy algorithm, in Progress in Combinatorial Optimization. Proceedings of the Silver Jubilee Conference on Combinatorics, Waterloo, June 1982, W. R. Pulleyblank, ed., Academic Press, New York, 1984, 221-243.

[KL3] _______________________, Shelling structures, convexity and a happy end, in Graph Theory and Combinatorics. Proceedings of the Cambridge Combinatorial Conference in honour of Paul Erdos, B. Bollobas (ed.), Academic Press, London (1984), 217-232.

[KL4] _______________________, Basis graphs of greedoids and two-connectivity, Report No. 84324-OR, Institute of Operations Research, University of Bonn, 1984.

[Ma] G. Markowsky, The representations of posets and lattices by sets, Alg. Univ., 11 (1980), 173-182.

[Mo] B. Monjardet, A use for frequently rediscovering a concept, Order 1 (1985), 415-417.

[Pf1] J. L. Pfaltz, Semi-homomorphisms of semimodular lattices, Proc. Amer. Math. Soc., 22 (1969), 418-425.

[Pf2] ____________, Convexity in directed graphs, J. Combin. Th., 10 (1971), 143-162.

[PP] H. Perfect and J. S. Pym, Submodular functions and independence structures, J. Math. Anal. Appl, 30 (1970), 1-31.

[Po] N. Polat, Treillis de separation des graphes, Canad. J. Math., 28 (1976), 725-752.

[Ro] G.-C. Rota. On the foundations of combinatorial theory: I. Theory of Möbius functions, Z. Wahrsch. Verw. Gebiet., 2(1964), 340-368.

[Sa] M. E. Saks, personal communication, 1984.

[St] R. P. Stanley, Combinatorial reciprocity theorems, Adv. in Math., 14(1974), 194-253.

[KL5] B. Korte and L. Lovasz, Homomorphisms and Ramsey properties of antimatroids, Report No. 85364-OR, Institute of Operations Research, University of Bonn, 1985.

DEPARTMENT OF MATHEMATICS
CARNEGIE-MELLON UNIVERSITY
PITTSBURGH, PA 15213 USA

Contemporary Mathematics
Volume 57, 1986

CORRELATION AND ORDER

Peter M. Winkler, Emory University[1]

Abstract. Suppose a finite set S is given with unknown linear order, and information about this order consists exactly of the order relation of some pairs in S. If P is the partial ordering determined by those pairs then all linear extensions of P become equally likely, and one may ask whether certain events in this probability space are necessarily correlated. For example, does it increase the probability that $x < y$ if it becomes known that x is less than some other element z?

Startling progress has been made on these questions in the last five years; the techniques and results are surveyed here, including recent theorems of Fishburn and Brightwell. Some familiarity with ordered sets and lattices, but none with correlation, is assumed on the part of the reader.

Contents:

1. Introduction

Consider the following three "situations."

(A) An academic job is advertised and n people apply. All of the information available to the prospective employers is in the form of letters from experts in the field, saying "Person A is better than person B" or "Person C is not as good as person D" etc.

1980 Mathematics Subject Classification: 06A10, 60E15

[1]Supported by NSF grant MCS 84-02054 and the Alexander von Humboldt Foundation.

How should the applicants be ranked? Is there a reasonable interpretation of "the probability that X is better than Y" when no direct or inferred comparison exists? How is this probability, or X's rank, affected when a new letter comes in saying that X is better than Z?

(B) A tournament is held among members of the newly-formed Arcata Tennis Club with the object of constructing a "tennis ladder." There is insufficient time for a complete round-robin, i.e. not all pairs of players meet.

How should the ladder be constructed? Can scores be reasonably taken into account? Is it possible to be certain that it is to each player's advantage (assuming he wishes to be high on the ladder) to play always to his best ability?

(C) At some point in the process of sorting a set S of n objects, some comparisons have been made and the computer thus "knows" for certain pairs (x,y) that $x < y$. Let U be a set of relations among members of S, e.g. $U = \{a < b,\ c < d,\ c > e\}$; then U has a certain probability of being "true" when the sorting is completed. If V is another such set, it may be "favorable to U" in the sense that if the relations in V are added to the present information, the probability of U will go up. When is V *always* favorable to U, that is, regardless of the initial information? When is it *un*favorable?

With certain assumptions --- reasonable in case (C), much less so in the other two cases --- these situations can be modelled by the following probability space. A finite (partially) ordered set P is given and the elements of the space are the linear extensions of P, each given equal probability. A typical "event" in this space is a Boolean combination of relations of the form "$x < y$" and the questions asked above are connected with the ways in which these events are correlated.

In what follows we define and explain correlation (for readers not familiar with probability theory) and raise some correlation questions for the model described above. We then review the tools of the theory and the remarkable results which have been obtained with them in recent years. Finally, we present five problems for further research.

Despite the youth of this subject there are already three excellent surveys (at least!) namely [9], [10] and [17]. We hope that this one will nevertheless contribute in several ways:

(1) by including some elementary discussion of correlation and justification of the linear extension probability space;

(2) by putting some new problems in print; and

(3) simply by being more up-to-date, in particular including the theorems of Fishburn and Brightwell.

We will take advantage of the existence of these surveys to omit technical proofs which can now be found in several places in the literature. Readers especially interested in the tools (e.g. the FKG inequality) are referred to [10] and the references therein, and those interested in various problems concerning linear extensions to [17].

2. Correlation

Readers familiar with the concept of correlation are advised to skim or skip this section. The mathematical content of what follows can be found in any text on probability theory, e.g. [19].

A *finite probability space* is a finite set X of "outcomes" x, each with probability $pr(x) > 0$, normalized so that the probabilities sum to 1. An *event* E is a subset of X, with $pr(E)$ defined to be the sum of the probabilities of the outcomes in E. The information that "E has occurred" reduces X to a new probability space $X|E$, obtained by throwing out the outcomes not in E and renormalizing.

If A and B are two events in X then the probability of B *given* A, denoted $pr(B|A)$, is defined as the probability of B in $X|A$, provided A is non-empty. It follows that $pr(B|A) = pr(B \cap A)/pr(A)$, an equation taken by many modern texts as the *definition* of conditional probability in order to make things easier for the author.

If $pr(A \cap B) = pr(A)pr(B)$, which occurs perforce when A or B has probability 1 or 0, then we say A and B are *independent* or *uncorrelated*; in that case $pr(B|A) = pr(B)$ and $pr(A|B) = pr(A)$ when defined.

If $pr(A \cap B) > pr(A)pr(B)$ then A and B are said to be *positively correlated*; in that case $pr(B|A) > pr(B)$ and $pr(A|B) > pr(A)$ so that the events A and B may be thought of as "favorable to one another". Putting it another way, if you have bet on B you should be happy to see A occur, and vice-versa.

Naturally if $pr(A \cap B) < pr(A)pr(B)$ then A and B are *negatively* correlated and the above inequalities are reversed.

The following facts are easily verified for arbitrary events A and B with nontrivial probability (not 0 or 1):

(C1) If A is contained in B or B in A then they are positively correlated;

(C2) If A and B are disjoint then they are negatively correlated;

(C3) A and B are positively correlated iff
A and $\neg B$ are negatively correlated iff
$\neg A$ and B are negatively correlated iff
$\neg A$ and $\neg B$ are positively correlated; ($\neg$ = complement)

(C4) If C is an event which contains A and B, and A and B are positively correlated in $X|C$, then A and B are positively correlated in X;

(C5) If $A \cap B \subset C \subset B$ and A and B are positively correlated, then A and C are positively correlated;

(C6) If A and B are positively correlated in X, and A and C are positively correlated in $X|B$, then A and $B \cap C$ are positively correlated in X.

Note that although positive correlation is reflexive and symmetric it is far from transitive. A more serious omission, however, is given in the following:

"Simpson's Paradox": If $E_1, E_2, \ldots, E_k$ constitute a partition of X, and A and B are positively correlated in $X|E_i$ for each i, it does *not* follow that A and B are positively correlated in X.

This fact is an insidious and omnipresent obstruction whenever one is trying to prove that two events are correlated. Since it might look harmless to the reader in the above form, we take the time to illustrate it with an example "from real life."

Suppose you have bet heavily that the Atlanta Falcons (an American football team) will win their upcoming match with the San Francisco Forty-Niners. At the moment you cannot recall whether the game is to be played in Atlanta or in San Francisco, but you do know that regardless of where the game is played, *rain favors Atlanta*: that is, Atlanta's probability of winning goes up if it rains at the game.

You turn the television set on to the game and see immediately that it is raining. Should you be happy? Not necessarily! Suppose rain is only a small factor, e.g. it improves Atlanta's chances by one or two percent; the big factor is location, each team being heavily favored in its own stadium. Furthermore, *rain is much more likely in San Francisco.* Thus, when rain appears on the television, there is an inference that the game is at San Francisco and this inference, though indirect, more than compensates for the favorable direct effect of the rain on Atlanta's chances. Without information as to the game's location, rain is actually unfavorable to Atlanta, even though Atlanta fans who know where the game is will be legitimately happy to see the rain.

Going back now to the abstract probability space X, a function f from X to the reals is called a *random variable* on X and its *expectation* $\bar{f}$ is defined by

$$\bar{f} = \sum_{x \in X} f(x)pr(x) \; .$$

If f and g are two random variables on X with pointwise product fg, they are said to be *positively correlated* if $(\overline{fg}) > \bar{f}\bar{g}$; intuitively, if high values of f tend to correspond with high values of g. The characteristic function $\chi(E)$ of an event E has expectation pr(E) and it is immediate that A and B are positively correlated iff $\chi(A)$ and $\chi(B)$ are. Warning: zero-correlation is strictly weaker than independence for random variables in general.

A simple but (for us) crucial fact about correlation of random variables is the following, usually attributed to Chebyshev:

(C7) Suppose the elements of X (of which there must be at least two) are *linearly ordered*, and that the random variables f and g are strictly increasing on X; then f and g are positively correlated.

Generalizations of this fact to certain *partial* orders are what will save us, in a sense, from being forever obstructed by Simpson's Paradox.

3. The Linear Extension Model

Henceforth we fix a set X of n elements (job applicants, tennis players, objects to be sorted) with *unknown linear order*. Thus our underlying probability space LO consists of all linear orderings of S, each taken to have probability 1/n!.

We now make the critical assumption: that we have information about this unknown linear order, and that the information consists precisely of answers to a set of questions of the form "is $x < y$ in the linear ordering?". The following consequent observations will make this assumption clear:

(i) Since the answers are assumed to be *reliable* (i.e. correct) they must be *consistent*, that is, there can be no cycle $x_1 < x_2 < \dots < x_k < x_1$. Since $x < z$ follows from $x < y$ and $y < z$, we may as well consider that the information consists of a *partial ordering* P of the elements of S.

(ii) Any other linear ordering consistent with P would have resulted in the same information. (This is what it means to say that P is the *only* information we have.) Thus we are now in the probability space $LO|P$ consisting of all linear extensions of P, where each linear extension L has probability

$$\begin{aligned} pr(L|P) &= pr(L \cap P)/pr(P) = pr(L)/pr(P) \\ &= (1/n!)/(\ell(P)/n!) = 1/\ell(P) \end{aligned}$$

where $\ell(P) = |L(P)|$ = number of linear extensions of the ordered set P.

Consequence (i) is normal in a sorting situation and not unreasonable for the job applicant comparisons, but implies that the assumption is untenable in a sports tournament where cycles are inevitable. Consequence (ii) is again normal in computer sorting; note that it is perfectly permissible for the selection of pairs submitted for comparison to depend on results of previous comparisons. In the job applicant situation, however, (ii) cannot account for the increased likelihood of a letter attesting to y's superiority to x when y is *much* better; and in the tennis model, different winning scores cannot be distinguished. In part 6 below we will consider the effect of relaxing these restrictions in hope of fitting the tennis and job-applicant models better.

Events in the probability space $L(P) = LO|P$ are sets of linear extensions of P and therefore representable as Boolean combinations of relations "$x < y$". It is natural to think of a *conjunction* of such relations as another partial ordering A on S; then $pr(A) = \ell(P \text{ and } A)/\ell(P)$ where $\ell(P \text{ and } A)$ is the number of linear orderings of S which are linear extensions of both P and A. We avoid writing $\ell(P \cup A)$ because unions of orderings correspond to intersections of sets of linear extensions, and thus to intersections of events in LO or L(P).

The events of greatest interest are of course single relations "$x < y$"; here we write $pr(x < y)$ $\ell(P$ and $x < y)/\ell(P)$ so that $pr(x < y) = 1$ if $x < y$ in P, 0 if $x > y$ in P, and strictly between otherwise.

The first correlation result for these events concerns the case where S is partitioned into sets S_1 and S_2 each of which is *linearly ordered* by P; one might think of a merge sort, or a match between two tennis teams each of which is already ranked. Note that additional relations in P *between* the two parts are permitted. Then:

Theorem 1 (Graham, Yao and Yao [11]). Let A and B be partial orderings of S each consisting entirely of pairs $\{x_1 < x_2\}$ with $x_1 \in S_1$ and $x_2 \in S_2$. Then A and B are non-negatively correlated, i.e. $pr(A \text{ and } B) \geq pr(A)pr(B)$.

This corresponds to the intuitively reasonable statement that victories by team S_2 tend to make further victories by S_2 more likely. The original proof of Theorem 1 used a rather complex combinatorial argument, but subsequent proofs using the FKG inequality (about which much more later) were found independently by Shepp [21] and by Kleitman and Shearer [16].

In [21] Shepp showed that Theorem 1 holds also without the restriction that S_1 and S_2 be chains, provided that there are *no* relations in P between S_1 and S_2. A simple example (see e.g. [10]) shows that one cannot drop *both* restrictions, and raises the question of whether any correlations can be deduced in the absence of restrictions on P. This brings us to the following very elegant (and farsighted) conjecture made by Rival and Sands in 1980:

Conjecture ([18]): For any P and for any x, y and z in S, $pr(x < y \text{ and } x < z) \geq pr(x < y)pr(x < z)$.

This says that the events "$x < y$" and "$x < z$" are *always* non-negatively correlated; such results, not depending on the information P, are said to be *universal*.

Consider the content of the conjecture --- usually called the "xyz conjecture" --- in the case of the job applicants. If applicants x and

y are to be compared (perhaps they are coming together for an interview) it would be natural to compute $pr(x < y)$. Surely if a new letter now comes in saying "$x < z$" then $pr(x < y)$ cannot go *down*; that is, $pr(x < y | x < z) \geq pr(x < y)$. Of course, if the pair (x,y) or the pair (x,z) is already comparable in P then the events "$x < y$" and "$x < z$" are trivially independent; otherwise it seems plausible that $pr(x < y)$ must actually go *up*. This is the substance of the

"Strict" xyz conjecture: If the pairs (x,y) and (x,z) are incomparable in P, then the events "$x < y$" and "$x < z$" are positively correlated i.e. $pr(x < y \text{ and } x < z) > pr(x < y)pr(x < z)$.

The xyz conjecture has a consequence in *ranking* that also has intuitive appeal. The *height* $h_L(x)$ of an element x in a linear ordering L is $|\{y: y \leq_L x\}|$; thus $1 \leq h_L(x) \leq n$. Thus $h_L(x)$ may be regarded as a random variable on $L(P)$, and the *average height* $\bar{h}(x) = \sum \{h_L(x): L \in L(P)\}/\ell(P)$ is its expectation. Average height is an obvious way to rank the job applicants or tennis players, and it is reasonable to ask how new information affects average height. In fact, the xyz conjecture implies that $\bar{h}(x)|(x > z)$ (that is, the average height of x in $L(P$ and $x > z))$ is *strictly* greater than $\bar{h}(x)$, assuming x and z are incomparable in P. (Interpretation: tennis player x always improves his rank by beating z.) This follows from the observation that $\bar{h}(x) = \sum \{pr(y \leq x): y \in P\}$; see [24] for details. In fact, from the strict form of the xyz conjecture (written "$pr(x > y \text{ and } x > z) > pr(x > y)pr(x > z)$", equivalent by property (C3)) we get that $\bar{h}(x)|(x > z) - \bar{h}(x)|(x < z) \geq 1$, with equality only in trivial cases.

Although no one doubted its truth the xyz conjecture sat around for two years before becoming the "xyz inequality", and the strict form for two more. They yielded finally to some clever arguments using the powerful tools which we describe in the next section.

4. Tools

Around 1980 two useful tools began to influence the state of knowledge concerning the probability space $L(P)$. One of these is the "Aleksandrov-Fenchel inequalities" [6] concerning mixed volumes, used by Stanley [23] to show that the height function $h_L(x)$ described above has unimodal distribution. More precisely, for any element x of any ordered set P, and any integer i with $1 < i < n = |P|$, $pr(h_L(x) = i-1)pr(h_L(x) = i+1) \leq (pr(h_L(x) = i))^2$ in the space $L(P)$.

Later, Kahn and Saks [13] used the Aleksandrov-Fenchel inequalities in a similar way in their proof that in any P which is not *linearly* ordered, there is a pair x, y of elements such that $pr(x < y)$ lies strictly between $3/11$ and $8/11$; this shows that the rest of the sorting of P can be done with at most $(2.17)\log_2(\ell(P))$ comparisons, which is best possible up to change of constant. It is conjectured by Fredman (see [13]) and roundly believed that in fact one can always get $1/3 \leq pr(x < y) \leq 2/3$.

The other tool has been used for correlation results and hence is the one of greatest interest here: it consists of variations of the statement that *increasing functions on a distributive lattice are non-negatively correlated.*

The central (but not most general) result of this kind is the *FKG inequality*, proved in 1971 as part of the research program of the Foundation for Fundamental Research on Matter, and applied to problems in statistical mechanics.

Theorem 2 (Fortuin, Kasteleyn and Ginibre [8]): Let D be a finite distributive lattice with measure μ satisfying $\mu(x)\mu(y) \leq \mu(x \vee y)\mu(x \wedge y)$ for every x, y in D. If f and g: $D \to \mathbf{R}$ are increasing functions (i.e. $x \leq y$ implies $f(x) \leq f(y)$ and $g(x) \leq g(y)$) then

$$\Big(\sum_{x\in D} f(x)g(x)\mu(x)\Big)\Big(\sum_{x\in D} \mu(x)\Big) \geq \Big(\sum_{x\in D} f(x)\mu(x)\Big)\Big(\sum_{x\in D} g(x)\mu(x)\Big).$$

This means, when μ is normalized to be a probability measure, that f and g are non-negatively correlated; thus it generalizes the inequality of Chebyshev mentioned earlier. The FKG inequality also generalizes two results concerning the Boolean algebra $\underline{2}^n$ of subsets of an n-element set, with inclusion. Call $A \subset \underline{2}^n$ a *down-set* if it is closed downward, i.e. under inclusion, and its complement an *up-set*. Then:

Theorem 3 (Kleitman [15]) If A is a down-set and B an up-set $\underline{2}^n$, then $|A \cap B|\cdot 2^n \leq |A|\cdot|B|$: and

Theorem 4 (Seymour [20]) If A and B are both up-sets of $\underline{2}^n$ then $|A \cap B| \cdot 2^n \geq |A| \cdot |B|$.

Theorems 3 and 4 are equivalent by (C3) and the latter follows immediately from Theorem 2 by taking $D = \underline{2}^n$, μ the uniform probability measure, and f and g the characteristic functions of A and B. Theorems 3 and 4 are of course true for distributive lattices in general, and in that form suffice for proving some of the results in the next section; however, we don't see a way to generalize them directly.

Theorem 2 was itself generalized by Holley [12], one of whose results (misstated in [10]) is the following:

Theorem 5 (Holley): Let D be a finite distributive lattice with probability measures μ_1 and μ_2 satisfying $\mu_1(x)\mu_2(y) \leq \mu_1(x \vee y)\mu_2(x \wedge y)$ for all x, y in D. Then

$$\sum_{x \in D} f(x)\mu_1(x) \geq \sum_{x \in D} f(x)\mu_2(x)$$

for any increasing function $f\colon D \to \mathbf{R}$.

To get the FKG inequality from this, assume g is everywhere positive (adding a constant if necessary) and define

$$\mu_1(x) = g(x)\mu(x)/(\sum \{g(z)\mu(z)\colon z \in D\})$$

and $\mu_2 = \mu$.

Holley's results were followed by an elegant and even more general theorem of Ahlswede and Daykin [1]. To state the theorem the following notation is useful: if X and Y are subsets of the lattice D and f is a function from D to $\mathbf{R}$, then $X \vee Y = \{x \vee y\colon\ x \in X,\ y \in Y\}$ (dually for $\wedge$), and $f(X) = \sum\{f(x)\colon\ x \in X\}$.

Theorem 6 (Ahlswede and Daykin): Let D be a finite distributive lattice and let a, b, c and d be four functions from D to the non-negative reals such that

for any x, $y \in D$, $a(x)b(y) \leq c(x \vee y)d(x \wedge y)$.

Then

for any $X, Y \subset D$, $a(X)b(Y) \leq c(X \vee Y)d(X \wedge Y)$.

One of the nice things about Theorem 6 is that unlike Theorems 2 and 5 it need only be proved for $D = \underline{2}^n$, since otherwise D can be represented as a sublattice of a Boolean algebra and the functions a, b, c, and d defined to be zero outside D. Thus Theorem 6 can be proved by a straightforward induction on the dimension n of the Boolean algebra; we refer the reader to [1], [9] or [10] for details.

The derivation of the FKG inequality from Theorem 6 is immediate by putting $X = Y = D$, $a(x) = f(x)\mu(x)$, $b(x) = g(x)\mu(x)$, $c(x) = f(x)g(x)\mu(x)$ and $d(x) = \mu(x)$, after boosting f and g by constants if necessary to make them everywhere positive. Interestingly, the FKG inequality can be obtained also in a quite different way, setting $a(x) = b(x) = c(x) = d(x) = \mu(x)$ and varying X and Y so as to build f and g from characteristic functions of up-sets, as is done in [10].

Theorem 5 is obtained simply by setting $a(x) = \mu_1(x)$, $b(x) = f(x)\mu_2(x)$, $c(x) = f(x)\mu_1(x)$ and $d(x) = \mu_2(x)$, again after boosting f suitably, and $X = Y = D$.

It is perhaps worth noting that the hypothesis of distributivity in Theorems 2, 5 and 6 is necessary, in a very strong sense: all three theorems *characterize* distributive lattices. In fact setting the four functions of Theorem 6 all identically 1 gives the inequality $|X||Y| \leq |X \vee Y||X \wedge Y|$ which, as is apparent from the title of [3], was shown by Daykin to imply distributivity. Theorems 3 and 4, however, hold also for sufficiently "skinny" lattices (or, more generally, ordered sets) like the one pictured below.

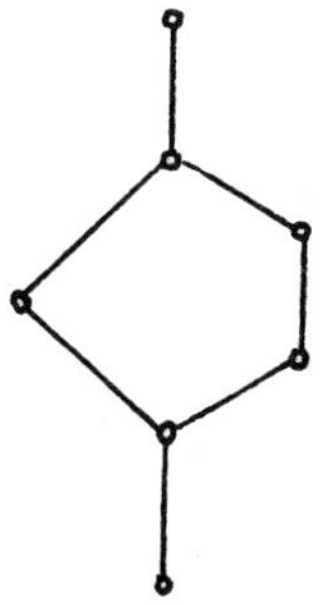

5. Results

The time has finally come to apply the tools of Section 4 to the probability space L(P), but the reader may well wonder: where is the distributive lattice? In fact L(P) is *asymptotically* equivalent to a sequence of probability spaces which *are* distributive lattices.

Let $\underline{N}$ be the chain $\{1 < 2 < 3 < \ldots < N\}$ for a given (large) positive integer N, and let $\underline{N}^P$ be the distributive lattice of all increasing (i.e., order-preserving) maps $f: P \to \underline{N}$. Assigning all maps equal probability makes $\underline{N}^P$ a probability space, and if A is another ordering of the underlying set of P, it follows that the probability of A in $\underline{N}$ --- which we denote $\mathrm{pr}_N(A)$ --- is exactly $|\underline{N}^{P \cup A}| / |\underline{N}^P|$. Letting $\mathrm{pr}(A)$ denote the probability of A in $L(P)$, we have

Lemma: For any P and any A, $\mathrm{pr}(A) = \lim_{N \to \infty} \mathrm{pr}_N(A)$.

Proof: Let $\underline{N}_+^Q = \{f \in \underline{N}^Q : f \text{ is injective}\}$, where Q is any ordering of the n-element underlying set S of P. Then for large N,

$$|\underline{N}_+^Q| = \ell(Q) \cdot \binom{N}{n} > cN^n$$

but

$$|\underline{N}^Q| - |\underline{N}_+^Q| < \binom{N}{n-1} \cdot (n-1)^n < c'N^{n-1}$$

where c and c' do not depend on N; hence

$$\lim_{N \to \infty} |\underline{N}_+^Q| / |\underline{N}^Q| = 1 .$$

It follows that

$$\lim_{N \to \infty} |\underline{N}^{P \cup A}| / |\underline{N}^P| = \lim_{N \to \infty} |\underline{N}_+^{P \cup A}| / |\underline{N}_+^P|$$

$$= \lim \left(\binom{N}{n} \ell(P \text{ and } A)\right) / \left(\binom{N}{n} \ell(P)\right) = \mathrm{pr}(A) \quad \text{as desired.}$$

Corollary: If orderings A and B are non-negatively correlated in $\underline{N}^P$ for sufficiently large N, then they are non-negatively correlated in $L(P)$.

Note that "non-negatively" cannot be replaced by "positively" in both hypothesis and conclusion since a strict inequality may disappear in the limit.

So far we have made no use of the lattice structure of $\underline{N}$, *normally* given by the relation $f \le g$ iff $f(u) \le g(u)$ for every $u \in P$. Since up- and down-sets in this ordering do not correspond to events in $L(P)$, however, this ordering must be changed in order to apply the tools of Section 4. This was in fact the approach of [16] and [21] in proving both versions of Theorem 1, and a very clever re-ordering --- given below --- enabled Shepp to prove the xyz inequality.

Theorem 7 (Shepp [22]) Let x, y and z be arbitrary elements of a finite ordered set P. Then in the space $L(P)$, $pr(x < y \text{ and } x < z) \geq pr(x < y)pr(x < z)$.

Proof. Define an ordering on $\underline{N}$ by $f \leq g$ iff $f(x) \geq g(x)$, and $f(u) - f(x) \leq g(u) - g(x)$ for every $u \neq x$. Then we have

$$(f \wedge g)(u) = \min(f(u) - f(x), g(u) - g(x)) + \max(f(x), g(x))$$

and

$$(f \vee g)(u) = \max(f(u) - f(x), g(u) - g(x)) + \min(f(x), g(x))$$

provided $\underline{N}^P$ is closed under these operations. To show this assume $u < v$ in P; then $f(u) \leq f(v)$ and $g(u) \leq g(v)$ for f and g in $\underline{N}$; then it is easily checked that $(f \wedge g)(u) \leq (f \wedge g)(v)$ and $(f \vee g)(u) \leq (f \vee g)(v)$ in all three cases: that is when $u = x$, when $v = x$, and when u, v and x are distinct.

The easiest way to see that under these operations $\underline{N}$ is distributive is to note that the map $f \to f'$ given by $f'(x) = -f(x)$, $f'(u) = f(u) - f(x)$ is a lattice embedding of $\underline{N}^P$ into the distributive lattice Z^n.

The object of this sleight-of-hand is to make the events "$x < y$" and "$x < z$" both up-sets in $\underline{N}^P$, and this is indeed now immediate: if $f(x) \leq f(u)$ and $f \leq g$ in our new ordering, then $g(x) \leq g(x)$ and $g(u) \geq f(u)$ so $g(x) \leq g(u)$ also. The characteristic functions of the events "$x < y$" and "$x < z$" are thus increasing on $\underline{N}$ and an application of the FKG inequality, with uniform probability measure, proves that the events are non-negatively correlated. The corollary above allows us to transfer this correlation to $L(P)$, completing the proof.

We may now ask whether there are further universal correlations to be found. For example, is the event "$x < u$ and $x < v$" always non-negatively correlated with the event "$y < u$ and $y < v$"? How about "$x < u$ and $y < v$" versus "$x < v$ and $y < u$"? For a while it was feared (hoped?) that arguments of ever-increasing subtlety were required, but then it turned out that *all* such questions could be answered by a combination of the xyz inequality, elementary properties (C1) through (C6) of correlation, and some simple constructions. Thus the beauty of the xyz inequality --- both the proof and the conjecture --- becomes even more apparent.

For convenience we will henceforth say that two orderings A and B of a finite set T are *universally* non-negatively correlated if for every ordering P defined on a finite set containing T, pr(A and B) $\geq$ pr(A)pr(B) in the probability space L(P). Further, for any ordered set Q, we let $\Delta(Q)$ stand for the set of *covering pairs* of Q, i.e. $\Delta(Q) = \{(u,v)\colon u < v$ in Q but for no $w \in Q$ is it the case that $u < w$ and $w < v\}$.

Theorem 8 (Winkler [25]). Two finite ordered sets A and B are universally non-negatively correlated if and only if

(i) they are consistent, i.e. their union is acyclic; and

(ii) for every pair $(x,y) \in \Delta(A \cup B) - \Delta(B)$ and every pair $(u,v) \in \Delta(A \cup B) - \Delta(A)$, either $x = u$ or $y = v$.

It follows that the answer to the first of the two new correlation questions asked above is "no", the second "yes".

We sketch the proof of Theorem 8 below; for details the reader is referred to [25] or [2].

To show that the conditions are sufficient, we use primarily property (C4) plus the fact that if the event E_i is non-negatively correlated with with the event F for i = 1, 2 and *for all P*, then the same is true for the event "E_1 and E_2" and F. This holds because the universality of the correlation allows us to deduce the conditional correlation in the hypothesis of (C6). Since "$x < y$" and "$x < v$" are universally non-negatively correlated by Theorem 7, and likewise "$x < y$" and "$u < y$", we can build up the correlation by successive conjunctions.

Necessity is obvious when A and B are inconsistent, by property (C2); one need only ensure that P is separately consistent with A and with B. Thus we may assume that there are pairs (x,y) and (u,v) failing to satisfy the second condition of the theorem. The ordered set P is now carefully selected so that (1) the events "$u < v$" and "$x < y$" are *negatively* correlated in L(P), and (2) all the other relations in $A \cup B$ are already true in P. It turns out that one of the four orders pictured below will always do the trick; the point labelled "z" is new but all the others lie already in the union of the underlying sets of A and B.

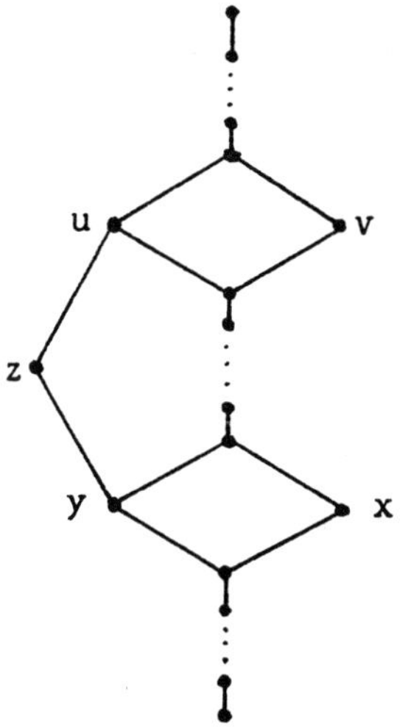
u
v
z
y
x

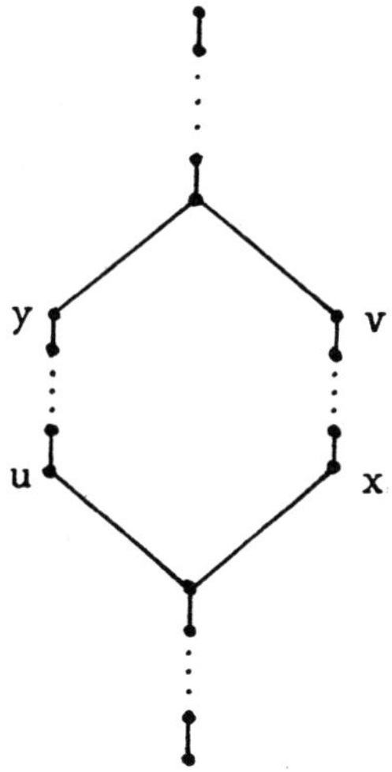
y
v
u
x

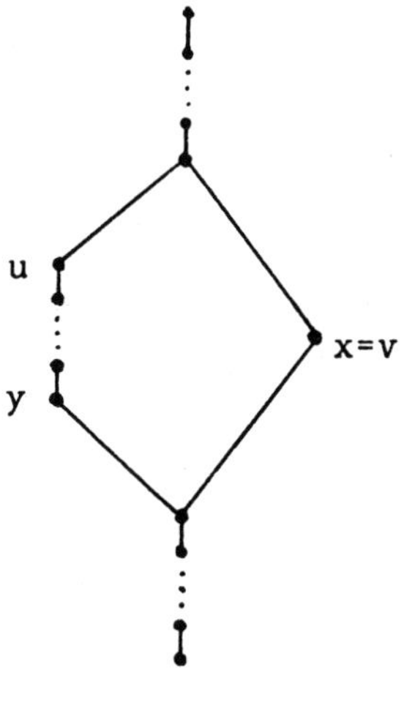
u
x=v
y

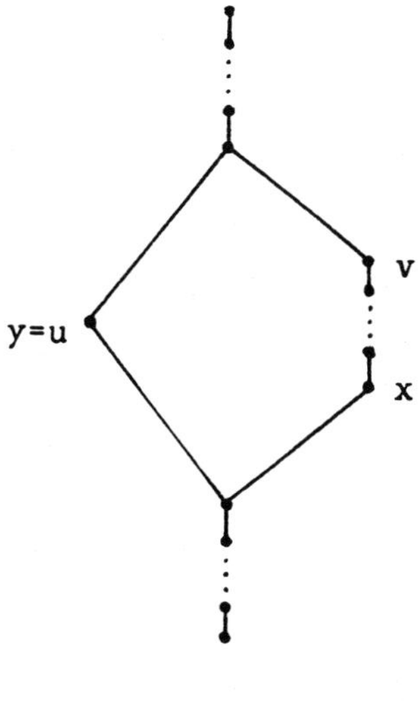
v
y=u
x

The xyz inequality can, of course, also be put in a negative form: the events "$x < y$" and "$x > z$" are universally *non-positively* correlated. The question of which pairs A, B of ordered sets have this property was brought up and investigated by Daykin(s) in [5] and [4], and recently Brightwell characterized them as follows.

Theorem 9 (Brightwell [2]). Let A and B be orderings of a finite set T. Then the following are equivalent:

(i) For every ordered set P whose underlying set contains T, $pr(A \text{ and } B) \le pr(A)pr(B)$;

(ii) $A \cap B = \emptyset$ and for any $(x,y) \in \Delta(A)$ and $(u,v) \in \Delta(B)$, $v = x$ or $u = y$; *or*, A and B are inconsistent.

Notice that Theorems 8 and 9 are *not* dual, nor do we know of any way to derive one easily from the other. The proof of Theorem 9, for which we refer the reader to [2], is similar in style to the proof of Theorem 8 but differs in some significant ways.

We return now not to non-negative correlation, but to *positive* correlation: what about the *strict* xyz conjecture? Since, as noted, strict inequalities can disappear in the limit, a more subtle argument than Shepp's seems to be required. This was finally provided by Fishburn, who managed, amazingly, to get exact bounds.

We note first that if $x||y$ (read: x is comparable to y) and $x||z$ but y and z *are* comparable in P, then "$x < y$" and "$x < z$" are automatically positively correlated by property (C1); thus the only interesting case of the strict xyz conjecture is where $x||y||z||x$.

Notice also that if we multiply both sides of the inequality "$pr(x < y$ and $x < z) > pr(x < y)pr(x < z)$" by $(\ell(P))^2$ and cancel like terms, we obtain the equivalent inequality "$\ell(P$ and $x < y$ and $x < z) \cdot \ell(P$ and $x > y$ and $x > z) > \ell(P$ and $y < x < z) \cdot \ell(P$ and $z < x < y)$". Thus the following implies the strict xyz inequality:

Theorem 10 (Fishburn [7]). Let P be an ordered set with n elements, in which x, y and z are pairwise incomparable. Then

$$\frac{\ell(P \text{ and } y < x < z) \cdot \ell(P \text{ and } z < x < y)}{\ell(P \text{ and } x < y \text{ and } x < z) \cdot \ell(P \text{ and } x > y \text{ and } x > z)} \le \lambda_n ,$$

with equality a possibility, where $\lambda_n = (n-1)^2/(n+1)^2$ for n odd and

$\lambda_n = (n-2)/(n+2)$ for n even.

The proof, which features *two* applications of the AD inequality (Theorem 6), is too complex to include here; but we cannot resist giving (a slightly modified version of) the delightful end of the proof.

Let U and V be subsets of $S - \{x\}$, where S is the underlying set of P, and define

$$\ell_U = |\{L \in L(P): \{u \in S: u < x \text{ in } L\} = U\}|$$

and the same for ℓ_V. Fishburn shows (using a Shepp-like argument and some fancy manipulation of inequalities) that if U contains y but not z, and V contains z but not y, then $\ell_U \cdot \ell_V \le \lambda_n \cdot \ell_{U \cup V} \cdot \ell_{U \cap V}$.

Now let D be the Boolean algebra $\underline{2}^{S-\{x\}}$ and define

$a(U) = \ell_U$ if $y \in U$ but $z \notin U$, and 0 otherwise;

$b(U) = \ell_U$ if $z \in U$ but $y \notin U$, 0 otherwise;

$c(U) = \lambda_n^{1/2} \cdot \ell_U$ if $y, z \in U$, 0 otherwise; and

$d(U) = \lambda_n^{1/2} \cdot \ell_U$ if $y, z \notin U$, 0 otherwise.

Then the hypothesis of Theorem 6 is fulfilled, and setting $X = Y = D$ produces exactly the required inequality.

Now that we know the "master" xyz inequality is strict in non-trivial cases, the proofs of Theorems 8 and 9 easily convert to produce the following final result:

Theorem 11. Suppose the ordered sets A and B are universally non-negatively (respectively, non-positively) correlated. Then they are positively (resp. negatively) correlated in L(P) exactly when neither pr(A) nor pr(B) has value 0 or 1.

6. Problems

Although vastly more has been done in the area of correlation and order than was imagined (at least by this author) only five years ago, some old problems remain and many new ones have arisen. We give a sampling below.

(1) Correlation of arbitrary events

Every event in $L(P)$ is a disjunction of conjunctions of simple relations (i.e. events of the form "$x < y$"); in fact, a disjunction of linear extensions of P. Theorems 8 and 9 tell us about universal correlation between conjunctions of simple relations, but since a disjunction of simple relations is the negation of the conjunction of the reversed relations, property (C3) permits us also to solve all universal correlation questions between two disjunctions of simple relations or between a conjunction and a disjunction. Thus it would seem that we are not very far from being able to answer *all* universal correlation questions. Specifically,

Problem 1: Let U and V be two subsets of the set of linear orderings of the finite set T. When is it the case that for any ordering P of a finite set containing T, $pr(L \cap T \in U \cap V) \geq pr(L \cap T \in U)pr(L \cap T \in V)$ in $L(P)$?

(2) Correlation among random variables

Colin McDiarmid of the Institute of Economics and Statistics, University of Oxford, brought the following question to the attention of the author at the British Combinatorial Conference in Glasgow, July 1985.

Problem 2: Suppose that $X_1, X_2, \ldots, X_n$ are independent real-valued random variables and P is a list of inequalities involving pairs of the X_i's, e.g. $P = \{X_3 < X_5, X_4 \geq X_7\}$. Then <u>conjecture:</u>

$$Pr(X_1 < X_2 | P) \leq Pr(X_1 < X_2 | P, X_1 < X_3) \quad \text{when defined.}$$

This would generalize Shepp's xyz inequality.

(3) Correlation in the set of all linear orderings

One ordering P of the finite set S in which correlation qestions are particularly intersting is the empty ordering; putting it another way, what correlations can be found in the probability space $L(S)$ of all linear orderings of the set S?

The following pretty conjecture is due to Kahn and Saks [14].

Problem 3: Let A and B be two orderings of S such that $A \cap B = \emptyset$ (that is, they have no relation in common) and $A \cup B$ is again a partial order (already closed under transitivity). Is $pr(A \text{ and } B) \geq pr(A)pr(B)$ in $L(S)$?

[Graham Brightwell has just discovered that the answer, unfortunately, is "no". Let S be the set $\{x, y, z, w\}$, let A be the ordering of S consisting of the three pairs $z < w$, $x < w$, $x < y$, and let B be the ordering $x < z$, $y < z$, $y < w$. Then A and B have empty intersection and their union is the total ordering given by $x < y < z < w$. Since A and B each have five linear extensions, we have

$$1/4! = \Pr(A \text{ and } B) < \Pr(A) \cdot \Pr(B) = (5/4!)^2 = (25/24)/4!$$

which constitutes a counterexample.

We must therefore restate Problem 3 as follows: Find conditions under which two orderings are non-negatively correlated over the empty ordering.]

(4) Correlation in Boolean algebras

The following problem was given by M. Aigner at the meeting on Combinatorics of Ordered Sets, Oberwolfach, Jan. 27 - Feb. 2, 1985. Aigner attributes it to Fürstenberg.

Let $\underline{2}^S$ be the Boolean algebra of subsets of some finite set S, and for $A \subset B \subset S$ let $[A, B]$ be the <u>interval</u> $\{U \subset S\colon\ A \subset U \subset B\}$ in $\underline{2}^S$. The "box" of two intervals is given by

$$[A_1, B_1] \mathbin{\square} [A_2, B_2] = \begin{cases} [A_1 \cup A_2, B_1 \cap B_2] & \text{if } (A_1 \cup B_1^c) \cap (A_2 \cup B_2^c) = \emptyset \\ \emptyset & \text{otherwise.} \end{cases}$$

Notice that for any B and C, $[\emptyset, B] \mathbin{\square} [C, S]$ is the same as $[\emptyset, B] \cap [C, S]$.

If $\mathcal{E} = \{[A_i, B_i]\}_{i \in I}$ and $\mathcal{F} = \{[C_j, D_j]\}_{j \in J}$ are two *collections* of intervals, then we define

$$\mathcal{E} \mathbin{\square} \mathcal{F} = \cup \{[A_i, B_i] \mathbin{\square} [C_j, D_j]\colon\ i \in I,\ j \in J\}.$$

Problem 4 is to prove or disprove the following conjecture:

For any two collections $\mathcal{E}$ and $\mathcal{F}$ of intervals of $\underline{2}^S$,

$$|\cup\mathcal{E}| \cdot |\cup\mathcal{F}| \geq 2^{|S|} \cdot |\mathcal{E} \mathbin{\square} \mathcal{F}|.$$

We point out that if E is a down-set and F an up-set of $\underline{2}^S$, and we define $\mathcal{E} = \{[\emptyset, B]\colon\ B \in E\}$ and $\mathcal{F} = \{[C, S]\colon\ C \in F\}$ then $\cup\mathcal{E} = E$ and $\cup\mathcal{F} = F$ and $\mathcal{E} \mathbin{\square} \mathcal{F} = E \cap F$; thus the conjecture generalizes

Kleitman's inequality (Theorem 3), but apparently not in the same way as does the FKG inequality.

(5) Unreliable information

In view of the remarks made earlier regarding the job-applicant and tennis-player models, it would be nice to be able to relax the *reliability* assumption and still get the results of Section 4; that is, to allow information of the sort "x is *probably* less than y".

The model we have in mind works something like this:

With each 2-element subset $\{x, y\}$ of S we associate an *oracle* $O(x, y)$ which has a certain *reliability* $r(x, y)$, taken to be a number between 0 and 1. If L^* is the *actual* linear ordering of S and $r(x, y) = 1$, then $O(x, y)$ reports "$x < y$" when $x < y$ in L^*, "$y < x$" when $y < x$ in L^*; if $r(x, y) = 0$ then $O(x, y)$ does the opposite. If $r(x, y) = 1/2$ then the oracle flips a coin to decide what to say, thus its information is worthless. Of course oracles with reliability below 1/2, e.g. certain weather forecasters, can be valuable.

One may equally well consider that there is only one oracle which has different reliabilities for different pairs, but it is crucial that the reports be *independent* otherwise there is no hope to derive correlations such as the xyz inequality. The model we have been working with --- where the various correlation results *do* hold --- corresponds to the case where all the reliabilities are in the set $\{0, 1, 1/2\}$.

We have not yet said how $O(x, y)$ reports when $r(x, y)$ is other than 0, 1 or 1/2. The obvious and simplest mechanism is that $O(x, y)$ tells the truth with probability $r(x, y)$ and lies with probability $1 - r(x, y)$. The result is that to each potential relation $x < y$ we can assign an *a priori* probability $ap(x < y)$ which is equal to $r(x, y)$ if the oracle has reported "$x < y$", and $1 - r(x, y)$ otherwise. Then, in the probability space $L(S)$ of all linear orderings of the underlying set, each linear ordering L will be *weighted* with measure μ given by

$$\mu(L) = \Pi\{ap(x < y): x < y \text{ in } L\}$$

and therefore has probability

$$pr(L) = \mu(L)/\Sigma\{\mu(L'): L' \in L(S)\}.$$

This probability assignment on $L(S)$ is thus determined by (1) a function r from the two-element subsets of S to the unit interval, and (2) the reports of the oracles, which together constitute a *tournament* (i.e. oriented complete graph) on S. When all the reliabilities are 0, 1 or 1/2 the arcs of the tournament with reliability 1, and the *reverses* of the arcs with reliability 0, determine (with probability 1)) an acyclic directed graph H; linear orderings consistent with H will receive weight $(1/2)^i$ where i is the number of pairs having reliability 1/2, and all other linear orderings get weight 0. This puts us back in L(P), where P is the transitive closure of H.

The question is, does the xyz inequality hold in general on the probability space L(S)? The unfortunate (and to us, slightly surprising) answer is "no". A counterexample is illustrated below, where arcs are labelled with their reliabilities and missing arcs all have reliability 1/2 (and are thus irrelevant).

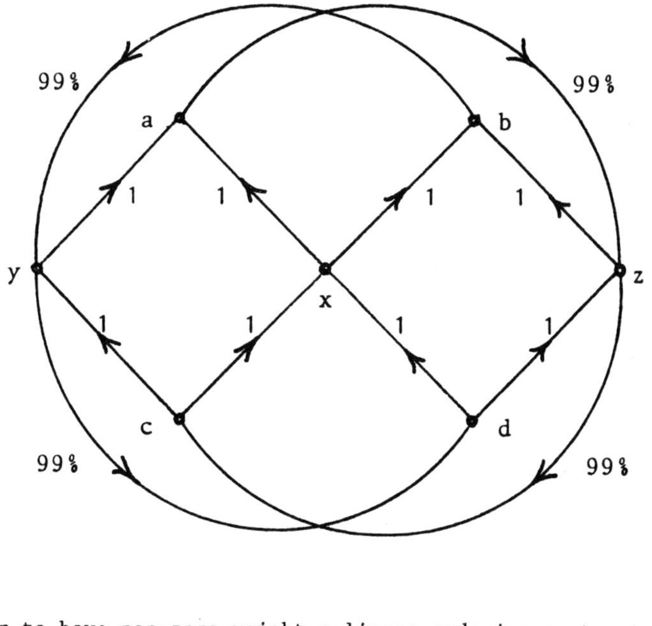

In order to have non-zero weight a linear ordering must extend the ordered set whose diagram is given by the straight lines above; and most of the weight goes to the extensions which reverse only two of the curved lines, namely the linear orderings $d < z < c < x < b < y < a$ and $c < y < d < x < a < z < b$. Thus knowing x is less than z makes it *less* likely that x is under y, showing that the xyz inequality does not hold here.

The source of this misbehavior seems to lie in the failure of the weight μ described above to satisfy the hypothesis of the FKG inequality in the distributive lattice $\underline{N}^S$. For example, let $S = \{x,y\}$ where $x < y$ with reliability 75% ; put $N = 4$ and let $f(x) = 4$, $f(y) = 1$, $g(x) = 2$, and $g(y) = 3$. Taking max and min for join and meet we see that $\mu(f) = 25\%$ and $\mu(g) = 75\%$ but $\mu(f \vee g) = \mu(f \wedge g) = 25\%$.

Study of this example suggests that a linear ordering must be heavily penalized for putting y *far* below x when there is information that x is probably below y. Thus, an oracle with high reliability is more likely to give an accurate report on a pair if they are widely separated in L^*.

We propose the following mechanism. Suppose $x < y$ in L^* (the reverse situation is dual) and in fact that $h_{L^*}(y) - h_{L^*}(x) = k > 0$. If oracle $O(x,y)$ has reliability $r(x,y) = 1 - 2s \geq 1/2$, then it says "$x < y$" with probability $1 - 2^k/2$, and "$y < x$" with probability $s^k/2$. An oracle with reliability $r(x,y) = 2s \leq 1/2$ does the reverse.

Now it is tedious but straightforward to verify that the resulting new weight function on $L(S)$ does satisfy the hypothesis of the FKG inequality when transferred in a reasonable way to $\underline{N}^S$. Alas, there are still complications in trying to generalize Shepp's proof. Hence we conclude with the following final question:

Problem 5: Does the xyz inequality hold in $L(S)$ for the weight determined by the mechanism described above?

REFERENCES

1. R. Ahlswede and D. E. Daykin (1978) An inequality for the weights of two families of sets, their unions and intersections, *Z. Wahrscheinlichkeitstheorie und Verw. Gebiete* **43**, 183-185.

2. G. Brightwell (1985) Universal correlations in finite posets, *Order*, to appear.

3. D. E. Daykin (1977) A lattice is distributive iff $|A|\cdot|B| \leq |A \vee B|\cdot|A \wedge B|$, *Nanta Math.* **10**, 58-60.

4. D. E. Daykin (1983) Addendum to Winkler's paper "Correlation among partial orders," preprint.

5. D. E. Daykin and J. W. Daykin (1983) Order preserving maps and linear extensions of a finite poset, preprint.

6. W. Fenchel (1936) Inégalités quadratique entre les volumes mixtes des corps convexes, *C. R. Acad. Sci. Paris* **203**, 647-650.

7. P. C. Fishburn (1984) A correlational inequality for linear extensions or a poset, *Order* **1** #2, 127-138.

8. C. M. Fortuin, J. Ginibre and P. N. Kasteleyn (1971) Correlation inequalities on some partially ordered sets, *Commun. Math. Phys.* **22**, 89-103.

9. R. L. Graham (1982) Linear extensions of partial orders and the FKG inequality, in *Ordered Sets*, I. Rival, ed., D. Reidel, Dordrecht, Holland, 213-236.

10. R. L. Graham (1983) Applications of the FKG inequality and its relatives, *Conference Proceedings*, 12th Intern. Symp. Math. Programming, Springer-Verlag, 115-131.

11. R. L. Graham, A. C. Yao and F. F. Yao (1980) Some monotonicity properties of partial orders, *SIAM J. Alg. Disc. Meth.* **1**, 251-258.

12. R. Holley (1974) Remarks on the FKG inequalities, *Commun. Math. Phys* **36**, 227-231.

13. J. Kahn and M. Saks (1984) Balancing poset extensions, *Order* **1** #2, 113-126.

14. J. Kahn and M. Saks, in a talk given by Kahn at Emory University, April 1984.

15. D. J. Kleitman (1966) Families of non-disjoint sets, *J. Comb. Thy* **1**, 153-155.

16. D. J. Kleitman and J. B. Shearer (1981) Some monotonicity properties of partial orders, *Studies in Appl. Math.* **65**, 81-83.

17. I. Rival (1984) Linear extensions of finite ordered sets, *Annals of Disc. Math.* **23**, 355-370.

18. I. Rival and B. Sands (1982) Problem 3.1, in *Ordered Sets*, I. Rival, ed., D. Reidel, Dordrecht, Holland, 806.

19. V. K. Rohatgi (1976) *An Introduction to Probability Theory and Mathematical Statistics*, J. Wiley and Sons, New York.

20. P. D. Seymour (1973) On incomparable collections of sets, *Mathematika* **20**, 208-209.

21. L. A. Shepp (1980) The FKG inequality and some monotonicity properties of partial orders, *SIAM J. Alg. Disc. Meth.* **1**, 295-299.

22. L. A. Shepp (1982) The XYZ conjecture and the FKG inequality, *Ann. Prob.* **10**, 824-827.

23. R. P. Stanley (1981) Two combinatorial applications of the Aleksandrov-Fenchel inequalities, *J. Comb. Thy* (A) **31**, 56-65.

24. P. M. Winkler (1982) Average height in a partially ordered set, *Disc. Math.* **39**, 337-341.

25. P. M. Winkler (1983) Correlation among partial orders, *SIAM J. Alg. Meth.* **4**, 1-7.

DEPARTMENT OF MATHEMATICS AND COMPUTER SCIENCE
EMORY UNIVERSITY
ATLANTA, GA 30322

Contemporary Mathematics
Volume 57, 1986

RETRACTS :
GRAPHS AND ORDERED SETS FROM THE METRIC POINT OF VIEW

El Mostapha JAWHARI
Département de Mathématiques
Ecole Normale Supérieure
MARRAKECH
MAROC

Maurice POUZET *
Laboratoire d'Algèbre ordinale
et d'algorithmique
Université Claude Bernard
VILLEURBANNE
FRANCE

Driss MISANE
Département de Mathématiques
Faculté des Sciences
RABAT
MAROC

ABSTRACT

Inspired by the work of A. Quilliot (1983), we propose here to consider graphs and ordered sets as a kind of a metric space, where - instead of real numbers - the values of the distance function d belong to an ordered semigroup equipped with an involution. In this frame, graph - or order - preserving maps of various sorts are exactly the non-expansive mappings(that is the maps f such that $d(f(x), f(y)) \leq d(x, y)$ for all x, y). Many known results on retraction and fixed-point property for classical metric spaces (whose morphisms are the non-expansive mappings) extend to these spaces. This is the case, for example, for the characterization of absolute retracts, by N. Aronszajn and P. Panitchpakdi (1956), the construction of the injective envelope by J. Isbell (1965) and the fixed-point theorem of R. Sine and P.M. Soardi (1979). For ordered sets these results immediately translate into the Banaschewski-Bruns theorem (1967), the MacNeille completion (1933) and the famous Tarski fixed-point theorem (1955). More importantly, they can be used to study various categories of posets and graphs along the lines of the classification scheme based upon retracts and products proposed by D. Duffus and I. Rival (1981). In particular, they give, or even improve, recent results concerning the variety generated by the fences (considered by D. Duffus and I. Rival (1981) and A. Quilliot (1983)), the variety generated by the undirected paths (R. Nowakowski, I. Rival (1983)) and the variety of directed graphs (A. Quilliot (1983)). Conversely, the study of holes initiated by I. Rival (1982) can be performed in this frame leading to new results. This metric point of view also establishes a link between the theory of precompact spaces and the theory of partially well ordered sets, initiated by M. Fréchet (1905) and G. Higman (1955). In conclusion this approach supports the idea that certain concepts of infinistic nature, like those which inspired metric spaces, can perfectly apply to the study of discrete, or even finite, structures.

* This work has been partially supported by the NATO Grant N° 339/84 .

0271-4132/86 $1.00 + $.25 per page

* * * * *

AMS (MOS) subject classifications (1980). Primary 05, 06, 08, Secondary 18, 03.

Key words. Ordered sets, graphs, metric spaces, retracts, absolute retracts, injectivity, injective hull, amalgamation, order-preserving maps, non-expansive mapping, essential morphism, fixed-point property.

* * * * *

*

INTRODUCTION

Retraction is an old notion. It was intensively studied for topological spaces, in relation with the fixed point property, but rather neglected for posets and graphs, except for the basic works of B. Banaschewski, G. Bruns [4] and P. Hell [16].

In 1981, D. Duffus and I. Rival [12] introduced retraction as one of the main ingredients of a structure theory for ordered sets. They proposed a classification scheme based upon the classes of posets closed under retracts and products, that they called *varieties,* and started the description of the lattice of varieties, characterizing some of the most important varieties. Since then new results have emerged (see for instance [47], [42], [34], [13]) and the same program for graphs, initiated by R. Nowakowski and I. Rival [36], has been developped (see for instance [19], [24], [37]).

In the meantime independently, A. Quilliot considered the retracts of posets and graphs for the purpose of pursuit games. In his thesis [44] (see also [45]), he pointed out a striking similarity between the properties of retractions of such objects and those of metric spaces, and he introduced very suggestives examples in the theory of posets and graphs.

This development called for a better understanding of the notion of retraction, led one to ask whether there was more than the similarity, observed by Quilliot, between ordered sets, graphs and metric spaces and it suggested that there is, may be, a general frame in which these studies can fit together.

The purpose of this paper is to report on our work done in this direction. This includes the results which already appeared in the theses of the first two authors [25] 1983, [32] 1984, obtained in collaboration with the third author, and new results as well.

We present a study of the retracts and fixed-point properties of graphs and posets in the frame of generalized metric spaces, with the non-expansive mappings as morphisms. Such spaces are equipped with a distance function d, whose values need not be real but belong to an ordered semigroup $\underset{\sim}{V} = \langle V, \leq, + \rangle$, not necessarily commutative, having a least element 0. They satisfy two of the usual properties of a metric:

$$d1 : d(x, y) = 0 \quad \text{iff} \quad x = y \; ;$$

$$d2 : d(x, y) \leq d(x, z) + d(z, y) \quad \text{for all } x, y, z.$$

The symmetry - not necessarily assumed - is replaced by:

$$d3 : d(x, y) = \overline{d(y, x)} \; ,$$

where the "bar" is a fixed order-preserving involution on the semigroup. As in the classical case, the non-expansive mappings are the maps f such that $d(f(x), f(y)) \leq d(x, y)$ for all x, y.

Many generalizations of metric spaces have been considered, very early and in various areas: geometry (K. Menger [30], [31], L.M. Blumenthal [5], [6]), logic and toposes, [20], [28], fuzzy sets [8], [9], for others, see [32], [33] and [10]. Our approach is closer to the current studies of classical metric spaces and particularly to the studies of the fixed-point property for non-expansive mappings, for which there is a very rich litterature, see for example the books of Istratescu [23] and R. Sine [50]. Indeed, it turns out that the basic results about retraction and fixed-point property for classical metric spaces extend to ours and translate into known or new results on posets and graphs.

For example, if we assume that our space of values $\underset{\sim}{V}$ is *Heyting* (that is $\underset{\sim}{V}$ is order complete and the addition distributes over the meets) then it can be equipped with a distance satisfiying d1 - d3 above; with this distance $\underset{\sim}{V}$ is hyperconvex and every metric over $\underset{\sim}{V}$ embeds isometrically into a power of $\underset{\sim}{V}$ (Proposition II-2.7). *Moreover, the hyperconvex spaces, the absolute retracts and the injective objects, (with respect to isometries), then coincide with the retracts of powers of* $\underset{\sim}{V}$ *and thus form a variety.* (Theorem 1). This is the straightforward extension of the characterization of absolute retracts, due to N. Aronszajn and P. Panitchpakdi [1]. Under the same assumption, we extend the construction of the injective envelope, done by J. Isbell [22] (Theorem 2). We also extend a recent theorem of R. Sine [49] and P.M. Soardi [52]; namely that *every non-expansive mapping on a bounded hyperconvex space has a fixed-point* (Theorem 4). These results immediately translate for posets into the Banaschewski-Bruns theorem [4], the MacNeille completion [29] and the Tarski fixed-point theorem [53]. But, more importantly, they apply to many other categories of posets and graphs. For example, they can be used in posets with the "fence"-distance and in undirected graphs, with all loops, equipped with the graph-distance. Theorem 1 above characterizes respectively the variety generated by fences (this extending a previous result of A. Quilliot [45]) and the variety generated by the paths (result due independently to A. Quilliot [44], R. Nowakowski and I. Rival [35]). Theorem 2 asserts that every poset isometrically embeds into a minimal retract of product of fences and also that every graph isometrically embeds into a minimal retract of product of paths (this last fact has been obtained independently by E. Pesh [37]). According to theorem 4 every order-preserving map on a retract of a power of a finite fence has a fixed point (this also follows from a more general resul, due to I. Rival [46] for finite posets and K. Baclawski and A. Bjőrner [2] for infinite posets) whereas every graph-preserving map on a retract of a power of a finite path (e.g. a finite tree) leaves invariant a complete subgraph (see also R. Nowakowski and I. Rival [35]). This applies in the same way to directed graphs (reflexive and antisymmetric) equipped with the "zig-zag"-distance and substantially completes the results of A. Quilliot [44].

On the other hand one can develop in this frame the study of "holes" and hole-preserving-maps, initiated by I. Rival first for posets under the name of "gaps" and gap-preserving-maps [12], and then for graphs [19]. For example, we show that for the hole preserving maps, *the absolute retracts and the injectives coincide, that every "metric" space embeds in one of them - by an hole-preserving map - and consequently that they form a variety* (Theorem 3). In the case of posets those objects are the posets having the "strong-selection property" (notion introduced by I. Rival and R. Wille [47] for lattices, an extended to posets by P. Nevermann and R. Wille [34]. For posets and graphs considered with the "fence"-distance and the graph-distance, Theorem 3 is due to P. Nevermann, I. Rival [33] and P. Hell, I. Rival [19], respectively. Of course it applies to directed graphs equipped with the "zig-zag"-distance and to classical metric spaces as well.

This metric approach also allows to consider convergence. We establish a link between the theory of precompact spaces and the theory of partially well-ordered sets, observing that the basic characterization of these two notions, due to M. Fréchet and G. Higman [21], are the same and, in fact, are reformulation of the Ramsey theorem. This observation, which was the other motivation for our approach, suggests further developments like cofinality, finite basis property, almost-periodic-sequences, etc.

In conclusion, this approach seems to support the idea that concepts originally of infinistic nature, like those behind metric spaces, can perfectly apply to the study of discrete - or even finite - structures.

This paper is divided into 5 sections. The first contains the basic categorical facts about retraction and introduces the metric approach. The second summarizes our results on generalized metric spaces and their absolute retracts and the third is on the fixed-point property. The examples are given in the fourth section . The convergence notions are briefly discussed in the fifth section. Since the variety of examples is the only justification of the present approach the reader can go directly to section 4, and then back to sections 2, 3.

This paper is based upon a preliminary version by the third author "Retracts: recent and old results on graphs, ordered sets and metric spaces", presented at the Bielefeld meeting on Combinatorics in November 1983, a French summary [41] presented at the "Infinistic Days" in October 1984 in Lyon, and also in a series of lectures given at the University of Calgary in January 1984.

The third author is pleased to thank the members of the Department of Mathematics of the University of Calgary for their hospitality during his stay. He his grateful to I. Rival, who introduced him to this subject, for the very stimulating discussions they had during the last three years. The authors thank I.G. Rosenberg for his help during the preparation of the manuscript.

I - RETRACTION: From the categorical approach to the metric point of view.

I - 1. Retraction and fixed point property.

Retraction is a categorical notion. We have objects, say P, Q, ..., and morphisms f, g, ... We say that the object P is a *retract* of the object Q and we note $P \lhd Q$ if there are morphisms $f : P \longrightarrow Q$ and $g : Q \longrightarrow P$ such that $g \circ f = 1_P$.

Examples:

a) The objects of the category are the posets and the morphisms are the order-preserving maps (i.e. the maps f such that $x \prec y$ implies $f(x) \prec f(y)$).

P is a retract of Q

b) The objects of the category are all reflexive graphs (which are the undirected graphs with a loop at every vertex, or, equivalently, the reflexive and symmetric binary relations) and the morphisms are all edge-preserving maps (note that an edge joining two vertices can be mapped on a loop).

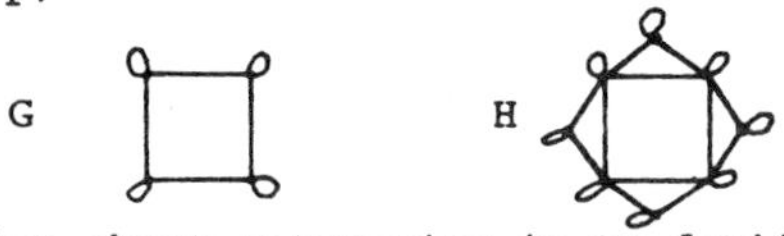

G is a retract of H

The central question about retraction is to decide, for two given objects P and Q, whether P is a retract of Q or not. A related question is to decide whether a given morphism $f : P \longrightarrow Q$ has a companion $g : Q \longrightarrow P$ such that $g \circ f = 1_P$; if this is the case f is said to be a *corectraction* and its companion is a *retraction,* but terminology does not provide an answer.

In fact, these questions are still largely unsolved, even for very simple categories like those of posets and graphs. Nevertheless a fruitful approach of a solution is this:
Identify a general property, say (p), that the coretractions enjoy in the category considered; for example, in the above category of posets each coretraction is an order-embedding (that is a map f such that $x \prec y$ is equivalent to $f(x) \prec f(y)$). Now looking at (p) as an approximation of the coretractions, then characterize the objects P for which this approximation is accurate, that is for which every morphism of source P and with property (p) is a coretraction. These P are commonly called the *absolute retracts* (briefly AR); (a terminology not perfectly adequate, since these objects depend upon the approximation, but commonly used in the field), we will rather say *AR with respect to the approximation (p).* There are at least two reasons for such an approach. First, in many instances the absolute retracts can be characterized very neatly. Second, they are the natural candidates for the fixed-point property. Let us precise this point: An object P (assumed to be concrete !) has the *fixed-point property* (f.p.p.) provided that every morphism $f : P \longrightarrow P$ has a fixed point, that is an x, $x \in P$, such that $f(x) = x$. Everybody has observed that if Q has the f.p.p. then every retract P has f.p.p. too (indeed, to a morphism $h : P \longrightarrow P$ associate the morphism $k = f \circ h \circ g : Q \longrightarrow Q$, where $f : P \longrightarrow Q$ and $g : Q \longrightarrow P$ are respectively the coretraction and the retraction maps; the image under g of every fixed point of k is a fixed point of h). But then, in order that an AR has the f.p.p., it suffices that it embeds by some approximation map into some object with the f.p.p. . This has a chance to happen, provided our category has enough objects with the f.p.p. .

This seems perfectly illustrated by the two following facts:

I-1.1. *Theorem.* (B. Banaschewski, G. Bruns, [4]).
In the category of posets and order-preserving maps, the following conditions are equivalent for a poset P:

1) *P is an absolute retract with respect to the order-embedding;*
2) *For every poset Q, every order-preserving map $f : A \longrightarrow P$ defined on a subset A of Q extends to an order-preserving map defined on Q;*
3) *P is a complete lattice;*
4) *P is a retract of a power of the two element chain $\underline{2}$.*

I-1.2. *Theorem.* (A. Tarski, [53]).
Every order-preserving map on a complete lattice has a fixed-point.

Characterizations of such absolute retracts like above abound in the litterature. Here we will gives 7 more. For others, see the survey paper "Categorical algebraic properties ..." by E.W. Kiss, L. Márki, P. Pröhle, and W. Tholen [27] and its 600 references.

These characterizations have in common: 1) a syntactical description of the AR (the third condition in the theorem above); 2) a categorical caracterization of the AR (the second condition); 3) a construction of the AR by some operations, usually products and retracts (the last condition). These two last parts use notions belonging to the folklore of category theory and universal algebra . Without proof we recall below the basic facts (for more see the above mentioned paper).

I - 2. Categorical basic facts on the absolute retracts and injective objects

Let $\mathfrak{C}$ be a category. We select a particular class $\mathfrak{D}$ of morphisms which will serve as approximations of the coretractions (thus we assume that $\mathfrak{D}$ contains all coretractions and is closed under composition); we denote the morphisms from $\mathfrak{D}$ by double arrows, $f : P \Longrightarrow Q$. In full generality the absolute retract with respect to $\mathfrak{D}$ are not preserved by any categorical construction: e.g. retracts and products of AR are not necessarily AR. There is a notable exception, namely when they enjoy the following property of extension of morphisms (see condition 2, in Theorem I-1.1.).

An object P is an *injective* with respect to $\mathfrak{D}$ if for all objects Q, R, each morphism $f : Q \longrightarrow P$, and every approximation $g : Q \Longrightarrow R$ there is a morphism $h : R \longrightarrow P$ such that $h \circ g = f$.

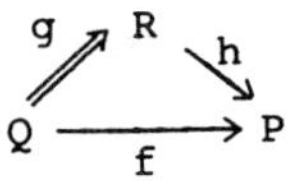

Fact 1. *Every injective is an absolute retract (with respect to $\mathfrak{D}$).*

Fact 2. *Every retract of an injective is an injective; moreover, if $\mathfrak{C}$ has products then every product of injectives is an injective.*

The category $\mathfrak{C}$ is said to have *enough injectives* if every object embeds by an approximation map into an injective. The approximations are *transferable*, or have the *transferability property*, (TP) if for all objects P, Q, R, morphism $f : P \longrightarrow Q$, approximation $g : P \Longrightarrow R$, there are an object P', a morphism $f' : R \longrightarrow P'$, an approximation $g' : Q \Longrightarrow P'$ such that $f' \circ g = g' \circ f$.

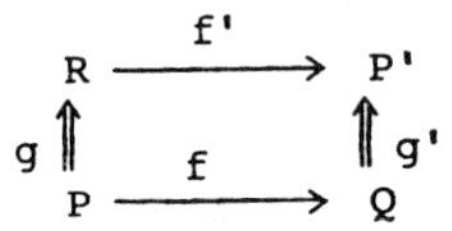

Fact 3. *If there are enough injectives then the approximation are transferable. If the approximations are transferable then the injectives and the absolute retracts are the same.*

If the above property holds for approximations f, f' (instead of morphisms) we have the *amalgamation property* (AP) introduced by R. Fraïssé in 1954 [14].

Fact 4. *Under the assumption that every morphism f is an approximation whenever for some morphism g the composition $g \circ f$ is an approximation, the transferability of approximations and the existence of finite products in* $\mathfrak{C}$ *imply the amalgamation property.*

Categories of relational structures with the amalgamation property (with respect to the embeddings) have been intensively studied since 1954. The basic examples are the categories of posets and graphs. Fact 4 explains why there are good candidates for the study of retraction, and suggests many other examples.

For a class U of objects we denote by $\underline{R}.U$ the class of retracts of members of U, by $\underline{H}.U$ the class of objects which embed by some approximation map into a member of U and - assuming that products of members of $\mathfrak{C}$ exist - by $\underline{P}.U$ the class of products of members of U. We say that a subclass S is a *cogenerator* of U if $U = \underline{H}.\underline{P}.S$. According to D. Duffus and I. Rival [12] we say that U is a *variety* if the retracts and the products of its members belong to U, i.e. $U = \underline{R}.U$ and $U = \underline{P}.U$, we say that a variety U is *generated* by a subclass S if U is the smallest variety containing S (it turns out that $U = \underline{R}.\underline{P}.S$).

If U is a subclass of $\mathfrak{C}$, then considering it as a subcategory of $\mathfrak{C}$ one can define its absolute retracts. For the notion of injectivity we say that an object of U is injective if it is injective in the category $\underline{H}.U$.

Fact 5. *For every class U, the classes* $\underline{R}.U$ *and* $\underline{H}.\underline{R}.U$ *have the same absolute retracts and the same injectives.*

The facts above can be summarized as follows:

I - 2.1. *Theorem.*
If U is a variety and has a cogenerator S consisting of injectives, then the injectives, the absolute retracts and the retracts of products of members of S coincide; and thus they form a variety. (In particular, the approximations are transferable).

To apply theorem I-2.1. one generally searches first for a collection of injectives. Then one tries the usual trick, namely to show that for every P and for some suitable set S' of injectives, the canonical morphism

$P \longrightarrow \prod_{Q \in S'} Q^{\mathrm{Hom}(P,Q)}$ is an approximation; this depends upon some "separation" properties and is usually achieved using the injectivity of the Q's.

For example, to get the equivalence between 1), 2), 4), in Theorem I-1.1. above from Theorem I-2.1. it only needs to show that $\underline{2}$ is injective and that every poset P order-embeds into $\underline{2}^{\mathrm{Hom}(P,\underline{2})}$. The first fact is obvious: if $f : A \longrightarrow \underline{2}$ is any order-preserving map defined on a subset A of Q then we may extend it to the map g where $g(x) = \mathrm{Sup}\,\{f(y) \,/\, y \in A \text{ and } y \leq x\}$. The second fact is well known, too. It consists to show that if $x \nleq y$ in P then there is an $f : P \longrightarrow \underline{2}$ such that $f(x) \nleq f(y)$. For that send x to 1, y to 0 and, since $\underline{2}$ is injective, extends this map to P. Now to get the syntactical characterization, namely the equivalence between 3) and these three conditions, show that, like for the poset $\underline{2}$ above, the complete lattices are injectives and then that completeness is preserved under retraction (meaning 4) implies 3)).

I - 3. The metric point of view

There is a striking similarity between the Banaschewski-Bruns' theorem, the Tarski's theorem and the following theorems about the category of metric spaces, where the morphisms are the non-expansive maps (and the approximations maps the isometries).

I - 3.1. *Theorem.* (N. Aronszajn and P. Panitchpakdi, [1]).
The following conditions are equivalent for a metric space E:

1) *E is an absolute retract with respect to the isometries;*
2) *For every metric space F, every non-expansive mapping* $f : A \longrightarrow E$, *defined on a subset A of F extends to a non-expansive mapping* $\tilde{f} : F \longrightarrow E$;
3) *E is hyperconvex* (that is the intersection of every family of closed balls $B(x_i, r_i)$ is non empty provided that $d(x_i, x_j) \leq r_i + r_j$ for all i, j);
4) *E is a retract of some* $\ell^{\infty}(I)$.

I - 3.2. *Theorem.* (R. Sine, [49], P.M. Soardi, [52]).
Every non-expansive mapping on a bounded hyperconvex space has a fixed point.

Theorems I-1.1. and I-3.1. have the same categorical content; as we will see below, it turns out that they have also the same syntactical content, and that theorems I-1.2. and I-3.2. are, despite of the appearance, the same.

At a first glance one might ask why there is a such similarity and, what is the common structure of posets and metric spaces ? This is not quite the right question. Absolute retractness and f.p.p. depend upon the morphisms (and the approximations) no matter what the structures are: For example, the f.p.p. for posets and metric spaces is a property of their semigroup of endomorphisms: P has the f.p.p. iff for every $f \in End(P)$ there is a $g \in End(P)$ such that $f \circ g = g$ (simply because the constant maps are in End(P)). So the right question rather concerns the common structure of the order-preserving maps and the non-expansive mappings.

That is common can be seen both ways: these maps are the maps preserving some binary relational structures, or the non-expansive maps associated to some natural generalizations of metric spaces. Let us be more explicit: On one hand, consider the category whose objects are binary relational structures of a given type I, that is the $R = (E, (R_i)_{i \in I})$ where $R_i \subseteq E \times E$, for every $i \in I$, and morphisms are the relational homomorphisms, that is the maps $f : R \longrightarrow R'$ such that $x R_i y$ imply $f(x) R'_i f(y)$ for all $i \in I$, and $x, y \in E$. On the other hand, consider the category whose objects are the structures, that we call *binary spaces,* consisting of pairs (E, d), where d maps $E \times E$ into a fixed ordered set $\underset{\sim}{V}$, and whose morphisms are $f : (E, d) \longrightarrow (E', d')$ such that $d'(f(x), f(y)) \leq d(x, y)$ for all $x, y \in E$; according to the metric space terminology we will call these maps *non-expansive* (although *contracting* would be better).

We can transform each category into the other. Indeed, if the first category is given, take for $\underset{\sim}{V}$ the set $\underline{2}^I$ ordered componentwise, and associate to every $R = (E, (R_i)_{i \in I})$ the binary space (E, d) over $\underset{\sim}{V}$, where $d(x, y)(i)$ is 0 if $(x, y) \in R_i$ and 1 otherwise. Usually this is done the other way, giving the value 1 if $(x, y) \in R_i$, but because of our choice a map $f : E \longrightarrow E'$ is a relational homomorphism from $R = (E, (R_i)_{i \in I})$ to

$R' = (E', (R'_i)_{i \in I})$ iff it is a non-expansive map from (E, d) to (E', d').
On the other hand if the second category is given, take for I the set $\underset{\sim}{V}$ and to (E, d) associate $(E, (R_i)_{i \in I})$ where $R_i = \{(x, y) / d(x, y) \leq i\}^{\sim}$; again $f : E \longrightarrow E'$ is a non expansive map from (E, d) to (E', d') iff it is a relational homomorphism from $(E, (R_i)_{i \in I})$ to $(E', (R'_i)_{i \in I})$.

From these trivial observations one can see the study of metric spaces and non-expansive mappings as a part of the study of the binary relational structures and relational homomorphisms, thus as a part of model theory. (And so, you will be less surprised that some results concerning these spaces, namely the Maurey's theorems (see [50]) on the fixed point property, which are among the deepest result in this field, have been obtained by some ultraproduct techniques, techniques which clearly belong to the model-theory area). Conversely, one can study the binary relational structures from the light of metric spaces, and one can at least expect that some standard techniques of metric spaces will extend enough to capture the most familiar relational structures, that is posets and graphs.

This is the point of view we develop here, (even though we belong to Fraïssé's relational school). Of course it could be argued that, in contrast to the positive reals, our set $\underset{\sim}{V}$ carries only an order, and thus our "distance" is very far from the traditionnal one. But we show, as observed recently by M. Pouzet and I.G. Rosenberg, that every reflexive binary relational structure $R = (E, (R_i)_{i \in I})$ has a very nice metric-like structure:

Let End(R) denote the set of all homomorphisms $f : R \longrightarrow R$, take for $\underset{\sim}{V}$ the set of all reflexive binary relations S (that is subsets S of $E \times E$ containing the set $\Delta_E = \{(x, x) / x \in E\}$) such that $f(S) \subseteq S$ for all $f \in End(R)$. Clearly $\underset{\sim}{V}$ ordered by inclusion, is a complete lattice, with a least element Δ_E and a largest element $E \times E$, (the meet of a subset W of $\underset{\sim}{V}$ is the set-theorical intersection $\cap W$, whereas the join of W is the least S containing the union $\cup W$ and preserved by all the $f \in End(R)$).

Now define $d(x, y) = \cap \{S \in \underset{\sim}{V} / (x, y) \in S\}$. As it is easy to check, End(R) is the set of non-expansive $f : (E, d) \longrightarrow (E, d)$, thus this translation is accurate. Now we come to the most interesting fact. Observe that if S and T belong to $\underset{\sim}{V}$ then their composition $S \circ T = \{(x, y) / (x, z) \in S$ and $(z, y) \in T$ for some z $\}$ also belongs to $\underset{\sim}{V}$, thus $\underset{\sim}{V}$ is a semigroup ; moreover, if $S \in \underset{\sim}{V}$ then the "inverse" $S^{-1} = \{(x, y) / (y, x) \in S\}$ belongs to $\underset{\sim}{V}$ too. Now if we denote + the semigroup operation, — ("bar") the inversion and 0 the least element of $\underset{\sim}{V}$ then our function d satisfies the following properties for all x, y, z :

d1) $d(x, y) = 0$ iff $x = y$

d2) $d(x, y) \leq d(x, z) + d(z, y)$

d3) $d(x, y) = \overline{d(y, x)}$.

Those properties are the basic axioms of the metric spaces that we will study all the rest of this paper.

NOTE: The above construction is reminiscent of universal algebra. Indeed, on an algebra, the distance between x, y can be defined as the smallest congruence containing x, y. Such metric approach occurs implicitly in the Grätzer's book [15].

II - METRIC SPACES OVER AN ORDERED SEMIGROUP AND THEIR RETRACTS

II - 1. The basic definitions and examples

- In order to mimic the above example let us consider a complete lattice $\underset{\sim}{V}$, with a least element 0, greatest element 1, equipped with a semigroup operation + and an involution satisfying the following properties:
  - . the semigroup operation is compatible with the ordering, that is $p \leq p'$ and $q \leq q'$ imply $p + q \leq p' + q'$, (i.e. $\underset{\sim}{V}$ is an ordered semigroup) and 0 is its neutral element;
  - . the involution (which satisfies $\bar{\bar{p}} = p$ for all $p \in \underset{\sim}{V}$) is order-preserving and reverses the semigroup operation, that is $\overline{p+q} = \bar{q} + \bar{p}$ holds for all p, q.
- Let E be a set. A *distance* on E is a map $d : E \times E \longrightarrow \underset{\sim}{V}$ satisfying d1), d2), d3) from I-3.

The pair (E, d) is a *metric space*; if there is no confusion we will denote it E. If we replace the semigroup operation + by its reverse, that is the operation $(x, y) \longmapsto y + x$, and leave unchanged the ordering and the involution, then the new structure $\underset{\sim}{V}'$ satisfies the same properties as $\underset{\sim}{V}$ and we can define distances over $\underset{\sim}{V}'$. For example if $d : E \times E \longrightarrow \underset{\sim}{V}$ is a distance then $\bar{d} : E \times E \longrightarrow \underset{\sim}{V}'$ defined by $\bar{d}(x, y) = d(y, x)$ is also a distance, the *dual distance*. We denote $\overline{(E, d)}$ or simply $\bar{E}$ the corresponding space.

These notions have non trivial instances: the map $d : \underset{\sim}{V} \times \underset{\sim}{V} \longrightarrow \underset{\sim}{V}$ defined by setting $d(x, x) = 0$ and $d(x, y) = \bar{x} + y$ if $x \neq y$, is a distance, so $\underset{\sim}{V}$ is metrizable.

Let (E, d) be a metric space over $\underset{\sim}{V}$. For all $x \in E$, $r \in \underset{\sim}{V}$, the *right* and *left ball* with *center* x and *radius* r are the sets $B_E(x, r) = \{y \in E \ / \ d(x, y) \leq r\}$ and $B_{\bar{E}}(x, r) = \{y \in E \ / \ d(y, x) \leq r\}$. Considered as sets, balls are both right and left since $B_E(x, r) = B_{\bar{E}}(x, \bar{r})$; thus we only need to consider right balls; if there is no confusion we will denote it $B(x, r)$ instead of $B_E(x, r)$.

If (E, d) and (E', d') are two metric spaces over $\underset{\sim}{V}$, then a map $f : E \longrightarrow E'$ is *non-expansive* provided

(1) $d'(f(x), f(y)) \leq d(x, y)$ for all $x, y \in E$

(or equivalently,

$f(B_E(x, r)) \subseteq B_{E'}(f(x), r)$ for all $x \in E$, $r \in \underset{\sim}{V}$);

If in (1) the equality holds for all $x, y \in E$, f is an *isometry*. If E is a subset of E' and the identity map is non-expansive we will say that (E, d) is a *subspace* of (E', d'), or that (E', d') is *an extension* of (E, d). If, moreover, this map is an isometry (that is d is the restriction of d' to $E \times E$), then we call (E, d) an *isometric subspace of (E', d') and (E', d') an isometric extension of (E, d)*.

If $\{(E_i, d_i) \ / \ i \in I\}$ is a family of metric spaces over $\underset{\sim}{V}$, the direct product $\prod_{i \in I} (E_i, d_i)$, is the cartesian product $E = \prod_{i \in I} E_i$, equipped with the "Sup" (or ℓ^{∞}) distance $d : E \times E \longrightarrow \underset{\sim}{V}$ defined by

$$d((x_i)_{i \in I}, (y_i)_{i \in I}) = \operatorname{Sup} \{d(x_i, y_i) \ / \ i \in I\} .$$

It is easy to check that this is the direct product in the category $\mathfrak{M}_{\underset{\sim}{V}}$ of metric

spaces over $\underset{\sim}{V}$ with the non-expansive mappings as morphisms.
For notational simplicity, whenever possible, we use d for all the distance functions.

Classical metric spaces: We take $\underset{\sim}{V} = \mathbb{R}^+ \cup \{+\infty\}$, and extend the addition to it in the obvious way, the spaces we get are just unions of disjoints copies of the classical metric spaces. We assume that $\underset{\sim}{V}$ is a complete poset only to have infinite products and to avoid ℓ^∞ type constructions.

Ordered set: Let $\underset{\sim}{V}$ be the complete lattice on four elements $\{0, a, b, 1\}$, with a incomparable to b, represented below:

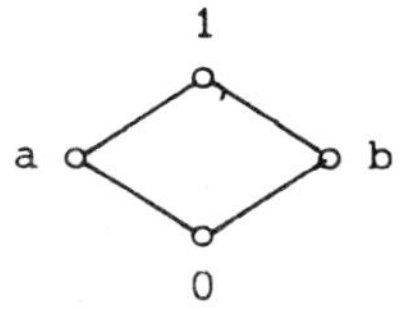

The semigroup operation is the join: $a + b = a \vee b$ and the involution exchanges a and b and fixes 0 and 1.

If $(P \leq)$ is any poset, the map $d : P \times P \longrightarrow \underset{\sim}{V}$, defined by $d(x, y) = 0$ if $x = y$, $d(x, y) = a$ if $x < y$, $d(x, y) = b$ if $y < x$ and $d(x, y) = 1$ if x and y are incomparable, is a metric over $\underset{\sim}{V}$. Conversely, if (E, d) is metric over $\underset{\sim}{V}$, then the relation defined by $x \underset{\sim}{\leq} y$ iff $d(x, y) \leq a$ is an ordering; the dual distance corresponds to the dual ordering; the non-expansive mappings are the order-preserving maps and the isometries the order-embeddings, whereas the direct product with the ℓ^∞ distance correspond to the direct product of posets.

Seven more examples are given in section IV. The above are enough to fully illustrate all the notions and results presented in this section and the next one.

II - 2. Absolute retracts with respect to the isometries

We can observe that in our category $\mathfrak{M}$ of metric spaces, the coretractions are isometric mappings. We use the isometries as approximations of the coretractions and thus we have absolute retracts and injective objects with respects to isometries (cf. I-2.).

Injectivity and Hyperconvexity.-

We start with the following observation:

II - 2.1. *Proposition.* A metric space E is injective (with respect to the isometries) if and only if for every metric space F, every non-expansive mapping $f : A \longrightarrow E$ defined on a proper subset A of F can be extended to a non-expansive mapping from $A \cup \{x\}$ to E, for some $x \in F \setminus A$.

Proof - If E is injective then, by definition, f can be extended to a non-expansive mapping defined on F. Conversely, let F be a metric space and let $f : A \longrightarrow E$ be a non-expansive mapping defined on a subset of F. Let Ext(f) denote the set of non-expansive $f' : A' \longrightarrow E$, with $A \subseteq A' \subseteq F$, which extend f.

We can order it in the obvious way and observing that this set is inductive, by the Zorn lemma, we get a maximal element, say $f_0 : A_0 \longrightarrow E$. Here A_0 must be F (otherwise f_0 would extend to some $x \in F \setminus A_0$ contradicting the maximality). □

We can translate the above categorical condition into a syntactical one that we borrow from the theory of classical metric spaces:

We say that a space E is *hyperconvex* if the intersection of every family of right balls $\{(B(x_i, r_i)) \mid i \in I\}$ is non empty whenever
$d(x_i, x_j) \preccurlyeq r_i + \overline{r_j}$ for all $i, j \in I$.

Hyperconvexity is equivalent to the conjunction of the following conditions H1 and H2, which are of independent interest:

H1) *Convexity :* for all $x, y \in E$, $p, q \in \underset{\sim}{V}$ such that $d(x, y) \preccurlyeq p + q$ there is z such that $d(x, z) \preccurlyeq p$ and $d(z, y) \preccurlyeq q$.
This condition, which is in fact a connectivity condition, expresses that the triangular inequality is best possible.

H2) *The 2-Helly property,* also called the *2-ball intersection property :* The intersection of every set (or, *equivalently, every family*) of balls is non empty provided the pairwise intersections are non empty.

II - 2.2. *Proposition.* *The injective objects in the category $\mathfrak{M}_{\underset{\sim}{V}}$ of metric spaces over $\underset{\sim}{V}$ are exactly the hyperconvex spaces.*

Proof - 1) <u>Hyperconvexity implies injectivity</u>. Let E be hyperconvex: According to the above proposition it is enough to show that if F is a metric space, $A \subseteq F$, $x_0 \in F$, $f : A \longrightarrow E$ is non expansive, then there is $y_0 \in E$ such that the map f' which coincides with f on A and sends x_0 on y_0 is non-expansive, that is - because of d3) - $d(f(x), y_0) \preccurlyeq d(x, x_0)$ for all $x \in A$. Thus all that we need is to choose y_0 in $\bigcap_{x \in A} B(f(x), d(x, x_0))$.
Since E is hyperconvex this intersection is non empty provided that for all $x, x' \in A$,

$$(2) \qquad d(f(x), f(x')) \preccurlyeq d(x, x_0) + \overline{d(x', x_0)} .$$

From the non-expansive character of f we have $d(f(x), f(x')) \preccurlyeq d(x, x')$ and from the triangular inequality
$d(x, x') \preccurlyeq d(x, x_0) + d(x_0, x') = d(x, x_0) + \overline{d(x', x_0)}$;
so (2) holds and we are done.

2) <u>Converse</u>. Let E be an injective space, and $\{(B(x_i, r_i)) \mid i \in I\}$ be a family of balls such that $d(x_i, x_j) \preccurlyeq r_i + \overline{r_j}$ for all $i, j \in I$. In order to prove that this intersection is non empty, we add a new element, say a, to I. On $F = I \cup \{a\}$ we can define a distance $d : F \times F \longrightarrow E$ by setting $d(i, j) = r_i + \overline{r_j}$ for all $i, j \in I$, $i \neq j$, $d(i, a) = r_i$ and $d(a, i) = \overline{r_i}$ for all $i \in I$, $d(x, x) = 0$ for all $x \in F$. The map $f : I \longrightarrow E$ defined by $f(i) = x_i$ is non-expansive by construction. Since E is injective, the map f extends to a and the image of a belongs to the intersection of the $B(x_i, r_i)$. □

With the Fact 2 of I-2., we get for free that *retracts and products of hyperconvex spaces are hyperconvex.*

This camouflaged definition of injectivity, even if it has some syntaxical flavour, is not internal as the index sets I can be big relatively to the space E. Can we replace these sets by subsets of E ? The answer depends upon the structure of $\underset{\sim}{V}$. Let us say that $\underset{\sim}{V}$ is an *Heyting involutive semigroup* (or briefly *Heyting*) if it satisfies the following distributivity condition

$\bigwedge_{\alpha \in A, \beta \in B} (p_\alpha + q_\beta) = \bigwedge_{\alpha \in A} p_\alpha + \bigwedge_{\beta \in B} q_\beta$ for every $p_\alpha, q_\beta \in \underset{\sim}{V}$ or, equivalently, (because of the involution) $\bigwedge_{\alpha \in A} p_\alpha + q = \bigwedge_{\alpha \in A} (p_\alpha + q)$ for all $p_\alpha, q \in \underset{\sim}{V}$.

From now and until the end of section II, we will assume this condition fulfilled.

II - 2.3. *Proposition.* Let E be a metric space over an Heyting involutive semigroup $\underset{\sim}{V}$; then E is hyperconvex iff for every map $f : E \longrightarrow \underset{\sim}{V}$ such that $d(x, y) \preccurlyeq f(x) + \overline{f(y)}$ for all $x, y \in E$, the intersection of balls $B(x, f(x))$ is non empty.

Proof - Let $\{(B(x_i, r_i)) \mid i \in I\}$ be a family of balls of E such that $d(x_i, x_j) \preccurlyeq r_i + \overline{r_j}$; define $f : E \longrightarrow \underset{\sim}{V}$ by $f(x) = \wedge \{r_i \ / \ x = x_i\}$. The distributivity insures that $d(x, y) \preccurlyeq f(x) + \overline{f(y)}$ □

Let E be a metric space over $\underset{\sim}{V}$ (not necessarily Heyting). For every isometric extension F of E and every $z \in F$, the map $f : E \longrightarrow \underset{\sim}{V}$, defined by $f(x) = d(x, z)$, satisfies $d(x, y) \preccurlyeq f(x) + \overline{f(y)}$ for all $x, y \in E$. We call *metric form* every map $f : E \longrightarrow \underset{\sim}{V}$ obtained by this process.

II - 2.4. *Proposition.* Let E be a metric space over an Heyting involutive semigroup $\underset{\sim}{V}$. Every $f : E \longrightarrow \underset{\sim}{V}$ such that $d(x, y) \preccurlyeq f(x) + \overline{f(y)}$ for all $x, y \in E$ majorizes a minimal metric form g (i.e. $g(x) \preccurlyeq f(x)$ for all $x \in E$).

Proof - Let E(f) be the set of $g : E \longrightarrow \underset{\sim}{V}$ such that $g(x) \preccurlyeq f(x)$ and $d(x, y) \preccurlyeq g(x) + \overline{g(y)}$ for all $x, y \in E$.
Observe first that E(f) has a minimal element. Indeed let $\{g_i \ / \ i \in I\}$ be a chain of elements of E(f) ; we have $d(x, y) \preccurlyeq g_i(x) + \overline{g}_j(y)$ for all $i, j \in I$; thus, from the distributivity, we have

$d(x, y) \preccurlyeq \bigwedge_{i,j} (g_i(x) + \overline{g}_j(y)) = \bigwedge_i g_i(x) + \bigwedge_j \overline{g}_j(y)$ and, consequently,

$\wedge g_i \in E(f)$. Zorn lemma insures our conclusion.

Secondly $g(x) \preccurlyeq d(x, y) + g(y)$ for all x, y . Indeed, assume this false, that is $g(x_0) \not\preccurlyeq d(x_0, y_0) + g(y_0)$ for some $x_0, y_0 \in E$; define $g' : E \longrightarrow \underset{\sim}{V}$ by $g' = g$ on $E \setminus \{x_0\}$ and $g'(x_0) = g(x_0) \wedge (d(x_0, y_0) + g(y_0))$. To check that $g' \in E(f)$, we have only to compare $d(x_0, y)$ and $g'(x_0) + \overline{g(y)}$; since

$$d(x_0, y) \preccurlyeq d(x_0, y_0) + d(y_0, y) \preccurlyeq d(x_0, y_0) + g(y_0) + \overline{g(y)} \quad \text{and}$$

$$d(x_0, y) \preccurlyeq g(x_0) + \overline{g(y)} \quad \text{we have}$$

$$d(x_0, y) \preccurlyeq (d(x_0, y_0) + g(y_0) + \overline{g(y)}) \wedge (g(x_0) + \overline{g(y)})$$

that is because of the distributivity,

$$d(x_0, y) \preccurlyeq (d(x_0, y_0) + g(y_0)) \wedge g(x_0) + \overline{g(y)} \quad \text{and we are done.}$$

But then we have g' in E(f) with $g' < g$; this contradicts the minimality.
Now we can extend E to a new element, in fact g, and extend the distance d on E to $F = E \cup \{g\}$, taking $d(x, g) = g(x)$. Because of our inequality this defines a distance, and g is a metric form. □

II - 2.5. *Corollary.* *E is hyperconvex iff for every minimal metric form* $f : E \longrightarrow \underset{\sim}{V}$, *the intersection* $\bigcap_{x \in E} B(x, f(x))$ *is non empty.*

In term of extension of homomorphisms this says:

II - 2.6. *Corollary.* Under the same assumptions the following are equivalent:

i) E is injective;
ii) E is an absolute retract;
iii) For every isometric extension F of E and $x \in F$, E is a retract of $E \cup \{x\}$.

But are there any such spaces ? We will see some in the next paragraph.

Hyperconvex metrization of the space of values.-

If we intend to define a distance d on $\underset{\sim}{V}$, the simplest minded one would have to satisfy $d(0, r) = r$ for all $r \in \underset{\sim}{V}$.
The distance d_M, defined by $d_M(p, q) = \overline{p} + q$ if $p \neq q$ and 0 otherwise is such an example, but in general, this space fails to be hyperconvex, the reason being that d_M is too big. So among the distances d satisfying $d(0, r) = r$ we have to search for one as small as possible.

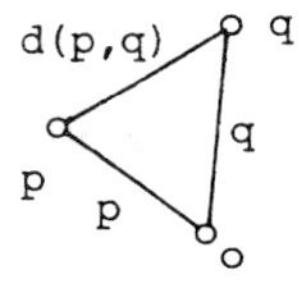

Because of the triangular inequality, such a distance satisfies

$$q \leq p + d(p, q) \quad \text{and} \quad p \leq q + \overline{d(p, q)}$$

for all $p, q \in \underset{\sim}{V}$, so $d(p, q)$ belongs to the set $D(p, q) = \{r / p \leq q + \overline{r} \text{ and } q \leq p + r\}$, whence the least element of $D(p, q)$, if there is any, is then a good candidate for $d(p, q)$.

II - 2.7. *Proposition.* Let $\underset{\sim}{V}$ be Heyting and $d_{\underset{\sim}{V}} : \underset{\sim}{V} \times \underset{\sim}{V} \longrightarrow \underset{\sim}{V}$ be the map defined by $d_{\underset{\sim}{V}}(p, q) = \wedge D(p, q)$ for all $p, q \in \underset{\sim}{V}$. Then

1) $(\underset{\sim}{V}, d_{\underset{\sim}{V}})$ is a hyperconvex metric space.
2) Every metric space over $\underset{\sim}{V}$ embeds isometrically into a power of $(\underset{\sim}{V}, d_{\underset{\sim}{V}})$.

Proof -

1 a) $d_{\underset{\sim}{V}}$ is a distance.

Let $p, q \in V$ and $a = \wedge D(p, q)$.
Observe <u>first</u> that a is the least element of $D(p, q)$, that is $p \leq q + \overline{a}$ and $q \leq p + a$. For the first inequality we compute $q + \overline{a}$; since "‾" is an order automorphism we have $q + \wedge \overline{D(p, q)}$, because of the distributivity condition $\wedge \{q + \overline{r} / r \in D(p, q)\}$, but from the definition of $D(p, q)$ we have $p \leq q + \overline{r}$ for all $r \in D(p, q)$ so $p \leq q + \overline{(\wedge D(p, q))}$. The second inequality comes along the same lines.

We check now that $d_{\underset{\sim}{V}}$ satisfies the axioms d1), d2), d3).

d1) $d_{\underset{\sim}{V}}(p, q) = 0$ means $p \prec q + \overline{0}$ and $q \prec p + 0$; but $\overline{0} = 0$ and $q + 0 = q$, $p + 0 = p$, thus this means $p \prec q$ and $q \prec p$, that is to say $p = q$.

d2) $d_{\underset{\sim}{V}}(p, q) \prec d_{\underset{\sim}{V}}(p, r) + d_{\underset{\sim}{V}}(r, q)$ for all $p, q, r \in \underset{\sim}{V}$.
Since $d_{\underset{\sim}{V}}(p, r)$ is the least element of $D(p, r)$ we have

(i) : $p \prec r + \overline{d_{\underset{\sim}{V}}(p, r)}$ and $r \prec p + d_{\underset{\sim}{V}}(p, r)$

(j) : $r \prec q + \overline{d_{\underset{\sim}{V}}(r, q)}$ and $q \prec r + d_{\underset{\sim}{V}}(r, q)$

(by a similar argument). Since + is compatible with the order and 0 is its neutral element, by combining (j_1) and (i_1) we get:
$p \prec q + \overline{d_{\underset{\sim}{V}}(r, q)} + \overline{d_{\underset{\sim}{V}}(p, r)}$, but since
$\overline{d_{\underset{\sim}{V}}(r, q)} + \overline{d_{\underset{\sim}{V}}(p, r)} = \overline{d_{\underset{\sim}{V}}(p, r) + d_{\underset{\sim}{V}}(r, q)}$ that says
$p \prec q + \overline{d_{\underset{\sim}{V}}(p, r) + d_{\underset{\sim}{V}}(r, q)}$. Along the same lines
$q \prec p + d_{\underset{\sim}{V}}(p, r) + d_{\underset{\sim}{V}}(r, q)$, given by (j_2) and (i_2) .
These two inequalities show that $d_{\underset{\sim}{V}}(p, r) + d_{\underset{\sim}{V}}(r, q) \in D(p, q)$.
Since $d_{\underset{\sim}{V}}(p, q)$ is the least element of $D(p, q)$ we get
$d_{\underset{\sim}{V}}(p, q) \prec d_{\underset{\sim}{V}}(p, r) + d_{\underset{\sim}{V}}(r, q)$.

d3) $d_{\underset{\sim}{V}}(p, q) = \overline{d_{\underset{\sim}{V}}(q, p)}$

For that it is enough to observe that

$D(p, q) = \overline{D(q, p)}$ (where $\overline{D(q, p)}$ stands for $\{ \overline{r} \,/\, r \in D(q, p)\}$

Finally, we have for every p, $D(o, p) = \{r \,/\, r \succ p\}$ and hence $d_{\underset{\sim}{V}}(o, p) = p$.
From the triangle inequality we get $d_{\underset{\sim}{V}}(p, q) \prec \overline{p} + q$.

1 b) $\underset{\sim}{V}$ is hyperconvex.

H1) $\underset{\sim}{V}$ is convex. Let $x, y, p, q \in \underset{\sim}{V}$ be such that $d_{\underset{\sim}{V}}(x, y) \prec p + q$.
We claim that $z = (x + p) \wedge (y + \overline{q})$ satisfies the desired

(*) $d_{\underset{\sim}{V}}(x, z) \prec p$ and $d_{\underset{\sim}{V}}(z, y) \prec q$.

First $d_{\underset{\sim}{V}}(x, y) \prec p + q$ means $p + q \in D(x, y)$ i.e.

(**) $x \prec y + \overline{q} + \overline{p}$ and $y \prec x + p + q$

The first inequality in (*) means $x \prec z + \overline{p}$ and $z \prec p$.
By distributivity,

$(x + p) \wedge (y + \overline{q}) + \overline{p} = (x + p + \overline{p}) \wedge (y + \overline{q} + \overline{p}) \succ x \wedge x = x$

For the second inequality in (*), we have to show that

$(x + p) \wedge (y + \overline{q}) \prec y + \overline{q}$, and $y \prec (x + p) \wedge (y + \overline{q}) + q$

The first one is immediate. For the second, by the distributivity and (**)

$(x + p) \wedge (y + \overline{q}) + q = (x + p + q) \wedge (y + \overline{q} + q) \succ y \wedge y = y$.

H2) $\underset{\sim}{V}$ has the 2-Helly property. First we show that the balls of $\underset{\sim}{V}$ are closed intervals (in the lattice $\underset{\sim}{V}$).

Let $B(x, r)$ be a (right) ball of $\underset{\sim}{V}$.

1) We claim that $p := \vee B(x, r)$ and $q := \wedge B(x, r)$ belongs to $B(x, r)$.

   For p this means $x \leq p + \overline{r}$ and $p \leq x + r$. Since $x \leq y + \overline{r}$ and $y \leq x + r$ for every $y \in B(x, r)$ these inequalities hold.

   For q this means $x \leq q + \overline{r}$ and $q \leq x + r$. Since $y \leq x + r$ for all $y \in B(x, r)$, particularly x, the second inequality holds. For the first the distributivity gives:

   $q + \overline{r} = \wedge \{y + \overline{r} \mid y \in B(x, r)\} \geq x$.

2) We claim that $B(x, r) = [q, p]$, that is $y \in B(x, r)$ whenever $q \leq y \leq p$. Indeed from $x \leq q + \overline{r}$ we get $x \leq y + \overline{r}$ and, similarly, from $p \leq x + r$ we get $y \leq x + r$, proving $y \in B(x, r)$.

Now in order to have the 2-Helly property, it suffices to show that the collection of closed intervals of a complete lattice has the 2-Helly property. This is immediate: let $([x_i, y_i])_{i \in I}$ be a collection of pairwise intersecting intervals. Then for all $i, j \in I$ we have $x_i \leq y_j$, thus $\bigvee_{i \in I} x_i \leq \bigwedge_{j \in I} y_j$. Since intervals are order-convex, the interval $[x, y]$ (where $x = \bigwedge_{i \in I} x_i$, $y = \bigwedge_{j \in I} y_j$) is included in all intervals $[x_i, y_j]$.

We base our proof of 2) upon the following.

II - 2.8. *Lemma.* Let E be a metric space over $\underset{\sim}{V}$. Then, for all $x, y \in E$, $d(x, y) = \text{Sup}\{d_{\underset{\sim}{V}}(d(z, x), d(z, y)) \mid z \in E\}$.

Proof- Let $x, y, z \in E$. The triangular inequality gives $d(z, x) \leq d(z, y) + \overline{d(x, y)}$ and $d(z, y) \leq d(z, x) + d(x, y)$, which just means $d(x, y) \in D(d(z, x), d(z, y))$, e.g. $d_{\underset{\sim}{V}}(d(z, x), d(z, y)) \leq d(x, y)$. For $z = x$ we get $d_{\underset{\sim}{V}}(d(x, x), d(x, y)) = d_{\underset{\sim}{V}}(0, d(x, y)) = d(x, y)$. This gives the above equality. □

Translated in more categorical terms:

1) For every $z \in E$ the map $\delta(z) : E \longrightarrow \underset{\sim}{V}$ defined by $\delta(z)(x) = d(z, x)$ is non-expansive, that is $\delta(z) \in \text{Hom}(E, \underset{\sim}{V})$;

2) If $\text{Hom}(\overline{E}, \underset{\sim}{V}')$ (shorthand for $\text{Hom}((E, \overline{d}), (\underset{\sim}{V}', d_{\underset{\sim}{V}'}))$ is considered as an isometric subspace of $\underset{\sim}{V}^E$, then the map $\overline{\delta} : E \longrightarrow \text{Hom}(\overline{E}, \underset{\sim}{V}')$, defined by $\overline{\delta}(x)(y) = d(y, x)$, is an isometry (from E onto its image).

   In particular $\overline{\delta}$ is an isometry from E into $\underset{\sim}{V}^E$ and 2) follows. □

The characterization of the absolute retracts.-

Our discussion can be summarized by the following:

II - 2.9. ***Theorem 1*** **-** *Let $\underset{\sim}{V}$ be Heyting. Then for a metric space E over $\underset{\sim}{V}$, the following conditions are equivalent:*

1) E is an absolute retract (with respect to the isometries),
2) E is injective (with respect to the isometries);
3) E is hyperconvex;
4) E is a retract of a power of $\underset{\sim}{V}$.

Proof- 1) $\Longrightarrow$ 4) : From Proposition II-2.7., E isometrically embeds into a power of $\underset{\sim}{V}$; since it is an absolute retract, it must be a retract of such a power.

4) $\Longrightarrow$ 3) : $\underset{\sim}{V}$ is hyperconvex, since the hyperconvex and the injective coïncide (Proposition II-2.2.), the retracts of powers of $\underset{\sim}{V}$ are hyperconvex.

3) $\Longrightarrow$ 2) : Proposition II-2.2.

2) $\Longrightarrow$ 1) : Fact 1 of section I-2. □

$\mathfrak{M}_{\underset{\sim}{V}}$ has other cogenerators than $\underset{\sim}{V}$, indeed:

II - 2.10. ***Proposition.*** Let D be a subset of $\underset{\sim}{V}$. The collection of subspaces $[0, r]$ $(r \in D)$ of $\underset{\sim}{V}$ is a cogenerating set of injectives provided that every element of $\underset{\sim}{V}$ is the supremum of elements of $D \cup \overline{D}$.

Proof - A set $\mathcal{S}$ of metric spaces is a cogenerating set provided that for every E the evaluation map $e : E \longrightarrow \prod_{S \in \mathcal{S}} S^{\mathrm{Hom}(E,S)}$ (defined by $e(x)(f) = f(x)$) is an isometric embedding, which simply means that for every $x, y \in E$, $d(x, y) = \mathrm{Sup}\,\{d(f(x), f(y)) \;/\; f \in \mathrm{Hom}(E, S),\ S \in \mathcal{S}\}$.
In our case, the intervals $[0, r]$ equipped with the induced distance by $d_{\underset{\sim}{V}}$ are injective (indeed, since they are balls in the hyperconvex space $\underset{\sim}{V}$ they are also hyperconvex and thus injective). Consequently, for every $r < d(x, y)$ the non expansive map which maps x to 0 and y to r extends on E to a non expansive map. So if every element of $\underset{\sim}{V}$ is the supremum of elements of some set, say A, it follows that
$d(x, y) = \mathrm{Sup}\,\{d(f(x), f(y)) \;/\; f \in \mathrm{Hom}(E, [0,r]) \;/\; r \in A\}$ for every $x, y \in E$, thus $\{[0, r] \;/\; r \in A\}$ is a cogenerating set. When $A = S \cup \overline{S}$, we only need $\{[0, r] \;/\; r \in S\}$. Indeed, if we take $\bar{r} \in \overline{S}$, $\bar{r} < d(x, y)$, then the map $f : E \longrightarrow [0, r]$ which sends y to 0 and x to r satisfies $d(x, y) = \bar{r}$.

□

Metric spaces.-

Theorem 1 applies to metric spaces over $[0, +\infty]$ and in this case, looks very similar to the Aronszajn-Panitchpakdi theorem's. The only difference is that the classical case deal with metric without infinite value, and consequently one has to replace the retracts of powers of $[0, +\infty]$ by the retracts of ℓ^{∞} spaces. The fact that every (classical) metric space E isometrically embed into some ℓ^{∞} follow trivially from the fact that it embeds into some power of $[0, +\infty]$ (Translate the connected component C of $[0, +\infty]^{I}$ in which E embeds into $\ell^{\infty}(I)$).

Ordered sets.-

For $\underset{\sim}{V} =$ 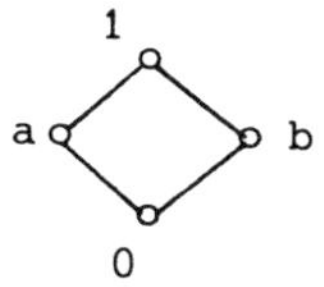

, the balls of a metric space E over $\underset{\sim}{V}$ are the set E, the singletons in E, the principal initial segments $(\leftarrow x]$ for all $x \in E$, and the principal final segments $[x \rightarrow)$ for all $x \in E$.

The space is *convex* if it is up - and down - directed (that is for every pair x, y there are z, t such that $z \leq x$, $y \leq t$). It has the *2-Helly* property iff every subset A of E has a supremum, resp. an infimum, whenever every two elements subset of A has a upper bound, resp. a lower bound. Thus *the hyperconvex spaces are the complete lattices.* Now the distance $d_{\underset{\sim}{V}}$ on $\underset{\sim}{V}$ defines an ordering for which b and a are respectively the least and last elements, and 0, 1 are incomparable, e.g. $\underset{\sim}{V}$ with is new ordering is order isomorphic to $\underset{\sim}{2} \times \underset{\sim}{2}$.

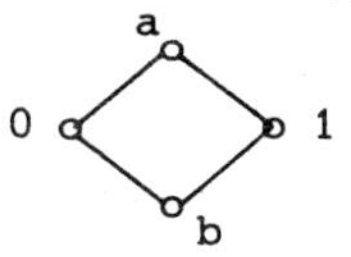

By II-2.10., the chain $\underset{\sim}{2}$ is a cogenerator; thus all that we get is the Banaschewski-Bruns theorem's.

II - 3. The injective envelope

In the categorical frame, the notion of *injective envelope* relies upon the notion of *essential morphism:*
A morphism $f : E \rightarrow F$ is *essential* with respect to the approximation $\mathfrak{D}$, if for every morphism $g : F \rightarrow G$, $g \circ f$ is an approximation iff g is an approximation (Note that, in particular, f is an approximation). An essential morphism f from E into an injective object F, (with respect to the same approximation $\mathfrak{D}$) is called an *injective envelope of E*. We will rather say that F (with f) is an injective envelope. It is uniquely determined, up to isomorphisms; for more details see [27]. In our case, we can view an injective envelope of a metric space E as an hyperconvex isometric extension F of E, which is minimal with respect to the inclusion (that is there is no proper hyperconvex subspace of F containing E). Since all those minimal extensions are isometric over E (in fact, as we shall see, by isometries which are the identity on E) we use a common symbol, $\underset{\sim}{N}(E)$ to denote them.

II - 3.1. ***Theorem 2*** - *Every metric space E over an Heyting involutive semigroup* $\underset{\sim}{V}$ *has an injective envelope.*

The proof is based upon the following property of metric form (defined in II-2.4).

II - 3.2. *Lemma.* The space $\underset{\sim}{L}(E)$ of metric forms is hyperconvex.

Proof - Observe first that $f \in \underset{\sim}{L}(E)$ iff $d(\overline{\delta}(y), f) = f(y)$ for all $y \in E$.
Indeed since $\overline{\delta}$ is an isometric embedding from E into $\underset{\sim}{V}^E$, each f satisfying this condition is a metric form. Conversely, assume f is of the form : $x \longmapsto d(x, u)$ for some u in an isometric extension of E. Then it satisfies

$$\begin{matrix} f(x) \leq d(x, y) + \underline{f}(y) \\ d(x, y) \leq f(x) + \overline{f}(y) \end{matrix} \quad \text{that is } d_{\underset{\sim}{V}}(d(x, y), f(x)) \leq f(y).$$

Writing this $d_{\underset{\sim}{V}}(\overline{\delta}(y)(x), f(x)) \leq f(y)$, we get

$d(\overline{\delta}(y), f)) = \text{Sup}\, \{d_{\underset{\sim}{V}}(\overline{\delta}(y)(x), f(x) \;/\; x \in E\} = f(y)$

From that it follows that $\underset{\sim}{L}(E)$ is the set $\text{Hom}(\overline{E}, \underset{\sim}{V}')$ equipped with the metric induced by $\underset{\sim}{V}^E$.

Now let F be an isometric extension of $\underset{\sim}{L}(E)$. For every $u \in F$ let $\tilde{u} : E \rightarrow \underset{\sim}{V}$ be defined by setting $\tilde{u}(x) = d(\overline{\delta}(x), u)$ for all $x \in E$. Since $\overline{\delta} : E \rightarrow \underset{\sim}{L}(E)$ is an isometry, $\tilde{u}$ is a metric form. We show that the map

$u \hookrightarrow \tilde{u}$ is a retraction, e.g. that $d(\tilde{u}, \tilde{v}) \prec d(u, v)$. For that we have

$$d(\overline{\delta}(x), u) \prec d(\overline{\delta}(x), v) + d(v, u) \qquad (1)$$

$$d(\overline{\delta}(x), v) \prec d(\overline{\delta}(x), u) + d(u, v) \qquad (2)$$

(1) and (2) give $\tilde{u}(x) \prec \tilde{v}(x) + \overline{d(u, v)}$ and $\tilde{v}(x) \prec \tilde{u}(x) + d(u, v)$ proving $d_{\underset{\sim}{V}}(\tilde{u}(x), \tilde{v}(x)) \prec d(u, v)$. Thus $\operatorname{Sup}\{d_{\underset{\sim}{V}}(\tilde{u}(x), \tilde{v}(x)) \,/\, x \in E\} \prec d(u, v)$. Whence $d(\tilde{u}, \tilde{v}) \prec d(u, v)$. □

Proof - of Theorem 2: Let $\underset{\sim}{N}(E)$ be the set of minimal metric forms. First $\underset{\sim}{N}(E)$ is hyperconvex : let F be an isometric extension of $\underset{\sim}{N}(E)$ and $u \in F$. We have to prove that there is $\omega \in \underset{\sim}{N}(E)$ such that $\omega \in \bigcap \{B(f, d(f, u)) \,/\, f \in \underset{\sim}{N}(E)\}$. For that let $\tilde{u} : E \longrightarrow \underset{\sim}{V}$ be defined by $\tilde{u}(x) = d(\overline{\delta}(x), u)$. This is a metric form and $d(\overline{\delta}(x), u) = d(\overline{\delta}(x), \tilde{u})$ for all x . Let ω be a minimal metric form , $\omega \prec \tilde{u}$. Consider the map $\varphi : \overline{\delta}(E) \cup \{u\} \subseteq F \longrightarrow \underset{\sim}{L}(E)$ sending u to ω and equal to the identity on $\overline{\delta}(E)$. Since $\underset{\sim}{L}(E)$ is hyperconvex, the map φ extends on F to a non expansive mapping $\overline{\varphi}$. We claim that $\overline{\varphi}$ is the identity on $\underset{\sim}{N}(E)$. Indeed, let $f \in \underset{\sim}{N}(E)$. Since $\overline{\varphi}$ is non expansive, for all $x \in E$, we have $d(\overline{\delta}(x), \overline{\varphi}(f)) \prec d(\overline{\delta}(x), f)$ meaning $\overline{\varphi}(f)(x) \prec f(x)$. Since f is a minimal metric form, it follows that $\overline{\varphi}(f) = f$. The facts that $\omega = \overline{\varphi}(u)$ and $\overline{\varphi}$ fixes $\underset{\sim}{N}(E)$, insures that ω has the required property. Second, $\underset{\sim}{N}(E)$ is a minimal hyperconvex isometric extension of $\overline{\delta}(E)$: If F is hyperconvex with $\overline{\delta}(E) \subseteq F \subseteq \underset{\sim}{N}(E)$ then, since F is hyperconvex, the identity map $i : \overline{\delta}(E) \subseteq \underset{\sim}{N}(E) \longrightarrow F$ extends on $\underset{\sim}{N}(E)$ to a non expansive map $\hat{i}$. As above we have $d(\overline{\delta}(x), \hat{i}(f)) \prec d(\overline{\delta}(x), f)$ for all $x \in E$, $f \in \underset{\sim}{N}(E)$; this gives $\hat{i}(f) \prec f$, thus $\hat{i}(f) = f$ by the minimality of f . Consequently $F = \underset{\sim}{N}(E)$.

Metric spaces.- This is a result of I. Isbell [22]. We own him the notion of minimal metric form (that he calls extremal). Although our proof is based upon the same ideas we avoid his computations, which are specific to the real line and do not extend to our case.

Ordered sets.-

II - 3.3. ***Theorem.*** (B. Banaschewski - G. Bruns [4]). *The injective envelope of a poset is its MacNeille completion.*

The illustration of the notions of metric form and minimal metric form provide a proof of this important result.

Let P be a poset; we consider it as a metric space over $\underset{\sim}{V}$ = a 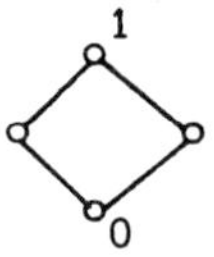 b .

For $f : P \longrightarrow \underset{\sim}{V}$ we put $A = f^{-1}(a)$, $B = f^{-1}(B)$.

If f is a metric form and $0 \in \operatorname{Im} f$ then $f(x) = d(x, u)$ for a unique u , so $A = (\leftarrow u[$, $B =]u \rightarrow)$. If $0 \notin \operatorname{Im} f$, then f is a metric form iff A is an initial segment, B is a final segment and every element of A is below every element of B. Indeed, the inequality $f(x) \prec d(x, y) + f(y)$ insures that A and B are an initial and a final segment respectively, and the inequality $d(x, y) \prec f(x) + \overline{f(y)}$ insures that every element of A is below every element of B.

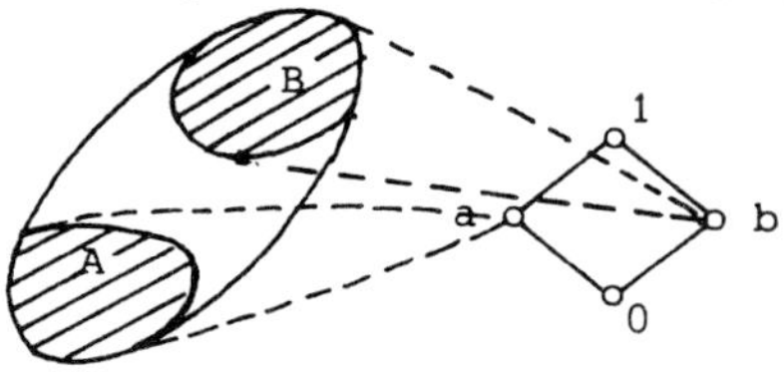

Clearly f is minimal if and only if $A^{+} = B$ and $B^{-} = A$ (As usual we denote $A^{+} = \{y \,/\, x \prec y \text{ for all } x \in A\}$ and $B^{-} = \{y \,/\, y \prec x \text{ for all } x \in B\}$).

In this case A and B define a *gap*, e.g. there is no x in between. Since $d(f, f') \prec a$ means $A \subseteq A'$ and $B' \subseteq B$, with obvious notations, the space $\underset{\sim}{N}(P)$, considered as a poset, is the MacNeille completion of P.

II - 4. Absolute retract with respect to the holes preserving maps

The definition of an *hole* is inspired by a similar notion introduced by R. Nowakowski and I. Rival [36] for graphs, see also [19],[24].

Let E be a metric space over $\underset{\sim}{V}$: An *hole* is a map $h : E \longrightarrow \underset{\sim}{V}$ such that the intersection of the balls $B(x, h(x))$ $(x \in E)$ is empty. Let f be a non expansive map from E into an other space F, and $h : E \longrightarrow \underset{\sim}{V}$; the *image* of h is the map $h_f = F \longrightarrow \underset{\sim}{V}$ defined by $h_f(y) = \wedge\{h(x) \,/\, f(x) = y\}$.

Clearly $\bigcap_{x \in E} B(f(x), h(x)) = \bigcap_{y \in F} B(y, h_f(y))$. The map f *preserves* h if h_f is a hole of F. Finally we say that *f preserves the holes* or is *hole-preserving* iff all the holes of E are preserved.

II - 4.1. *Proposition.* The coretractions preserve the holes, the hole-preserving maps are isometries.

Proof - Let $f : E \longrightarrow F$ be a coretraction ; let $h : E \longrightarrow \underset{\sim}{V}$ be a hole. If there is $z \in \bigcap_{y \in F} B(y, h_f(y))$ then, for every retraction g , $g(z) \in \bigcap_{x \in E} B(x, h(x))$. Let $f : E \longrightarrow F$ be hole-preserving and let $x, y \in E$. The map $h : E \longrightarrow V$ defined by setting $h(x) = 0$, $h(y) = d(f(y), f(x))$ and $h(z) = 1$ for $z \neq x, y$, cannot be a hole since $f(x) \in \bigcap_{z \in F} B(z, h_f(z))$; thus $x \in B(y, d(f(y), f(x))$ that is $d(y, x) \prec d(f(y), f(x))$. □

Using the hole-preserving maps as approximations of the coretractions we get:

II - 4.2. *Theorem 3* - *The absolute retracts and the injective objects, with respect to the hole-preserving maps, coincide; thus they form a variety $\mathcal{H}$ of metric spaces. Moreover every metric space embeds into a member of $\mathcal{H}$ by a hole-preserving map.*

Proof - To every metric space E we associate a metric space $\mathcal{H}(E)$ the *replete-space*. We prove that E embeds into $\mathcal{H}(E)$ by the hole-preserving map $\overline{\delta}$ and $\mathcal{H}(E)$ is an absolute retract (Lemma II-4.3.). Then we prove, using $\mathcal{H}(E)$, that the hole-preserving maps are transferable (Lemma II-4.6.), namely, that for every $f : F \longrightarrow E$, every $g : F \Longrightarrow G$, there is $f : G \longrightarrow \mathcal{H}(E)$ such that the diagram commutes:

$$\begin{array}{ccc} F & \overset{g}{\Longrightarrow} & G \\ f \big\downarrow & & \big\downarrow \hat{f} \\ E & \overset{\overline{\delta}}{\Longrightarrow} & \mathcal{H}(E) \end{array}$$

From fact 3 of I-2. follows that the absolute retracts and the injectives coïncide. □

II - 4.3. *Lemma.* Let $\mathcal{H}(E) = \{h \in \underset{\sim}{L}(E) \ / \cap \{B(x, h(x)) \ / \ x \in E\} \neq \emptyset\}$ and $\overline{\delta} : E \longrightarrow \mathcal{H}(E)$, defined by setting $\overline{\delta}(x)(y) = d(y, x)$ for all $x, y \in E$. Then $\overline{\delta}$ is hole-preserving and $\mathcal{H}(E)$ is an absolute retract.

Proof- For every $x \in E$, $\overline{\delta}(x) \in \mathcal{H}(E)$. Indeed $\cap \{B(t, \delta(x)(t)) \ / \ t \in E\} = \{x\}$. The map $\overline{\delta}$ is hole-preserving: Let $h : E \longrightarrow \underset{\sim}{V}$ be a hole. If $h_{\overline{\delta}} : \mathcal{H}(E) \longrightarrow \underset{\sim}{V}$ (where $h_{\overline{\delta}}(u) = \wedge \{h(x) \ / \ \overline{\delta}(x) = u\}$) is not a hole, choose $u \in \cap \{B(y, h_{\overline{\delta}}(y)) \ / \ y \in \mathcal{H}(E)\} = \cap \{B(\overline{\delta}(x), h(x)) \ / \ x \in E\}$. Now, since $u \in \mathcal{H}(E)$, we have $\cap \{B(x, u(x)) \ / \ x \in E\} \neq \emptyset$. Since $u(x) = d(\overline{\delta}(x), u) \prec h(x)$ we have $B(x, u(x)) \subseteq B(x, h(x))$ for all $x \in E$. Thus $\cap \{B(x, h(x)) \ / \ x \in E\} \neq \emptyset$. Contradiction. Finally $\mathcal{H}(E)$ is an absolute retract. Indeed, let F be a hole-preserving extension of $\mathcal{H}(E)$. For $u \in F$, let $\tilde{u} : E \longrightarrow \underset{\sim}{V}$ defined by setting $\tilde{u}(x) = d(\overline{\delta}(x), u)$ for all $x \in E$. By construction, $\tilde{u}$ is a metric form. Moreover $u \in \mathcal{H}(E)$. Indeed, assume the contrary, that is $\cap \{B(x, \tilde{u}(x)) \ / \ x \in E\} = \emptyset$. Since $\mathcal{H}(E)$ is a hole-preserving extension of E (through $\overline{\delta}$) and F is a hole-preserving extension of $\mathcal{H}(E)$, this imposes $\cap \{B(\overline{\delta}(x), \tilde{u}(x)) \ / \ x \in E\} = \emptyset$. But u belongs to this intersection. Contradiction. The map $u \longrightarrow \tilde{u}$ is a retraction from F onto $\mathcal{H}(E)$. Indeed it fixes $\mathcal{H}(E)$ (since $d(\overline{\delta}(x), u) = u(x)$ for every metric form) and, as we have already seen in the proof of Lemma II-3.2., it is non-expansive. □

II - 4.4. *Lemma.* Let $\underset{\sim}{C}(E) = \{v \in \underset{\sim}{V}^E \ / \ d(x, y) \prec v(x) + \overline{v(y)}$ for all $x, y \in E\}$. For $v \in \underset{\sim}{C}(E)$, let $v_M : E \longrightarrow \underset{\sim}{V}$ defined by setting $v_M(x) = \wedge \{d(x, y) + v(y) \ / \ y \in E\}$. Then v_M is the largest metric form below v and $\cap \{B(x, v(x)) \ / \ x \in E\} = \cap \{B(x), v_M(x)) \ / \ x \in E\}$. The map $v \longrightarrow v_M$ is a retraction from $\underset{\sim}{C}(E)$ onto $\underset{\sim}{L}(E)$.

Proof- Let $\omega \in \underset{\sim}{L}(E)$, $\omega \prec v$, then $\omega \prec v_M$. Indeed, since ω is a metric form, we have $\omega(x) \prec d(x, y) + \omega(y)$ and, since $\omega \prec v$, we have $\omega(y) \prec v(y)$, thus $\omega(x) \prec d(x, y) + v(y)$. It follows $\omega(x) \prec \wedge \{d(x, y) + v(y) \ / \ y \in E\} = v_M(x)$. Now v_M is a metric form, that is $d(x, y) \prec v_M(x) + \overline{v_M(y)}$ and $v_M(x) \prec d(x, y) + v_M(y)$ for all $x, y \in E$. For the first inequality, we compute

$v_M(x) + \overline{v_M(y)} = \wedge\{d(x, z) + v(z) / z \in E\} + \wedge\{\overline{v(t)} + d(t, y) / t \in E\}$. From the distributivity we get $\wedge\{d(x, z) + v(z) + \overline{v}(t) + d(t, y) / z, t \in E\}$. From the triangular inequality and $v(z) + \overline{v}(t) \geq d(z, t)$, we get $d(x, z) + v(z) + \overline{v}(t) + d(t, y) \geq d(x, y)$, thus $v_M(x) + \overline{v_M(y)} \geq d(x, y)$. For the second inequality, we have $d(x, z) + v(z) \leq d(x, y) + d(y, z) + v(z)$ for all $z \in E$, thus

$\wedge\{d(x,z) + v(z) / z \in E\} \leq \wedge\{d(x,y) + d(y,z) + v(z) / z \in E\} =$

$d(x,y) + \wedge\{d(y,z) + v(z) / z \in E\}$ (from the distributivity). That is $v_M(x) \leq d(x, y) + v_M(y)$.

Since $v_M \leq v$ we have $\cap\{B(x,v_M(x)) / x \in E\} \subseteq \cap\{B(x,v(x)) / x \in E\}$. For the reverse inclusion let $t \in \cap\{B(x, v(x)) / x \in E\}$. This amounts to say that $\overline{\delta}(t) \leq v$. Since $\overline{\delta}(t)$ is a metric form and v_M is the largest metric form below v, it follows that $\overline{\delta}(t) \leq v_M$, which amounts to say that $t \in \cap\{B(x, v_M(x)) / x \in E\}$.

The map $v \mapsto v_M$ is a retraction. First this map is non-expansive. Indeed let $v, \omega \in \underset{\sim}{C}(E)$; since $v(y) \leq \omega(y) + \overline{d(v,\omega)}$ we have $d(x,y) + v(y) \leq d(x,y) + \omega(y) + \overline{d(v,\omega)}$, thus from the distributivity we have

$\wedge\{d(x,y) + v(y) / y \in E\} \leq \wedge\{d(x,y) + \omega(y) + \overline{d(v,\omega)} / y \in E\} =$

$\wedge\{d(x,y) + \omega(y) / y \in E\} + \overline{d(v,\omega)}$ that is $v_M(x) \leq \omega_M(x) + \overline{d(v,\omega)}$. The same argument shows that $\omega_M(x) \leq v_M(x) + d(v,\omega)$. Consequently $d_{\underset{\sim}{V}}(v_M(x), \omega_M(x)) \leq d(v,\omega)$ for every $x \in E$; that is $d(v_M, \omega_M) \leq d(v, \omega)$. Finally this map fixes $\underset{\sim}{L}(E)$. Indeed, since v_M is the largest metric form below v , we have $v_M = v$ whenever v is a metric form. □

II - 4.5. *Lemma.* Let E, F be two metric spaces over $\underset{\sim}{V}$; then every non-expansive map $f : F \longrightarrow E$ extends to a non expansive map $\mathbb{H}_f : \mathbb{H}(F) \longrightarrow \mathbb{H}(E)$.

Proof - Let $\alpha_f : \underset{\sim}{V}^F \longrightarrow \underset{\sim}{V}^E$ defined by setting $\alpha_f(v) = v_f$ for $v \in \underset{\sim}{V}^F$ ($v_f(x) = \wedge\{v(x') / x' \in F$ and $f(x') = x\}$, in particular $v_f(x) = 1$ for all $x \in E \setminus f(F)$) . This map is non-expansive. Indeed let $v, \omega \in \underset{\sim}{V}^F$. The inequality $v(x') \leq \omega(x') + \overline{d(v, \omega)}$ for $x' \in F$ and the distributivity give $\wedge\{v(x')/x' \in F$ and $f(x')=x\} \leq \wedge\{\omega(x')/x' \in F$ and $f(x')=x\} + \overline{d(v,\omega)}$, that is $v_f(x) \leq \omega_f(x) + \overline{d(v,\omega)}$, for $x \in f(F)$. The same arguments give $\omega_f(x) \leq v_f(x) + d(v,\omega)$. Consequently $d_{\underset{\sim}{V}}(v_f(x), \omega_f(x)) \leq d(v, \omega)$. Since this also holds for $x \in E \setminus f(F)$ (since in this case $v_f(x) = \omega_f(x) = 1$) we get $d(v_f, \omega_f) \leq d(v, \omega)$.

Now, if $v \in \underset{\sim}{C}(F)$ then $\alpha_f(v) \in \underset{\sim}{C}(E)$. Indeed let $x, y \in E$. If x or y does not belong to f(F) then $v_f(x)$ or $v_f(y)$ equals 1 thus $d(x, y) \leq v_f(x) + \overline{v_f(y)}$. If $x, y \in f(F)$ then we have

$d(x, y) = d(f(x'), f(y')) \leq d(x', y') \leq v(x') + \overline{v(y')}$ for all $x', y' \in F$ such that $f(x') = x$, $f(y') = y$. Thus, from the distributivity, $d(x,y) \leq \wedge \{v(x') + \overline{v(y')}/f(x')=x, f(y')=y\} = \wedge \{v(x')/f(x')=x\} + \wedge \{\overline{v(y')}/f(y')=y\}$ that is $d(x, y) \leq v_f(x) + \overline{v_f(y)}$. Let us denote M the retraction map from $\underset{\sim}{C}(E)$ onto $\underset{\sim}{L}(E)$ defined by setting $M(v) = v_M$ for all $v \in \underset{\sim}{C}(E)$. The composition $M \circ \alpha_f : \underset{\sim}{C}(F) \longrightarrow \underset{\sim}{L}(E)$ is non-expansive.

It extends f; more precisely $M \circ \alpha_f(\overline{\delta}(x')) = \overline{\delta}(f(x'))$ for every $x' \in F$. Indeed, since $\alpha_f(\overline{\delta}(x'))(f(x')) = 0$ and $M \circ \alpha_f(\overline{\delta}(x')) \leq \alpha_f(\overline{\delta}(x'))$, it follows that $M \circ \alpha_f(\overline{\delta}(x'))(f(x')) = 0$. Since $M \circ \alpha_f(\overline{\delta}(x'))$ is a metric form this imposes $M \circ \alpha_f(\overline{\delta}(x')) = \overline{\delta}(f(x'))$. (Observe that $d(\overline{\delta}(f(x')), M \circ \alpha_f(\overline{\delta}(x'))) = 0$). Now for $v \in \underset{\sim}{C}(F)$, we have $\cap \{B(f(x'), v(x')) / x' \in F\} = \{B(x, \alpha_f(v)(x)) / x \in E\} = \cap \{B(x, M \circ \alpha_f(v)(x))/x \in E\}$ (from Lemma II-4.4.). Consequently if $v \in \mathcal{H}(F)$ then $M \circ \alpha_f(v) \in \mathcal{H}(E)$. The restriction $\mathcal{H}_f$ of $M \circ \alpha_f$ to $\mathcal{H}(F)$ has the required properties. □

II - 4.6. *Lemma.* The hole-preserving maps are transferable.

Proof - Let $f : F \longrightarrow E$ and $g : F \Longrightarrow G$ be a non-expansive map and a hole-preserving map respectively. We define $\hat{f} : G \longrightarrow \mathcal{H}(E)$ such that $\hat{f} \circ g = \overline{\delta} \circ f$. For that we define a map $\mathfrak{C}_g : G \longrightarrow \mathcal{H}(F)$.
For every $u \in G$, let $\tilde{u} : F \longrightarrow \underset{\sim}{V}$ defined by setting $\tilde{u}(x) = d(g(x), u)$ for all $u \in F$. Clearly $\tilde{u} \in \mathcal{H}(F)$. (Indeed $u \in \cap \{B(g(x), d(g(x),u)) / x \in F\}$ and since g is hole-preserving, $\cap \{B(x, d(g(x),u)) / x \in F\} = \cap \{B(x, \tilde{u}(x)) / x \in F\}$ is non empty).
The map $\mathfrak{C}_g : G \longrightarrow \mathcal{H}(F)$, where $\mathfrak{C}_g(u) = \tilde{u}$, is non expansive (Cf. Proof of Lemma II-3.2.) and $\mathfrak{C}_g(g(x)) = \overline{\delta}(x)$ for every $x \in F$ (indeed, since g is hole preserving, g is an isometry, thus $\mathfrak{C}_g(g(x))(y) = d(g(y), g(x)) = d(y, x) = \overline{\delta}(x)(y)$). Let $\hat{f} = \mathcal{H}_f \circ \mathfrak{C}_g$ where $\mathcal{H}_f$ is given by Lemma II-4.5. Then $\hat{f} \circ g = \overline{\delta} \circ f$. □

In the exemples considered in IV-1.2, 2.2., 3. and 5., we deal with a particular subcategory $\mathfrak{C}$ of metric space over $\underset{\sim}{V}$. And contrarily to the case of hyperconvexity it is not clear that theorem 3 will apply to $\mathfrak{C}$. Such a category has the following form:

Let C be a subset of $\underset{\sim}{V}$. A metric space E over $\underset{\sim}{V}$ is *C-convex* if for all $x, y \in E$, $p, q \in C$ such that $d(x, y) \leq p + q$ then there is $z \in E$ such that $d(x, z) \leq p$ and $d(z, y) \leq q$. Let $\mathfrak{C}$ be the subcategory of all C-convex metric spaces (and the non-expansive mappings). To extend Theorem 3 to $\mathfrak{C}$ it would be enough to show that $\mathcal{H}(E)$ belongs to $\mathfrak{C}$ whenever E belongs to $\mathfrak{C}$. We are unable to prove it, even with the help of additional properties that C satisfies in our examples. Nevertheless these properties allow us to extend Theorem 3. We describe it below.

Let D be a subset of $\underset{\sim}{V}$, D^* be the set of finite sequences $r_0 \cdot r_1 \ldots r_{m-1}$ of elements r_i of D, i.e; "words" over the alphabet D, and ΣD - our set C - be the set of finite sums of elements of D

(i.e. $\Sigma D = \{ r_0 + r_1 + \ldots + r_{m-1} / r_0 \, r_1 \cdot \ldots \cdot r_{m-1} \in D^* \}$). We assume that the following holds

    1) $D = \overline{D}$

    2) Every non zero element of $\underset{\sim}{V}$ is the meet of elements of ΣD

(i.e. $x = \wedge \{ p \in \Sigma D / x \leqslant p \}$ for every x, $x \neq 0$) .

    3) For all p, $q \in \Sigma D$, $X \subseteq D^*$, if

$\wedge \{ r_0 + r_1 + \ldots r_{m-1} / r_0 r_1 \ldots r_{m-1} \in X \} \leqslant p + q$ then there are $r_0 \ldots r_{m-1} \in X$, $j \in [0, m-1]$ such that $r_0 + \ldots + r_{j-1} \leqslant p$ and $r_j + \ldots + r_{m-1} \leqslant q$. In particular every $r \in D$ is *irreducible* that is for every p, $q \in \Sigma D$ if $r \leqslant p + q$ then $r \leqslant p$ ou $r \leqslant q$.

Let E be a metric space over $\underset{\sim}{V}$. From the distance d on E, we define a new distance d_D : $E \times E \longrightarrow \underset{\sim}{V}$ by setting $d_D(x, y) = \wedge \{ r$ / there are $r_0, \ldots r_{m-1}$ in D , $x_0 = x, x_1, \ldots, x_m = y$ in E such that $d(x_i, x_{i+1}) \leqslant r_i$ for all i = 0, ..., m-1 and $r_0 + \ldots r_{m-1} \leqslant r \}$. We denote E_D the space (E, d_D).

Let $\mathcal{H}_D(E)$ be the set $\mathcal{H}(E)$ previously defined, **endowed** with the distance d_D . We prove using $\mathcal{H}_D(E)$ the following

    II · 4.7. *Proposition.*   Let D satisfying the above conditions.
In the category of ΣD-convex metric spaces, the absolute retracts and the injective objects, with respect to the hole-preserving maps, coïncide and they form a variety $\mathcal{H}_D$ of metric spaces. Moreover every metric space of this category embeds into a member of $\mathcal{H}_D$ by a hole-preserving map.

Proof· Let E be a ΣD-convex metric space. We prove first that $\mathcal{H}_D(E)$ is ΣD-convex, the map $\overline{\delta}$ is a hole preserving map and $\mathcal{H}_D(E)$ is an absolute retract.

The first fact is obvious. Indeed by definition of the distance d_D and property 3, the spaces endowed with such a distance are ΣD-convex. For the second observe that, because of 2., the distance d and d_D coïncide on ΣD-convex sets, thus $E = E_D$ and moreover that

$d(\overline{\delta}(x), f) = d_D(\overline{\delta}(x), f)$ for all $f \in \mathcal{H}(E)$, $x \in E$.

To get this it is enough to show that *if $d(\overline{\delta}(x), f) \leqslant p + q$, where p, $q \in \Sigma D$ then there is $h \in \mathcal{H}(E)$ such that $d(\overline{\delta}(x), h) \leqslant p$ and $d(h, f) \leqslant q$* . For that let k : $E \longrightarrow \underset{\sim}{V}$ be the map defined by setting $k(t) = (d(t, x) + p) \wedge (f(x) + \overline{q})$. The argument given in 2) of Proposition II-2.7. shows that $d(\overline{\delta}(x), k) \leqslant p$ and $d(k, f) \leqslant q$. Let $y \in \cap \{ B(t, f(t)) / t \in E \}$.

Since $f(x) = d(\overline{\delta}(x), f) \prec p+q$ we have $d(x, y) \prec p+q$. Since E is ΣD-convex there is $z \in E$ such that $d(x, z) \prec p$ and $d(z, y) \prec q$. Clearly $z \in \cap \{B(t, k(t)) / t \in E\}$. Consequently the map $h = k_M$ (Cf Lemma II-4.4.) has the required property. (We do not know whether this property holds with arbitrary elements g of $\mathcal{H}(E)$ instead of $\overline{\delta}(x)$).

Now, since $d \prec d_D$ on $\mathcal{H}(E)$ and $\overline{\delta}$ is an hole-preserving map for d it is also a hole-preserving map for d_D. For the third fact observe that if E', F' are two metric spaces then a map from E' to F' is a non-expansive mapping from E'_D to F'_D if it satisfies the following property: *for every* $r \in D$, $d(f(x), f(y)) \prec r$ *whenever* $d(x, y) \prec r$. If F is an hole-preserving extension of $\mathcal{H}_D(E)$, and F is ΣD convex, then the map $u \longrightarrow \tilde{u}$ as defined in Lemma II-4.3. satisfies this property. Hence it is a retraction from F onto $\mathcal{H}_D(E)$.

Finally, we prove that the hole-preserving maps are transferable. This is straight-forward: The map $\hat{f}$ from G to $\mathcal{H}(E)$ is non-expansive thus satisfies the above property. Hence it is non-expansive mapping from $G = G_D$ to $\mathcal{H}_D(E)$. □

The replete-space of E is a witness of the absolute retractness of E, since E is an absolute retract iff it is a retract of $\mathcal{H}(E)$. The retraction maps have the following nice characterization (whose routine proof is omitted).

II - 4.8. *Proposition.* The map $f : \mathcal{H}(E) \longrightarrow E$ is a retraction (and $\overline{\delta}$ is the coretraction) iff f is non expansive and for every $h \in \mathcal{H}(E)$ we have $f(h) \in \cap \{B(x, h(x) / x \in E\}$.

We call such a map a *selection,* if it exists, the space has *the selection property*. In other words, *E is an A.R., or equivalently an injective, iff it has the selection property.*

Comment.-The notion of hole is a way, rather primitive, to express an obstruction to retraction. Indeed a hole $h : E \longrightarrow \underset{\sim}{V}$ is ment to capture the fact that there is an element u in some extension F such that $d(x, u) \prec h(u)$ and $\cap \{B(x, h(u)) / x \in E\}$ is empty in E. However not all holes have this property. The correct notion relies of course upon the notion of metric forms; but it use requires the metrization of the spaces of values, which appears first time in this paper. Nevertheless as it is defined it seems to be an useful notion. For example most of the results of [14], [19], [26], [28] are expressed in terms of holes. Let us rephrase briefly our previous results in these terms. For that we need the following definitions.

The *domain* of an hole $h : E \longrightarrow \underset{\sim}{V}$ is the set $D_f = \{x / f(x) \neq 1\}$. A hole h is *minimal* if $|D(h')| = |D(h)|$ for every hole h', with $h' \succ h$ ($h' \succ h$ means $h'(x) \succ h(x)$ for all $x \in E$), and we say that it is a $|D(h)|$*-hole*.

Clearly for every hole h there is a minimal hole h' such that $h' \succ h$, and a map f preserves all the holes iff it preserves the minimal one. This notion of minimality is very weak, except in one case:

II - 4.9. *Proposition.* *A space E has the 2-Helly property iff every minimal hole is a 2-hole*

This does not merit a formal proof, but it is a suggestive way to express the 2-Helly property.

The case of posets.-

Let P be a poset. If we consider only map $h : P \longrightarrow \underset{\sim}{V} \setminus \{0\}$, a hole is simply a pair of sets A, B, such that there is no element x in between, that is a *gap*. The replete space $\mathcal{H}(P)$ is the set of pairs (A, B) where A is an initial segment, B is a final segment and $A^+ \cap B^- \neq \phi$. The ordering on $\mathcal{H}(P)$ (obtained from the metric) is the following $(A, B) \prec (A', B')$ iff $A \subseteq A'$ on $B' \subseteq B$.

A selection is an increasing map from $\mathcal{H}(P)$ to P such that $f(A, B) \in A^+ \cap B^-$ for every A, B.

In the case of posets, the replete space has been previously introduced by P. Nevermann and R. Wille [34]. They showed that the selection property (that they called strong selection property) is preserved under retracts and products ans is equivalent to absolute retractness. The equivalence to injectivity is new. For further results on absolute retracts see [13], [42].

III - FIXED POINT PROPERTY FOR NON EXPANSIVE MAPPINGS

III - 1. Invariant subspace of an hyperconvex space

Let $\underset{\sim}{V}$ be our ordered semigroup (we do not assume distributivity). An element r of $\underset{\sim}{V}$ is *self-dual* if $\overline{r} = r$, *accessible* if for some $u \not\succ r$ $r \prec u + \overline{u}$, and *inaccessible* otherwise. Roughly, r is accessible if one can "reach" it from some element not already larger. Clearly 0 is inaccessible and every inaccessible element is self-dual. For example, in the poset $\underset{\sim}{V} = \{0, a, b, 1\}$, with a incomparable to b, 0 is the unique inaccessible element whereas in $\underset{\sim}{V} = [0, +\infty]$, 0 and $+\infty$ are the inaccessible elements.

III - 1.1. *Main Lemma.* Let E be a non empty hyperconvex space and $f : E \longrightarrow E$ be a non-expansive mapping: then there is a non empty hyperconvex subspace S of E such that $f(S) \subseteq S$ and its diameter $\underset{\sim}{\delta}(S) = \vee \{d(x, y) / x, y \in S\}$ is inaccessible.

Proof - Let $\mathcal{B}$ be the collection of non empty intersections of closed balls and let $\mathcal{B}_f = \{X \in \mathcal{B} / f(X) \subseteq X\}$.

The space E belongs to $\mathcal{B}_f$, (indeed, $E = B(x, 1)$ for all $x \in E$). The compactness property insures that $\mathcal{B}_f$ ordered by inclusion contains a minimal element S. Since S is an intersection of balls, it is hyperconvex. We show that $\underset{\sim}{\delta}(S)$ is inaccessible.

Let $u \in \underset{\sim}{V}$ such that $\underset{\sim}{\delta}(S) \prec u + \overline{u}$. Consider the family of balls of S, $B_S(x, u)$, $(u \in S)$. For all $x, y \in S$ we have $d(x, y) \prec \delta(S) \prec u + \overline{u}$. By the hyperconvexity of S, the set $S' = \cap \{B_S(x, u) / x \in S\}$ is non empty. Let $y \in S'$, then $S \subseteq B(y, \overline{u})$ and, since f is non expansive, $f(S) \subseteq B(f(y), \overline{u})$. Thus, if we define $\overline{A} = \cap \{B \in \mathcal{B} / A \subseteq B\}$ for every $A \subseteq E$, then we have $\overline{f(S)} \subseteq B(f(y), \overline{u})$. On an other hand, we have

$\overline{f(S)} = S$. Indeed $f(S) \subseteq S$ implies $f(S) \subseteq \overline{f(S)} \subseteq S$ thus $f(\overline{f(S)}) \subseteq f(S) \subseteq \overline{f(S)}$, that is $\overline{f(S)} \in \mathcal{B}_f$ and since S is minimal, $\overline{f(S)} = S$. Hence $S \subseteq B(f(y), \overline{u})$, proving $f(y) \in S'$. We have $f(S') \subseteq S'$ and by construction $S' \in \mathcal{B}$ (indeed $S' = \cap \{B(x, u) / x \in S\} \cap S$) thus $S' \in \mathcal{B}_f$. By minimality $S' = S$. But if $x, y \in S'$ then $d(x, y) \prec u$, hence $\underset{\sim}{\delta}(S') \prec u$. Since $S' = S$ this gives $\underset{\sim}{\delta}(S) \prec u$.

This reproduce an argument given by W. Kirk [26] p.246 (instead of the longer proof given by the second and third author in [32]).

For a metric space E, the *automorphism group,* Aut(E), consists of all bijective isometries $f : E \longrightarrow E$. A subspace X is *invariant* provided $f(X) = X$ for every $f \in \mathrm{Aut}(E)$. We have the following very easy result (observed by A. Quilliot for graphs [44]).

III - 1.2. *Lemma.* Let E be a non empty hyperconvex space. Then there is a non empty hyperconvex invariant subspace S whose diameter is inaccessible.

Proof - The proof, much easier, is an adaptation of the previous one. Let $\mathcal{B}_I = \{X \in \mathcal{B} \ / \ X \text{ is invariant}\}$. This set, ordered by inclusion has a minimal element S. Indeed, we have shown earlier that $\mathcal{B}$ is closed under intersections of descending chains and its known (and easy to show) that the same holds in the poset of invariant spaces.

As before the set S is hyperconvex. Let $p = \underset{\sim}{\delta}(S)$. To prove that p is inaccessible let u such that $p \prec u + \overline{u}$.

Consider the family $\{B(x, u) / x \in S\}$. For all $x, y \in S$ we have $d(x, y) \prec p \prec u + \overline{u}$. Put $S' = \{y \in S \ ; \ d(x, y) \prec u \text{ for all } x \in S\}$.

By the hyperconvexity of S, the set S' is non empty. Now S is invariant and so it is easy to check that $f(S') = S'$ for every $f \in \mathrm{Aut}(E)$.

Consequently $S' = S$, but since $\underset{\sim}{\delta}(S') \prec u$, this gives $p \prec u$. □

III - 2. Fixed point property for a non expansive mapping

A metric space E over $\underset{\sim}{V}$ is *bounded* if there are $x \in E$ and $r \in V$ such that

1) 0 is the unique inaccessible element below r
2) $E = B(x, r)$.

This amounts to the fact that 0 is the unique inaccessible element below the diameter $\underset{\sim}{\delta}(E)$ of E (observe that $\underset{\sim}{\delta}(E) \prec \overline{r} + r$, thus if p is inaccessible with $p \prec \delta(E) \prec \overline{r} + r$ then $p \prec \overline{r}$, that is $p = \overline{p} \prec r$).

III - 2.1. ***Theorem 4*** - *Let E be a non empty bounded hyperconvex space. Then every non-expansive mapping f has a fixed point. Moreover, the set* Fix(f) *of its fixed points is hyperconvex.*

Proof - Since 0 is the unique inaccessible element below $\underset{\sim}{\delta}(E)$, the diameter of the non empty set S given by the main lemma (III-1.1.) is 0, thus S reduces to a single element fixed by f.

Let $\{(B_F(x_i, r_i)) / i \in I\}$ be a family of balls of Fix(f) with $d(x_i, x_j) \prec r_i + \overline{r}_j$ for all $i, j \in I$. Since E is hyperconvex,

$T = \cap \{B(x_i, r_i) \ / \ i \in I\} \neq \phi$ and, as any intersection of balls of an hyperconvex space, T is hyperconvex and, of course, bounded. Now since f is non expansive and the x_i are fixed by f, we have $f(T) \subseteq T$. The f.p.p. applied to T gives an $x \in Fix(f) \cap T$ thus the above intersection is non empty, and Fix(f) is hyperconvex.

III - 2.2. *Corollary.* Let E be a bounded hyperconvex space. Among the subspaces of E, the retracts of E are the sets of fixed points of the non expansive mapping from E into itself.

Proof - If A is a retract of E, then $A = Fix(g)$ for every retraction. Conversely from the above result the set Fix(f) of fixed points of a map $f : E \longrightarrow E$ is hyperconvex. But the hyperconvex are absolute retracts, thus Fix(f) is a retract.

We do not know whether this result hold for unbounded spaces.
For classical metric space theorem 4 is the result of Sine [50] and Soardi [52], for ordered sets the Tarski fixed point theorem.

III - 3. Fixed point property for a commuting set of maps

The Tarski theorem is much better than ours in many respects. In fact, A. Tarski [53] proved that every set G of pairwise commuting maps on a complete lattice P has a common fixed point (an $x \in P$ s.t. $f(x) = x$ for every $f \in G$).

We do not know if such a result holds for bounded hyperconvex spaces. Worse, this seems unknown for the classical metric spaces, and yet, this is an interesting problem since in many instances what is really needed is a fixed point theorem for a set of functions, rather than a single one (A well known example is the proof of the existence of a Haar measure). We have only partial results that we summarize here.

III - 3.1. *Proposition.* Let E be a non empty bounded hyperconvex space, and G be a countable set of pairwise commuting non-expansive maps. Then the set of fixed point of G is hyperconvex.
This will follow from:

III - 3.2. *Proposition.* Let E be a bounded hyperconvex and $(E_n)_{n \in \mathbb{N}}$ be a descending sequence of non empty hyperconvex subspaces of E. Then their intersection $\cap \{E_n \ / \ n \in \mathbb{N}\}$ is hyperconvex and non empty.

Proof- Consider the product, say $F = \prod_{n \in \mathbb{N}} E_n$, of the E_n, equipped with the "sup"-distance.
This space is hyperconvex and bounded, (bounded because $\underset{\sim}{\delta}(F) = \underset{\sim}{\delta}(E_0)$; the shift operator S, defined by $S((x_n)_{n \in \mathbb{N}}) = (x_{n+1})_{n \in \mathbb{N}}$ is a non expansive map from E to F (as the E_n are decreasing). The set of its fixed points is the set of constant maps; the induced space is isometric to the intersection of the E_n. By our fixed point Theorem the set of fixed point of the shift is hyperconvex and the results follows. □

Proof- of Proposition III-3.1. : Let G be a set of commuting maps. Let $f, g \in G$; according to Theorem 4, Fix(f) is a non empty hyperconvex space; the fact that f and g commute implies that g transforms Fix(g) in itself, thus, according to Theorem 4, $\mathrm{Fix}(f) \cap \mathrm{Fix}(g)$ is again non empty and hyperconvex. More generally, for every finite subset F of G, the set $\mathrm{Fix}\,F = \{x \,/\, f(x) = x \text{ for all } f \in F\}$ is a non-empty hyperconvex space. Assuming G countable, it follows from the previous proposition that $\mathrm{Fix}\,G = \cap \{\mathrm{Fix}\,\{f_0, \ldots, f_n\} \,/\, n \in \mathbb{N}\}$ (where $(f_n)_{n \in \mathbb{N}}$ is an enumeration of G) is a non empty hyperconvex space.

To get a positive answer without any restriction on the size of G it would be enough to show that the intersection of any family of non empty hyperconvex spaces, totally ordered by inclusion, is non empty. This turns out to be an other way to ask the same question. Indeed one can assume that this family forms a descending sequence $(E_\alpha)_{\alpha<K}$, indexed by a regular ordinal K. The set of ordinals less than K (that is K) is equipped with the *Hessenberg sum* of ordinals:

For $\alpha = \sum^*_{\gamma<K} \omega^\gamma n_\gamma$, $\beta = \sum^*_{\gamma<K} \omega^\gamma m_\gamma$, where both n_γ and m_γ vanish almost everywhere, we set $\alpha \oplus \beta = \sum^*_{\gamma<K} \omega^\gamma (n_\gamma + m_\gamma)$. We define a commutative semigroup $\{S_\alpha : \alpha < K\}$ which acts on $E = \prod_{\alpha<K} E_\alpha$ as follows : to every α associate $S_\alpha : E \longrightarrow E$ defined by setting, $S_\alpha(f)(\beta) = f(\alpha \oplus \beta)$ for each $f \in E$. Then S_α is non-expansive and $S_\alpha \circ S_\beta = S_\beta \circ S_\alpha = S_{\alpha \oplus \beta}$ for all α, β. The common fixed points of S_α, $\alpha < K$, are the constants maps, and they form a space isometric to the intersection of the E_α's $(\alpha < K)$.
Consequently, from a fixed point theorem for sets of maps, we can deduce that this intersection is non empty. (This seems to be the shortest way to prove that for complete lattices, the intersection is non empty).

Even for the first uncountable cardinal, $\aleph_1$, we don't know whether the intersection of a descending sequence of $\aleph_1$ hyperconvex spaces is non empty (this question makes sens because most of the interesting semigroups are uncountable, in fact of size $2^{\aleph_0}$, so possibly very large !). The intersection question relates to the injective envelope. Indeed, if E is a subspace of a hyperconvex space F, the injective envelope is isometric to any of the minimal hyperconvex subspaces containing E. A positive answer to our question would probably gives an alternative proof for the existence of the injective envelope. Can we expect the converse to be true ?
On an other hand the formulation of the question in semigroup terms suggests a kind of duality between the semigroup and the set of values of the metric (for example for two elements x, y of E one can think about the set $\{f \,/\, f \in G \text{ and } f(x) = y\}$ as the "distance" from x to y).

Comment.- Theorem 4 is due to D. Misane and M. Pouzet ([25] without the observation that Fix(f) is hyperconvex). All results and proofs of this section are straightforward adaptations of well - known results on non expansive mappings on classical metric spaces.

IV - APPLICATIONS TO REFLEXIVE GRAPHS AND ORDERED SETS

The graphs we consider have a loop at every vertex, the morphisms are the edges-preserving maps; thus considered as binary relations they are simply the reflexive relations, with the relational homomorphisms. The product of graphs $G_i = (E_i, \mathcal{E}_i)$, $(i \in I)$, is the relational product, i.e. the graph on the cartesian product $\prod_{i \in I} E_i$ in which two vertices $x = (x_i)_{i \in I}$, $y = (y_i)_{i \in I}$ are connected by an edge if $(x_i, y_i) \in \mathcal{E}_i$ for all $i \in I$.

IV - 1. Undirected graphs

The graphs here are binary reflexive and symmetric relations.

IV - 1.1. Absolute retracts with respect to the graph-embedding

There is an easy way to convert graphs into metric spaces.
Let $\underset{\sim}{V} = \{0, \frac{1}{2}, 1\}$ with the usual ordering $0 < \frac{1}{2} < 1$; take for the semigroup operation on $\underset{\sim}{V}$ the truncated sum $x \oplus y = \mathrm{Min}\,\{x+y, 1\}$ take the identity for involution. For a graph $G = (E, \mathcal{E})$ define $d : E \times E \longrightarrow \underset{\sim}{V}$ by setting $d(x, y) = 0$ if $x = y$, $d(x, y) = \frac{1}{2}$ if $x \neq y$ and $\{x, y\} \in \mathcal{E}$, $d(x, y) = 1$ otherwise. Then d is a distance. Conversely if $d : E \times E \longrightarrow \underset{\sim}{V}$ is a distance, put $\mathcal{E}_{\frac{1}{2}} = \{(x, y) \in E \times E \,/\, d(x, y) < \frac{1}{2}\}$ and obtain a graph $G = (E, \mathcal{E}_{\frac{1}{2}})$.

Since $\underset{\sim}{V}$ is Heyting, it has itself a metric structure, the corresponding graph is the three elements path P_2 :

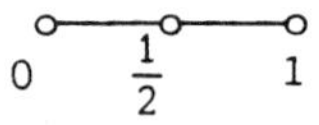

The non-expansive maps correspond to edge-preserving maps, and the isometries to graph-embeddings (that is maps f such that for all $x, y \in E$, the pair $\{x, y\}$ is an edge iff $\{f(x), f(y)\}$ is an edge).

Proposition II-2.7. et Theorem 1 translate into two simple facts:

IV - 1.1.1. ***Proposition.***
Every reflexive, unoriented graph embeds into a power of the path P_2 .

IV - 1.1.2. ***Theorem.***
For a reflexive unoriented graph G, the following are equivalent
1) G is an absolute retract (with respect to the graph-embeddings);
2) G is an injective object (with respect to the graph-embeddings);
3) G is central (that is it has a vertex connected to all vertices).
4) G is a retract of a power of P_2 .

Among the varieties of graphs, the variety $\{P_2\}^{\vee}$ generated by P_2 plays a similar role to the variety generated by the two element chain $\underline{2}$ among the varieties of posets. It covers the variety of complete graphs. For a description of varieties above $\{P_2\}^{\vee}$ see E. Jawhari, M. Pouzet, I. Rival [24].

Every graph can be made central by adding a new vertex joined to all the others, thus has an injective envelope, which by itself is not a very interesting fact ! The notion of hole and hole-preserving map are meaningful. In this special case, we can translate it as follows (even if our translation distorts proposition II-4.9.): a *hole* of $G = (E, \mathcal{E})$ is a subset H of E having no common neighbour (i.E. to every $x \in E$ there is $h \in H$ such that $(x, h) \notin \mathcal{E}$) it is minimal if every subset of H of strictly smaller size has a common neighbour. With this definition, central graphs have no holes. We say that a graph is *locally central* if it has no finite holes. The typical locally central graphs are the graphs M_ω, J_ω, I_ω, L_ω defined as follows : They have the same set of vertices $E = A \cup B$, where $A = \{a_n \mid n < \omega\}$ and $B = \{b_n \mid n < \omega\}$ are two countable disjoint sets. Their edge sets are

$\mathcal{E}_M = \{\{b_n, b_m\} / n, m < \omega\} \cup \{\{a_n, b_m\} / n < m < \omega\} \cup \{\{a_n\} / n < \omega\}$

$\mathcal{E}_J = \mathcal{E}_M \cup \{\{a_n, b_m\} / m < n < \omega\}$, $\mathcal{E}_I = \mathcal{E}_M \cup \{\{a_n, a_m\} / n, m < \omega\}$,

$\mathcal{E}_L = \mathcal{E}_J \cup \mathcal{E}_I$ respectively.

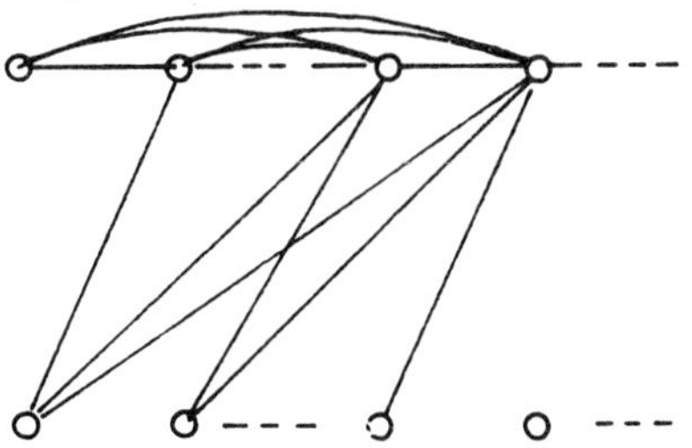

The graph M_ω

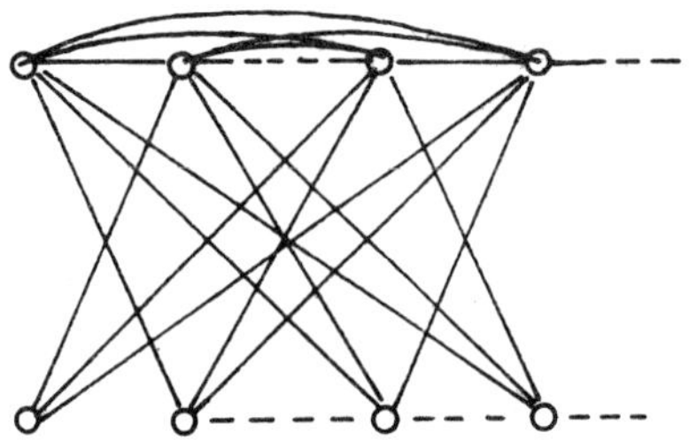

The graph J_ω

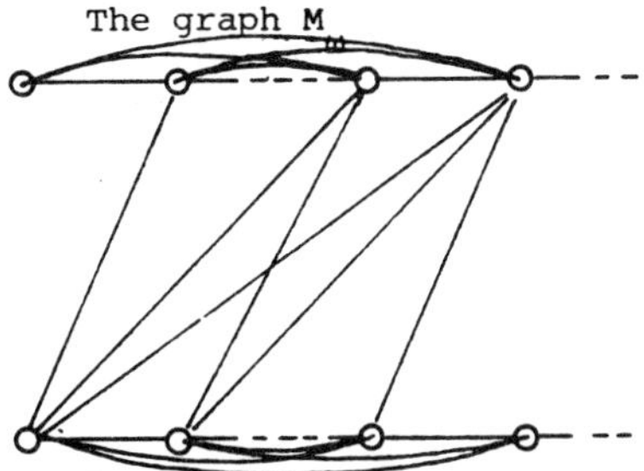

The graph I_ω

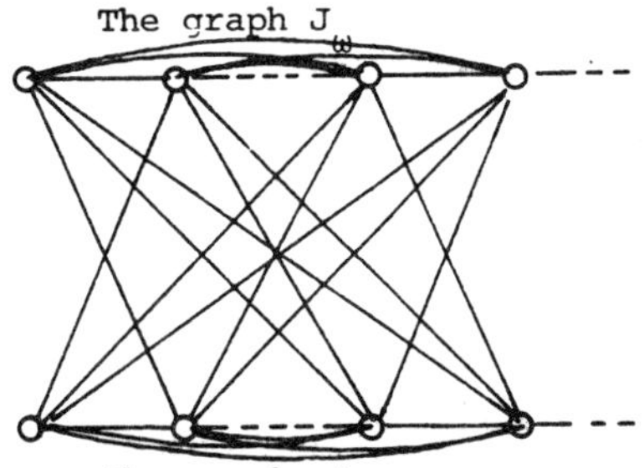

The graph L_ω

IV - 1.1.3. *Theorem.*

Every locally central graph with at least an infinite hole has an induced subgraph isomorphic to one of the graphs M_ω , J_ω , I_ω , L_ω . *Moreover in the lattice of varieties of graphs, the varieties* $\{J_\omega\}^\vee$, $\{I_\omega\}^\vee$ *and* $\{L_\omega\}^\vee$ *are pairwise incomparable and contain the variety* $\{M_\omega\}^\vee$.

These results have been obtained by the first and third author and are exposed in [25] .

IV - 1.2. Absolute retracts with respect to the isometric embedding

The coretractions of graphs are isometries for the standard graph-distance (defined below). Thus it make sense to ask for a description of the absolute retracts with respect to those isometries, which is within the frame of metric spaces. Indeed let $\underset{\sim}{V} = \mathbb{N} \cup \{+\infty\}$; the semigroup operation is the usual addition on $\mathbb{N}$ extended in an obvious way to $\underset{\sim}{V}$ and the involution is the identity. Every graph $G = (E, \mathcal{E})$ is endowed with the graph distance d_G (where $d_G(x, y) = \mathrm{Min}\,\{n \,/\, \text{there is a path } x_0 = x, \ldots, x_n = y\}$ if x, y are connected, and $d_G(x, y) = +\infty$ otherwise). This defines a well known metric space. A map $f : G \longrightarrow G'$ is edge-preserving iff it is non expansive. Thus the category of graphs is a full subcategory of the category of metric spaces over $\underset{\sim}{V}$ (in particular the graph-distance of a product of graphs is the sup-distance of the graph-distances of its components). Some distance on $\mathbb{N} \cup \{+\infty\}$ are not graph-distances. However it is easy to see that a distance on $\mathbb{N} \cup \{+\infty\}$ is a graph-distance iff it is *convex* (see II-2.), thus hyperconvex metric spaces come from graphs for which the collection of balls has the 2-Helly property. Since the distance $d_{\underset{\sim}{V}}$ on $\underset{\sim}{V}$, as defined in II-2., is the usual distance on $\mathbb{N} \cup \{+\infty\}$, it is the graph distance for the graph on $\underset{\sim}{V}$ whose edges are the pairs $\{x, y\}$ such that $|y - x| \leq 1$; that is the infinite path:

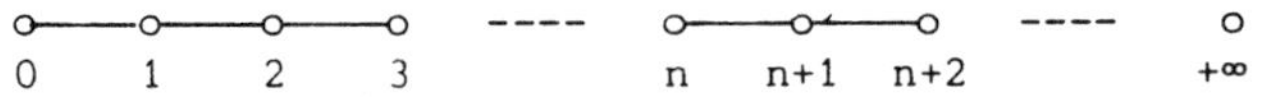

with the point at infinity disconnected from all the others. Despite the fact that some distances are not graph-distances we get from Propositions II-2.7. and II-2.11.

IV - 1.2.1. ***Proposition.***

Every metric space over $\mathbb{N} \cup \{+\infty\}$ and in particular every graph, embeds isometrically into a power of the infinite path (or equivalently into a product of finite paths) equipped with the graph-distance.

Because of IV-1.2.1., Theorem 1 applies to graphs, we state it in full details:

IV - 1.2.2. ***Theorem.***

The following are equivalent for a reflexive symetric graph G

1) *G is an absolute retract with respect to the isometries* (i.e. is a retract of every isometric extension);
2) *G is injective with respect to the isometries* (i.e. for every graph H, every $A \subseteq H$, each $f : A \longrightarrow G$, non-expansive for the graph distance induced on H, extends to a graph-preserving map on H);
3) *The balls of G have the 2-Helly property;*
4) *G is a retract of a power of an infinite path* (or, equivalently, G is a retract of a product of path).

Proposition IV-1.2.1. and Theorem IV-1.2.2. are substantially due to R. Nowakowski and I. Rival [35] and, independently A. Quilliot [44] . The present formulation is in [25].

From Theorem 2 we get:

IV - 1.2.3. ***Theorem.*** *Every graph has an injective envelope (with respect to the isometric embedding), i.e. up to isomorphism, a minimal isometric extension to a retract of product of paths.*

This result has been obtained independently by E. Pesh [37] and M. Pouzet [41], [69].

IV - 1.3. Absolute retracts with respect to the hole-preserving maps. Fixed point property

The distance is the graph-distance and the holes are defined accordingly. The adaptation of Theorem 3 given in II-4.7. gives the following:

IV - 1.3.1. *Theorem.* (P. Hell, I. Rival [19])
Absolute retracts with respect to the hole- preserving maps form a variety of graphs.

Only the singleton graph has the fixed point property, nevertheless, since the inaccessible elements of V are exactly 0, 1, $+\infty$, our main lemma and the caracterization of hyperconvexity (see Theorem IV-1.2.2.) give the following result:

IV - 1.3.2. *Theorem.* *In a retract G of a power of a finite path every edge-preserving self-map of G maps a complete subgraph into itself.*

This is substantially due to A. Quilliot [44].
Since the trees have the 2-Helly property, it follows that a tree without any infinite branch has a fixed edge (or loop). This is a particular instance of a much more general result due to R. Nowakowski, I. Rival [35].

IV - 2. Directed graphs

The graphs here are binary reflexive relations.

IV - 2.1. Absolute retracts with respect to the graph-embedding

We convert directed graphs into metric spaces as follows: Let $\underset{\sim}{V}$ be the five elements set $\{0, \frac{1}{2}, a, b, 1\}$; take for the ordering $0 < \frac{1}{2} < a, b < 1$ with a incomparable to b ; take for the semigroup

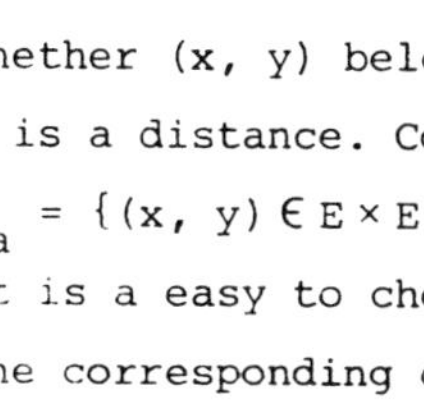

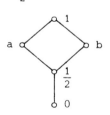

operation $x + y = 1$ if $x, y > \frac{1}{2}$ and $x + y = \mathrm{Max}\,\{x, y\}$ otherwise; take for the involution the map which exchange a and b and fixes $0, \frac{1}{2}, 1$. For a directed graph $G = (E, \mathcal{E})$ define $d : E \times E \longrightarrow \underset{\sim}{V}$ by setting $d(x, y) = 0$ if $x = y$ and, if $x \neq y$, $d(x, y)$ equals to $\frac{1}{2}$, a, b, 1 according to whether (x, y) belongs to $\mathcal{E} \cap \mathcal{E}^{-1}$, $\mathcal{E} \setminus \mathcal{E}^{-1}$, $\mathcal{E}^{-1} \setminus \mathcal{E}$, $E \times E \setminus \mathcal{E} \cup \mathcal{E}^{-1}$. Then d is a distance. Conversely if $d : E \times E \longrightarrow \underset{\sim}{V}$ is a distance, put $\mathcal{E}_a = \{(x, y) \in E \times E \;/\; d(x, y) \leq a\}$ and obtain a directed graph $G = (E, \mathcal{E}_a)$
It is a easy to check that $\underset{\sim}{V}$ is Heyting; thus it has a metric structure.
The corresponding directed graph is the following graph:

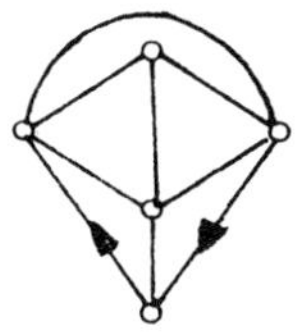

We will denote D_2 the induced graph on $\{0, \frac{1}{2}, a\}$, that is the graph:

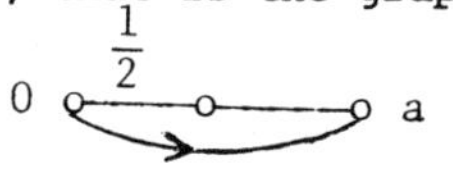

As before the non-expansive maps correspond to edge-preserving maps, and the isometries to graph-embeddings (but here edges are couples instead of pairs). With the help of Proposition II-2.10, Proposition II-2.7. and Theorem 1 translate to:

IV - 2.1.1. *Proposition.*
Every reflexive directed graph embeds into a power of the directed graph D_2 .

IV - 2.1.2. *Theorem.* *For a reflexive, directed graph G, the following are equivalent:*

1) G is an absolute retract (with respect to the graph-embeddings);

2) G is an injective object (with respect to the graph-embeddings);

3) G is central (that is it has a vertex connected by a double edge to all vertices);

4) G is a retract of a power of D_2 .

These results are substantially in [43] .

IV - 2.2. The zig-zag distance on directed graphs

The Heyting semigroup.-

Let A be a set consisting of two elements a and b, the "alphabet", and A^* be the set of all finite words $r = r_0 r_1 \dots r_{n-1}$ over A, including the empty word $\square$, (thus $A^* = \bigcup_{n \in \mathbb{N}} A^n$). We order A^* with the Higman ordering [21]: For two words $r = r_0 r_1 \dots r_{n-1}$ and $s = s_0 s_1 \dots s_{m-1}$, $r \leq s$ if there are integers $0 \leq j_0 < \dots < j_{n-1} \leq m-1$ such that $r_i = s_{j_i}$ for all $i \leq n-1$. In other "words", $r \leq s$ if r can be obtained from s by deleting some letters of s. Then A^* is an ordered semigroup with respect to the concatenation of words:

$$r_0 \dots r_{n-1} + s_0 \dots s_{m-1} = r_0 \dots r_{n-1} \cdot s_0 \dots s_{m-1} ,$$

the empty word $\square$ is the least element of A^* and the neutral element for the semigroup operation. The involution - on A which exchanges a and b (i.e. $\overline{a} = b$, $\overline{b} = a$), extends on A^* in such a way that $\overline{r+s} = \overline{s} + \overline{r}$ (i.e. for every word $r = r_0 \dots r_{n-1}$, $\overline{r} = \overline{r}_{n-1} \dots \overline{r}_0$) .

Except that A^* is not a complete lattice, A^* has all the property required for $\underset{\sim}{V}$.

In order to have a complete lattice, we choose $\underset{\sim}{V}$ to be the set $\underset{\sim}{F}(A^*)$ of final segments of A^* (that is the subsets F of A^* such that $x \in F$ and $x \leq y$ imply $y \in F$), ordered by the reverse inclusion $\supseteq$. This $\underset{\sim}{V}$ is a complete lattice with the least element A^* and the largest element ϕ which we denote 0 and 1. We extend to $\underset{\sim}{V}$ the semigroup operation on A^* and the involution in the obvious way : if $p, q \in \underset{\sim}{V}$ then $p + q = \{x + y \ / \ x \in p, y \in q\}$ and $\overline{p} = \{\overline{x} \ / \ x \in p\}$ (since in A^* clearly $x + y \leq z$ implies $z = x_1 + y_1$ for some $x_1 \geq x$, $y_1 \geq y$, we have $p + q \in \underset{\sim}{V}$ whenever $p, q \in \underset{\sim}{V}$). The semigroup $\underset{\sim}{V}$ is an Heyting involutive semigroup (the distributivity comes from the fact that the infimum in $\underset{\sim}{V}$ is the union). Moreover the map $A^* \longrightarrow \underset{\sim}{V}$ which associate to every $r \in A^*$ the set $[r] = \{y \; / \; y \in A^* \text{ and } r \prec y\}$ preserves both the ordering, the semigroup operation and the involution.

The distance.-

Let $G = (E, \mathcal{E})$ be a directed graph (i.e. a reflexive binary relation). A *valued zig-zag* from x_0 to x_m is a sequence $(x_0, \ldots, x_m \; ; \; r_0 \cdot \ldots \cdot r_{m-1})$ where $x_0, \ldots, x_m \in E$, $r_0 \cdot \ldots \cdot r_{m-1} \in A^*$ and $(x_i, x_{i+1}) \in \mathcal{E}$ if $r_i = a$, $(x_{i+1}, x_i) \in \mathcal{E}$ if $r_i = b$ $(i = 0, \ldots, m-1)$, (if $m = 0$ this is the sequence $(x_0, \square)$).

We define $d_G : E \times E \longrightarrow \underset{\sim}{V}$ by setting $d_G(x, y) = \{r_0 \ldots r_{m-1} \mid (x_0, \ldots, x_m \; ; \; r_0 \ldots r_{m-1})$ is a valued zig-zag from x to y$\}$. It is easy to check that d_G satisfies d1), d2), d3). Also $(x, y) \in \mathcal{E}$ if and only if $d_G(x, y) \prec [a]$, thus the graph can be reconstructed from the distance. More generally if $d : E \times E \longrightarrow \underset{\sim}{V}$ is a distance then d is of the form d_G if and only if the following weak from of convexity holds:
$r + s \in d(x, y)$ implies $r \in d(x, z)$ and $s \in d(z, y)$ for some z.
In this case put $\mathcal{E}_a = \{(x, y) \; / \; d(x, y) \prec [a]\}$ and obtain a directed graph $G = (E, \mathcal{E}_a)$. Now a space (E, d) is hyperconvex if and only if $d = d_G$ and the balls $B(x, [r])$ $(r \in A^*)$ satisfy the 2-Helly property (observe that every ball $B(x, p) = \bigcap \{B(x, [r]) \; / \; r \in p\}$). Since $\underset{\sim}{V}$ itself is hyperconvex for its metric structure, its distance comes from a graph $G_{\underset{\sim}{V}} = (\underset{\sim}{V}, \mathcal{D})$. The formal definition (see II-2.) is simple: $(p, q) \in \mathcal{D}$ iff $p \prec q + [b]$ and $q \prec p + [a]$, but not easy to handle.
On words of lenght at most 2 this give the following graph

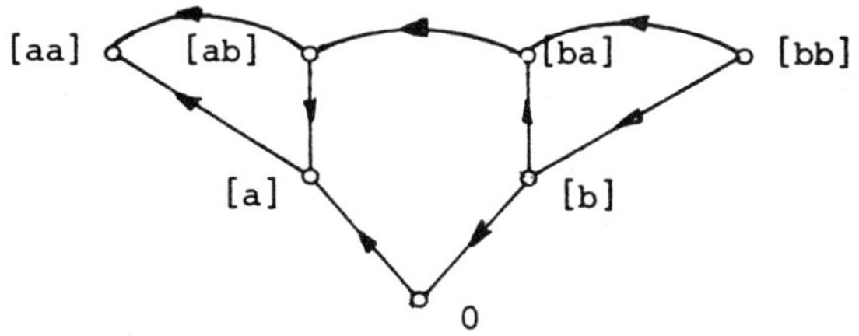

This graph has to be described. The Higman theorem [21] asserts that every $p \in \underset{\sim}{V}$ is of the form $p = [r_1] \wedge [r_2] \wedge \ldots \wedge [r_k]$, that is $p = \{s \in A^* \; / \; s \succ r_i \text{ for some } 1 \prec i \prec k\}$ for some $r_1, \ldots, r_k \in A^*$, thus $G_{\underset{\sim}{V}}$ is countable and between two elements p, q of $G_{\underset{\sim}{V}}$, one can find finitely many "zig-zags" $Z_1, \ldots, Z_k$ coded by $r_1, r_2, \ldots, r_k$ respectively such that $d(p, q) = [r_1] \wedge [r_2] \wedge \ldots \wedge [r_k]$ (see IV-4.) .

Here again, a map $f : G \longrightarrow G'$ is edge-preserving iff it is non-expansive for the associated distances. Thus we have a full subcategory of the category of metric spaces over $\underset{\sim}{V}$. From Proposition II-2.7. we get.

IV - 2.2.1. *Proposition*. *Every reflexive oriented graph isometrically embeds into a power of* $G_{\underset{\sim}{V}}$.

According to Fact 5, Theorem 1 applies and gives:

IV - 2.2.2. *Theorem*. *The following are equivalent for a reflexive directed graph* $G = (E, \mathcal{E})$

1) G is an absolute retract (with respect to the isometries);

2) G is injective (with respect to the isometries);

3) The balls $(B(x, [r])_{x \in E,\ r \in A^*}$ *have the 2-Helly property;*

4) G is a retract of a power of $G_{\underset{\sim}{V}}$.

IV - 2.2.3. *Theorem*. *Every reflexive directed graph has an injective envelope.*

IV - 2.3. Absolute retracts with respect to the hole preserving maps. Fixed point property

Proposition II-4.7. applies and gives:

IV - 2.3.1. *Theorem*. *With respect to the hole preserving maps, the absolute retracts are exactly the injective objects. Moreover they form a variety.*

Note that here the notion of a hole (see II-4.) cannot be replaced by the weaker notion of a map $f : G \longrightarrow A^*$ such that $\bigcap B(x, [f(x)]) = \phi$.

Concerning the fixed point property, one can notice that 0 , $[a] \wedge [b]$, (that is the set of all non empty words) and 1 are the inaccessible elements. Indeed, let $p \in \tilde{V} \setminus \{0, 1, [a] \wedge [b]\}$. Then $p = [r_1] \wedge [r_2] \wedge \ldots \wedge [r_k]$ where $r_1, \ldots, r_k$ are the minimal elements of p, i.e. p consists of words having at least one subword r_i . Assume that $r_i = s_0 \ldots s_{n-1}$ contains both a and b . Then for $u = [s_0 \cdot \ldots \cdot s_{n-2}]$ we have $p \not\prec u$ whereas $p \prec u + \bar{u}$. Thus suppose each r_i is of the form aaa... or bbb... . First suppose that $r_1 = s_0 \ldots s_{n-1}$ where $n > 1$, take $u = [s_0 \ldots s_{n-2} \bar{s}_{n-1}]$, again $p \not\prec u$, but $p \prec u + \bar{u}$. We have still the case $p = [a]$ or $p = [b]$. If $p = [a]$ then choosing $u = [b]$ we have $p = [a] \not\prec [b] = u$, but $p \prec u + \bar{u} = [ab]$. The case $p = [b]$ is similar.

We have then the:

IV - 2.3.2. *Theorem*. *Let G be a retract of a power of the graph* $G_{[0,r]}$ *where* $r \in \underset{\sim}{V} \setminus \{1\}$. *Then every edge preserving map f of G sends a complete subgraph of G into itself.*

The fact that the results concerning directed graphs are similar to those concerning undirected graph is not surprising. Let G be an unoriented reflexive graph. If $r \in A^*$ belongs to $d_G(x, y)$ then all $u \in A^*$ of the same length belong to $d_G(x, y)$, therefore $d_G(x, y)$ is determined by the length of the shortest path from x to y. In fact the set $\underset{\sim}{L} = \{[A^n \to) \,/\, n \in \mathbb{N}\} \cup \{\emptyset\}$ is a substructure of $\underset{\sim}{V}$ isomorphic to $\mathbb{N} \cup \{+\infty\}$, considered as an induced subgraph of $G_{\underset{\sim}{V}}$ it is an infinite path with an endpoint. The consideration of substructures of $\underset{\sim}{V}$ gives three other interesting subcases. We discuss them in the next paragrapns.

IV - 3. Antisymmetric graphs.

Let $\underset{\sim}{V}_a = \underset{\sim}{V} \setminus \{[a] \wedge [b]\}$. With the induced structure of $\underset{\sim}{V}$, this is an Heyting involutive semigroup (it is obtained by collapsing O and $[a] \wedge [b]$). The same study as before can be done. We summarize our results as follows:

IV - 3.1. *Theorem.*

1) Every reflexive antisymmetric graph determines a metric space over $\underset{\sim}{V}_a$;

2) The graph $G_{\underset{\sim}{V}_a} = (\underset{\sim}{V}_a, \mathfrak{D}_a)$ where $\mathfrak{D}_a = \{(x, y) \,/\, d_{\underset{\sim}{V}_a}(x, y) \prec [a]\}$ is a reflexive antisymmetric graph ;

3) Every reflexive antisymmetric graph embeds isometrically into a power of $G_{\underset{\sim}{V}_a}$.

4) In the variety of reflexive antisymmetric graphs, the absolute retracts (with respect to the isometries), the injective objects (with respect to the isometries), the graphs G whose the collection of balls $(B(x, [r]))_{x \in G,\, r \in A^*}$ have the 2-Helly property, the retracts of powers of $G_{\underset{\sim}{V}_a}$ coincide.

5) Every reflexive antisymmetric graph has an injective envelope (with respect to the isometries).

6) With respect to the hole-preserving maps, the absolute retracts and the injective are the same; moreover they form a variety.

Since 0 and 1 are exactly the inccessible element, a graph G is bounded (as a metric space) if there is an $r \in A^*$ such that $d_G(x, y) \prec [r]$ for all $x, y \in G$.

We have:

IV - 3.2. *Theorem.* *If G is a bounded reflexive antisymmetric graph, and if the balls of G have the 2-Helly property, then every edge-preserving map has a fixed point.*

IV - 4. W-Connected graphs

Let $\underset{\sim}{V}_z$ be the subset of $\underset{\sim}{V}$ consisting of p such that $(p_-)^+ = p$ (where p_- (resp. p^+) is the set of lower (resp. upper) bounds of the subset p of A^*, in $(A^* \prec)$). Clearly $(\underset{\sim}{V}_z, \supseteq)$ is order-isomorphic to the MacNeille completion of A^* .

It is preserved by the involution and, if one define the operation + by $p + q = [(p+q)_-]^+$, then it becomes an Heyting involutive ordered semigroup

(Indeed observe that $(\bigwedge_{\underset{\sim}{V}_Z} p_i) + q$ and $(\bigwedge_{\underset{\sim}{V}_Z} (p_i + q)$ are equal to $\bigwedge_{\underset{\sim}{V}_Z} \{[r] + [s] \,/\, r \in p_i,\ s \in q\}$).

Graphs with distance value in $\underset{\sim}{V}_Z$ will be called *W-connected.* They enjoy all the previous properties, so we do not recall them. In this particular case we have a little more satisfactory description of the absolute retracts. For that we say that a graph $G = (E, \mathcal{E})$ is a *zig-zag* if its symmetric hull $\tilde{G} = (E, \mathcal{E} \cup \mathcal{E}^{-1})$ is a path (see figure below). If G is finite, its *endpoints*, say x_0 and x_{n-1} are the endpoints of its symmetric hull.

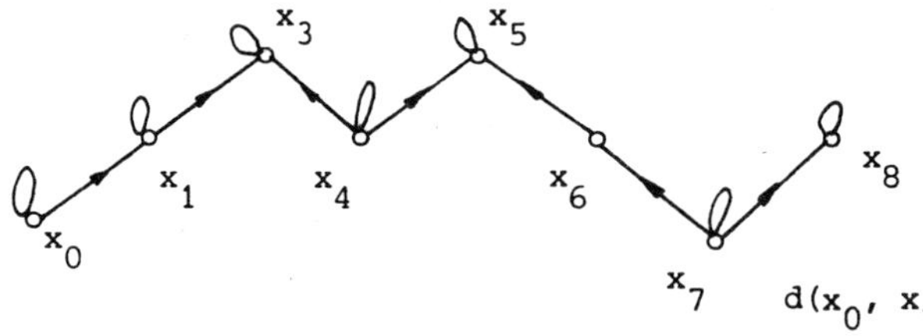

$d(x_0, x_7) = [aababba]$

The distance $d_G(x_0, x_{n-1})$ (in $\underset{\sim}{V}$) is of the form $[p]$ where $p \in A^*$, thus for a reflexive graph $d_G(x, y)$ measures the zig-zags with end points x, y which are subgraphs of G.

IV - 4.1. *Proposition.* A graph is W-connected if and only if it isometrically embeds into a product of zig-zags.

Proof- To get such an embedding it is enough to show that for every W-connected graph G, and $(x, y) \in G$, there is a zig-zag Z and a collection $(f_i)_{i \in I}$ of edge preserving maps $f_i : G \longrightarrow Z$ such that $d_G(x, y) = \text{Sup}\{d_Z(f_i(x), f_i(y)) \,/\, i \in I\}$. For that we take $I = d_G(x, y)_-$ and for every $[r] \in I$ we take a zig-zag Z_r with endpoints x', y' such that $d_{Z_r}(x', y') = [r]$. Using the fact that zig-zags are hyperconvex among the reflexive antisymetric graphs, we extend the map sending x to x', y to y', to an edge-preserving map. Finally we take a zig-zag Z containing all the Z_r. The converse is straight-forward. □

IV - 4.2. *Theorem. Among the W-connected graphs, a graph is an absolute retract (or, equivalently, an injective) if and only if it is a retract of a product of zig-zags.*

The W-connected graphs are acyclic (have no oriented cycles). We consider separately posets in this framework.

IV - 5. Ordered sets with the fence-distance

The space of value is derived from $(F(A^{*}), \supseteq)$ (see IV-2).

Because of the transitivity, the zig-zags (defined, according to IV-4., as posets whose comparability graph is a path) have a simple form (see below).

According to the standard terminology we call them *fences*.

One formal definition of the space of values is this: Let M be the set of non empty alternating words of a and b, that is the set of $r = r_0 r_1 \ldots r_{n-1}$ such that $n \geq 1$ and $r_i \neq r_{i+1}$ for every $i < n$; M is ordered as before; the semigroup operation on M is defined as follows: for $r = r_0 r_1 \ldots r_{n-1}$ and $s = s_0 s_1 \ldots s_{m-1}$, $r + s$ is $r_0 \ldots r_{n-1} s_0 s_1 \ldots s_{m-1}$ if $r_{n-1} \neq s_0$ and $r_{n-1} \ldots r_{n-1} s_1 \ldots s_{m-1}$ otherwise; the involution is the same as before. Let $\underset{\sim}{V} = (\underset{\sim}{F}(M), \supseteq)$ be the set of final segments of M ordered by reverse inclusion; $\underset{\sim}{V}$ is a complete lattice with least element M and greatest element $\emptyset$ that we denote 0 and 1. Equipped with the semigroup operation and the involution deduced from M, it becomes an Heyting involutive semigroup. It is easy to draw (see below).

The semigroup $\underset{\sim}{V}$ **The fences**

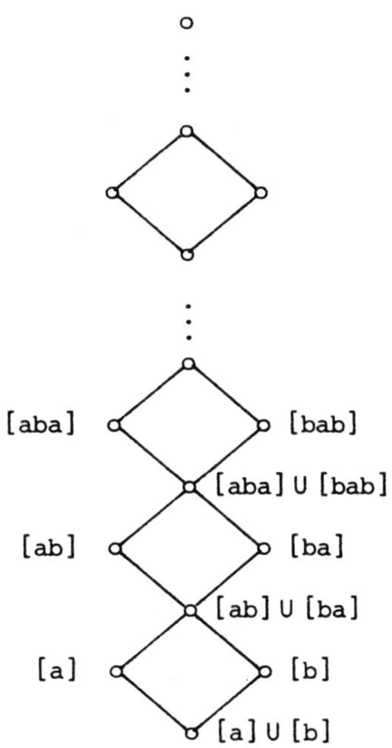

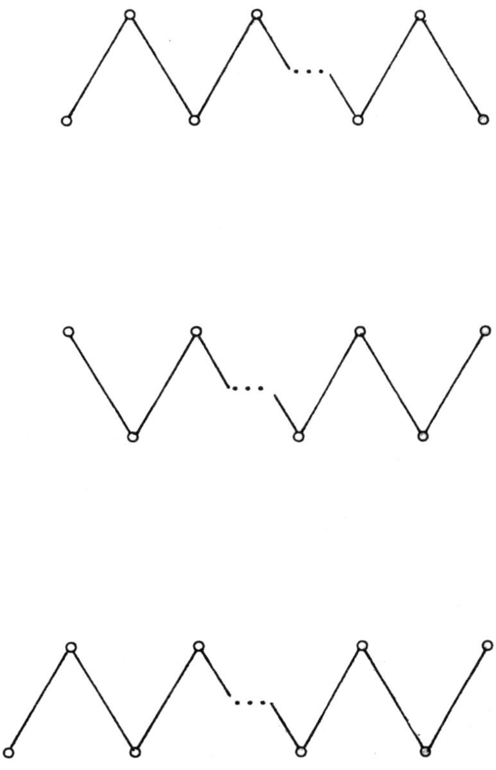

Let P a poset; for every x, y P let

$d_P(x, y) = \{r_0 \dots r_{m-1} \in M \mid (x_0 \dots x_m ; r_0 \dots r_{m-1})$ is a valued zig-zag from x to y$\}$.

Since an ordering is reflexive, $d_P(x, y)$ is a final segment of M ; its minimal elements correspond to fences joining x to y. It can be checked directly that the map $d_P : P \times P \longrightarrow \underset{\sim}{V}$ satisfies $d_1)$, $d_2)$, $d_3)$; also $x \prec y$ if and only if $d_P(x, y) < [\tilde{a}]$. As for graphs a distance $d : E \times E \longrightarrow \underset{\sim}{V}$ is of the form d_P if and only if for every $r + s \in d(x, y)$ there is some $z \in E$ such that $r \in d(x, z)$ and $s \in d(z, y)$. In particular E is hyperconvex if and only if $d = d_P$ and the collection of balls $(B(x, [r]))_{x \in E,\, r \in M}$ has the 2-Helly property. The poset $P_{\underset{\sim}{V}}$ is easy to draw:

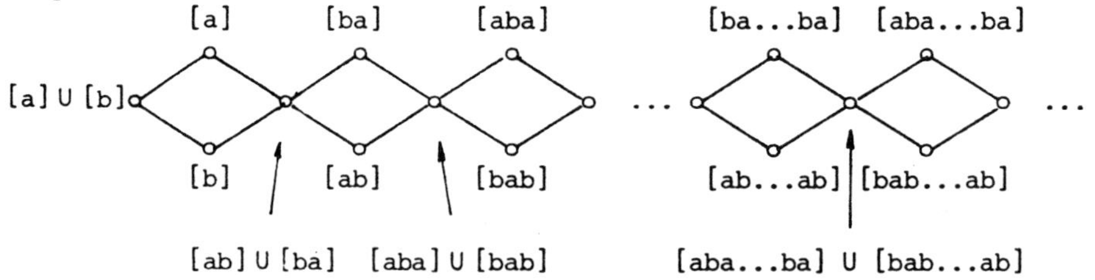

It look like a "double"-fence with an element incomparable to all the others. Every fence or "double"-fence, either finite or infinite on one side, is clearly a retract of $P_{\underset{\sim}{V}}$.

On the other hand since the fences are hyperconvex and since for every pair x, y, the distance $d_P(x, y)$ measures the collection of fences from x to y, one can find a fence F and two order preserving maps f_1, f_2 from P to F such that $d_P(x, y) = d_F(f_1(x), f_1(y)) \vee d_F(f_2(x), f_2(y))$.

This gives:

IV - 5.1. Proposition. *Every metric space over* $\underset{\sim}{V}$, *and in particular every poset, isometrically embeds into a power of* $P_{\underset{\sim}{V}}$ *and also into a product of double fences, or into a product of fences as well.*

The properties of posets with respect to the "fence"-distance can be summarized as follows.

IV - 5.2. *Theorem.* *The following holds in the variety of posets:*

1) The absolute retracts (with respect to the isometric embeddings), the injective posets (with respect to the isometric embeddings), the posets P whose the balls $(B(x, [r]))_{x \in P,\, r \in M}$ *have the 2-Helly property and the retracts of products of fences coïncide.*

2) Every poset has an injective envelope (i.e. embeds isometrically into a minimal retract of products of fences).

3) With respect to the holes preserving maps, the absolute retracts are exactly the injectives; moreover they form a variety. (P. Nevermann, I. Rival [33]).

4) Retracts of powers of finite fences have the fixed point property (for order preserving maps).

4) follows from a more general result due to K. Baclawski and A. Björner about "dismantability" [2]. (In a general setting, the injective are dismantable).

Comment.- The definition for a graph, or a poset, G of the set $d_G(x, y)$ of words coding zig-zags, or fences, joining x to y is due to A. Quilliot [33] . He uses the analogy between such functions and the usual distance functions and proves (see Theorem IV-2.2.2. and V-2.) that the notion of injectivity and 2-Helly property coïncide, but does not consider the order structure of the space of value nor its metric structure. The possibility of a characterization of the absolute retracts as Theorem 1 has been investigated first by E. Jawhari and M. Pouzet. In [25], Theorem IV-1.1.2. and the results of B. Banaschewski, G. Bruns [Theorem I-1.1.], R. Nowakowski, I. Rival [Theorem IV-1.2.2] are presented in this way, and the result of N. Aronszajn, P. Panitchpakdi [Theorem I-3.1.] is rediscovered. The equivalence between the notion of injectivity and absolute retractness, (in Theorem IV-3.1.) the full equivalence in 1) of Theorem IV-5.2. Theorem IV-3.2., Prop. IV-3.1., Theorem IV-3.2., Proposition IV-5.1. are due to D. Misane and M. Pouzet and have appeared in [32]. The other results are new.

V - FURTHER DEVELOPMENTS

V - 1. The metric point of view revisited

Our development leaves two questions unanswered. Does every category of binary relational structures, with the relational homomorphisms, can be viewed as a category of metric spaces over an involutive ordered semigroup (or, better, over an Heyting semigroup), with the non expansive mappings ? Does our axiomatic appropriate - neither too general nor too restrictive - and meaningful ?

Let us attempt to answer to these questions. First, our point of view was designed for *reflexive* binary relational structures. The case of irreflexive structures, even the simplest case of irreflexive undirected graphs, is considerably more complicated. This, in spite of the fact that the study of retracts of graphs originates in his case and that the fundamental results (due to P. Hell [16], [17], [18]) look very similar to those for reflexive graphs. We do not know whether the methods used in this paper can cope with this situation. Next, there are many ways to deal with reflexive binary relational structures from a metric point of view. One could translate the relational notions in terms of binary spaces, this in the simplest possible way, and then uses in this frame the notions borrowed to the theory of metric spaces. For example, one could only assume that the set of value $\underset{\sim}{V}$ has a least element 0 and the binary spaces (E, d) over $\underset{\sim}{V}$ satisfy

$d'_1)$ $\quad d(x, x) = 0 \quad$ for every $\quad x \in E$.

This was our first approach, in the years 1982-1983, and at that time we were convinced that additional structure on $\underset{\sim}{V}$, e.g. semigroup or involution, were useless. After a lecture on *generalized semimetrics*, given by Montserrat

Pons Valles to the First international symposium on ordered algebraic structures (Luminy, June 11-16, 1984), we prefered the presentation in terms of metric spaces over an involutive semigroup, realizing that it make more transparent the connection with classical metric spaces and also allows to deal with the current studies on their generalizations.

According to M. Pons Valles [28], [39], and her terminology and notations, a generalized semimetric is defined as follows : Let $(\Sigma, +, \preccurlyeq)$ be an ordered semigroup with a zero element, which is also the least element of the ordering. A *generalized semimetric* on a set E is a map $m : E \times E \longrightarrow \Sigma$ such that

1) $m(x, x) = 0$ for all $x \in E$
2) $m(x, y) \preccurlyeq m(x, y) + m(y, z)$ for all $x, y \in E$

The semimetric is *separating* if

3) $m(x, y) = m(y, x) = 0$ implies $x = y$.

Let us explain why our presentation includes both the reflexive binary relational structures and the generalized semimetric spaces.

First, in categorical terms, there is no difference between binary spaces (E, d) satisfying

$d'_1)$: $d(x, x) = 0$ for all $x \in E$,

and binary spaces (E, d) satisfying

$d''_1)$: $d(x, y) = d(y, x) = 0$ iff $x = y$, (i.e. 1) and 3) above).

Indeed let $\underset{\sim}{V}$ be a poset. Adding a new element, let 0_*, below all the elements of $\underset{\sim}{V}$ we get a poset $\underset{\sim}{V}_*$; if + is a semigroup operation on $\underset{\sim}{V}$ we extend it to $\underset{\sim}{V}_*$ so that 0_* is the zero. To every binary space (E, d) over $\underset{\sim}{V}$ correspond the binary space (E, d_*) over $\underset{\sim}{V}_*$ where $d_*(x, y) = d(x, y)$ for all $x \neq y$, $x, y \in E$, and $d_*(x, x) = 0_*$. Clearly if $\underset{\sim}{V}$ has a least element 0, then this correspondance establishes a functorial bijection between the categories of binary spaces (E, d) over $\underset{\sim}{V}$ satisfying

$d'_1)$: $d(x, x) = 0$ for all $x \in E$, and binary spaces (E, d) over $\underset{\sim}{V}_*$ satisfying (d''_1) : $d(x, y) = d(y, x) = 0_*$ iff $x = y$. In particular, for the study of non-expansive maps the semimetrics or separating semimetrics play the same role.

The involution introduces no more restriction. Indeed let $\underset{\sim}{V}$ be a poset and let $\underset{\approx}{V} = \underset{\sim}{V} \times \underset{\sim}{V}$ ordered componentwise. Define an involution on $\underset{\approx}{V}$ by setting $\overline{(p, q)} = (q, p)$ for all $p, q \in \underset{\sim}{V}$. If + is a semigroup operation on $\underset{\sim}{V}$, then the product with its reverse defines a semigroup operation on $\underset{\approx}{V}$, i.e. $(p, q) + (p', q') = (p+p', q'+q)$ for all $p, p', q, q' \in \underset{\sim}{V}$. If $\underset{\sim}{V}$ is an ordered semigroup then $\underset{\approx}{V}$ is an involutive ordered semigroup. To every binary space (E, d) over $\underset{\sim}{V}$ correspond a binary space $(E, \underline{d})$ over $\underset{\approx}{V}$, where $\underline{d}(x, y) = (d(x, y), d(y, x))$. Such space satisfies

$d_3)$: $\overline{\underline{d}(x, y)} = \underline{d}(y, x)$ for all $x, y \in E$.

Conversely for every binary space (E, d') over $\underset{\approx}{V}$ satisfying $d_3)$, d' is of the form $\underline{d}$. In particular (E, d') is a metric space over $\underset{\approx}{V}$, i.e. d' satisfies $d_1)$, $d_2)$, $d_3)$ if d is a separating semimetric over $\underset{\sim}{V}$.

For the study of reflexive relational structures the semigroup operation is no more restrictive. Indeed we convert the reflexive binary relations $R = (E, (R_i)\ i \in I)$ of a given type I into binary spaces (E, d), satisfying d'_1), over the complete ordered set $\underset{\sim}{V} = 2^I$ ordered componentwise. Since $\underset{\sim}{V}$ has least and largest elements 0 and 1, we can define a semigroup operation by setting $p+q = 1$ if p and q are both distincts from 0, $p + q = \text{Max}\ \{p, q\}$ otherwise. The triangular inequality then is satisfied. Applying the two previous construction we get an Heyting involution ordered semigroup $\underset{\sim}{V}_I = \{0_*\} + 2^I \times 2^I$, thus we can convert the reflexive binary relations of type I into metric spaces over $\underset{\sim}{V}_I$. By this way one can obtain the same results as those obtained for reflexive graphs in IV-1.1. and IV-2.1., thus, in particular, a characterization of the absolute retracts with respect to the embeddings (for general relational structures see [43]).

This answers the first question and shows that our axiomatic is appropriate, but this does not say that it is meaningful. On the contrary one could argue that our axiomatic is appropriate because both the involution and the semigroup operation are irrelevant ! In our opinion a justification of our presentation sits in the constructions initiated by A. Quilliot that we have developped in IV. These constructions can be done in a general setting, and can be applied to reflexive binary relational structures as follows:

Let $\underset{\sim}{V}$ be a complete poset, with least and largest element 0 and 1, equiped with an order-preserving involution "$\_$". Let $D \subseteq \underset{\sim}{V} \setminus \{0, 1\}$ such that 1) $D = \overline{D}$, 2) Every non zero element of $\underset{\sim}{V}$ is the infimum of elements of D. Let D^* be the set of finite words over D. We order D^* with the Higman ordering: for two words $r = r_0 \ldots r_{n-1}$, $s = s_0 \cdot \ldots \cdot s_{m-1}$, we put $r \leq s$ if there are integers $0 \leq j_0 < \ldots < j_{n-1} \leq m-1$ such that $r_i \leq s_{ij}$ for all $i = 0, \ldots, n-1$. Let $\underset{\sim}{V}_D^*$ be the set of final segments of D^*, $\underset{\sim}{V}_D^* = \underset{\sim}{F}(D)$; As in IV-2.2. we order $\underset{\sim}{V}_D^*$ by the reverse inclusion $\supseteq$, we extend to $\underset{\sim}{V}_D^*$ the semigroup operation on D^* and the involution and we get an Heyting involutive ordered semigroup.

Now let us consider the binary spaces (E, d) over $\underset{\sim}{V}$ satisfying d_1) and d_3). For such a space (E, d) a *D-valued path* from x_0 to x_m is a sequence $(x_0, \ldots, x_m ; r_0 \ldots r_{m-1})$ where $x_0, \ldots, x_m \in E$, $r_0 \ldots r_{m-1} \in D^*$, and $d(x_i, x_{i+1}) \leq r_i$ for all $i = 0 \ldots m-1$. We define $d^* : E \times E \longrightarrow \underset{\sim}{V}_D^*$ by setting

$d^*(x, y) = \{r_0 \ldots r_{m-1} / (x_0, \ldots, x_m ; r_0 \ldots r_{m-1})$ is a D-valued path from x to y$\}$. As in IV-2.2. it is easy to check that d^* satisfies d_1), d_2), d_3). Thus (E, d^*) is a metric space over $\underset{\sim}{V}_D^*$. Also d can be reconstructed from d^* $(d(x, y) = \wedge \{r \in d^*(x, y) \cap D^*\})$ thus for every map $f : E \longrightarrow E'$, f is non-expansive from (E, d) to (E', d') if and only if f is non-expansive from (E, d^*) to (E', d'^*). As in IV-2.2. a distance $m : E \times E \longrightarrow \underset{\sim}{V}_D^*$ is of the forme d^* if and only if for every

$r + s \in m(x, y)$, there is some $z \in E$ such that $r \in m(x, z)$ and $s \in m(z, y)$; this space (E, m) is hyperconvex if and only if $m = d^*$ and the collection of balls $B(x, [r])$ for $x \in E$, $r \in D^*$ has the 2-Helly property; with its metric structure $\underset{\sim}{V}_D^*$ is hyperconvex and every space over $\underset{\sim}{V}_D^*$ embed into a power of $\underset{\sim}{V}_D^*$. It follows that the absolute retracts (with respect to the d^*-isometries) over $\underset{\sim}{V}$ correspond exactly to the absolute retracts (with respect to the isometries) over $\underset{\sim}{V}_D^*$. Formally we get the same result as in IV-2.2. . If we take $\underset{\sim}{V} = \{0, \frac{1}{2}, 1\}$ and $D = \{\frac{1}{2}\}$ we get the results of IV-1.2. ; if we take $\underset{\sim}{V} = \{0, \frac{1}{2}, a, b, 1\}$ and $D = \{a, b\}$ we get those of IV-2.2. .

For reflexive binary relational structures of type I, take $\underset{\sim}{V} = \{0_*\} + 2^I \times 2^I$, D be the set of element of $2^I \times 2^I$ such that all the components exept one are equal to 1 . This construction also make sense for classical metric spaces and relates to the classical notion of ε-connexity. This type of construction can be used for various purposes. For example, let us define the "ℓ_1-distance" on a product $E = \prod \{E_i \mid i \in I\}$ of metric spaces (E_i, d_i) over a involutive ordered semigroup $\underset{\sim}{V}$. If the semigroup operation + is commutative then for every $\vec{x}, \vec{y} \in E$ we can define $\delta(\vec{x}, \vec{y})$ by setting $\delta(\vec{x}, \vec{y}) = \vee \{ \sum_{i \in F} d_i(\vec{x}(i), \vec{y}(i)) \ / \ F \subseteq I , |F| < \omega\}$.

Since + commutes the summations over the finite sets F are defined unambisguously and more importantly δ satisfies the triangular inequality and finally is a distance over $\underset{\sim}{V}$. But, if the semigroup operation is not commutative, this does not work, even in the case of a product of two spaces E_1, E_2 . We do that as follows : Let $\underset{\sim}{W} = \underset{\sim}{V} \times \underset{\sim}{V}'$ ordered with the order product; the semigroup operation is the product of the semigroup operation on $\underset{\sim}{V}$ and its reverse ; the involution is defined by setting $\overline{(p, q)} = (\bar{q}, \bar{p})$ for all $p, q \in \underset{\sim}{V}$. Let $D = \{(p, 1), (1, p) \ / \ p \in V \setminus \{0, 1\}\}$ and $\underset{\sim}{W}_D^*$ as defined above. Let $\Delta : E \times E \longrightarrow \underset{\sim}{W}$ defined by setting $\Delta(\vec{x}, \vec{y}) = (\delta(\vec{x}, \vec{y}), \delta(\vec{y}, \vec{x}))$ for all $\vec{x}, \vec{y} \in E$, where

$$\delta(\vec{x}, \vec{y}) = \vee \{ \sum_{i=i_1}^{i_k} d_i (\vec{x}(i), \vec{y}(i)) \ / \ \{i_1, \ldots, i_k\} \subseteq I , \ k \in \mathbb{N} \} .$$

This map satisfies $d_1), d_3)$ thus it corresponds a map $\Delta_* : E \times E \longrightarrow \underset{\sim}{W}_D^*$ satisfying $d_1), d_2), d_3)$. To get a distance over $\underset{\sim}{V}$ it is enough to "represent" $\underset{\sim}{W}_D^*$ in $\underset{\sim}{V}$. We do that first for D^* : we set $\varphi((1,p)) = \bar{p}$, $\varphi((p, 1)) = p$ for all $p \in D$, and then $\varphi(r_0 \ldots r_{n-1}) = \varphi(r_0) + \ldots + \varphi(r_{n-1})$ for every $r_0 \ldots r_{n-1} \in D^*$; then we extend φ to $\underset{\sim}{W}_D^*$ by setting $\bar{\varphi}(F) = \wedge \{\varphi(r) \ / \ r \in F\}$ for all $F \in \underset{\sim}{W}_D^*$. Finally the map $\hat{\delta} : E \times E \longrightarrow \underset{\sim}{V}$ defined by $\hat{\delta}(\vec{x}, \vec{y}) = \bar{\varphi} \circ \Delta_*(\vec{x}, \vec{y})$ is a distance. One can easily check that if $\vec{x}, \vec{y}$ only differ from the ith coordinate then $\hat{\delta}(\vec{x}, \vec{y}) = d_i (\vec{x}(i), \vec{y}(i))$.

(Note in view of δ that if $\vec{x}, \vec{y}$ differ only on a finite set F then

$\hat{\delta}(\vec{x}, \vec{y}) < \wedge \{ \sum_{i=i_i}^{i_k} d(\vec{x}(i), \vec{y}(i)) / i_1 \ldots i_k$ is an arbitrary permutation of $F\}$. Similar ideas can be used to define the length of a "curve" as well.

We have seen that every metric over an Heyting involutive semigroup $\underset{\sim}{V}$ isometrically embeds into a power of $(\underset{\sim}{V}, d_{\underset{\sim}{V}})$ equiped with the ℓ^{∞} distance. It seems to be an interesting problem to find conditions which insures that that a metric space embeds into such a power, equiped with the ℓ^1 distance. For symmetric graphs, the graph to corresponding to the ℓ^1 distance of the product of graphs G_i is the "cartesian" product (two distincts elements $\vec{x}, \vec{y}$ of G are connected iff they differ exactly on one component i, and x_i, y_i are connected). Thus depending upon the distance on graphs one considers, a solution of this problem will include the caracterization of graphs embeddable, or isometrically embeddable, into the hypercube (see [59] [71] for results in this direction).

V - 2. The use of injectivity

One can illustrate the use of injectivity with some translations of results borrowed to topology or metric spaces (e.g. the connection between contractibility and injectivity). We choose here a notion borrowed to Universal algebra, the notion of majority function.

Let P be a poset, a *majority function* on P is an order-preserving map $f : P^3 \longrightarrow P$ such that $f(x, y, y) = f(y, x, y) = f(y, y, x) = y$ for all $x, y \in E$. Quite recently (June 1985), I. Rival proved the following striking result.

V - 2.1. *Theorem.* *A finite poset P has a majority function if and only if P is a retract of a product of fences.*

His proof is based upon the following observations:

1) Every fence has a majority function. Products and retracts of posets with a majority function have a majority function. Thus every retract of a product of fences has a majority function.

2) If a finite posed P has a majority function then the collection of balls $B(x, [r])$, $x \in P$, $r \in M$ (Cf. IV-4) has the 2-Helly property, thus from Theorem IV-4.1. P is a retract of a product of fences.

It seems to us that the present point of view is appropriate to deal with such questions. Indeed, let E be a metric space over V. A *majority* function on E is a non-expansive map $f : E^3 \longrightarrow E$ such that $f(x, y, y) = f(y, x, y) = f(y, y, x) = y$ for all $x, y \in E$. More generally, for an integer n $(n \neq 2)$, an n-ary *near-unanimity* function is a non-expansive map $f : E^n \longrightarrow E$ such that $f(x_1, \ldots, x_n) = y$ whenever $x_i = y$ for all $i = 1, \ldots, n$, except at most one, (in these definitions E^n is equipped with the "Sup" distance).

Clearly, the existence of an n+1-ary near-unanimity function implies the existence of an n-ary near-unanimity function. Also products and retracts of spaces with a n-ary near-unanimity function have a n-ary near-unanimity function.

This relates with the Helly-property has follows: Let n be an integer, the collection of balls of a space E has the *finite n-Helly property* if for every *finite* family $\mathcal{F}$ of balls, the intersection is non empty provided all the intersection of the $<n$ element subfamilies $\mathcal{F}'$ are non empty. The key observation of I. Rival translates to the following

V - 2.2. *Lemma.* If E has an n+1-ary near-unanimity function then the collection of balls has the finite n-Helly property.

Proof- Let $B(x_i, r_i)$, $i \in I = \{1, \ldots, m\}$ be a finite family of balls such that $\cap \{B(x_i, r_i) / i \in I\} \neq \emptyset$ for every $I' \subseteq I$, $|I'| < n$. We show by induction on $m = |I|$ that $\cap \{B(x_i, r_i) / i \in I\} \neq \emptyset$. For every $j \in I$, let $C_j = \cap \{B(x_i, r_i) / i \in I$ and $i \neq j\}$. We can assume $m > n+1$ and $C_j \neq \emptyset$ for every $j \in I$. We choose $y_j \in C_j$, for $j = 1, \ldots, n+1$. We claim that $f(y_1, \ldots, y_{n+1}) \in \cap \{B(x_i, r_i) / i \in I\}$, that is $d(x_i, f(y_1, \ldots, y_{n+1})) < r_i$ for all $i \in I$. Since $d(x_i, y_j) < r_i$ for all $j \neq i$, it is enought to show that

$$d(x_i, f(y_1, \ldots, y_{n+1})) < \mathrm{Sup}\{d(x_i, y_j) / 1 < j < n+1, i \neq j\} .$$

This follows from the fact that f is non-expansive, indeed if $i \notin \{1, \ldots, n+1\}$ we have $x_i = f(x_i, \ldots, x_i)$, thus

$$d(x_i, f(y_1, \ldots, y_{n+1})) = d(f(x_i, \ldots, x_i), f(y_1, \ldots, y_{n+1})) < d((x_i, \ldots, x_i), (y_1, \ldots, y_{n+1})) = \mathrm{Sup}\ \{d(x_i, y_j) / 1 < j < n+1\}$$

if $i \in \{1, \ldots, n+1\}$ we have $x_i = f(x_i, \ldots, x_i, y_i, \ldots, x_i)$ thus

$$d(x_i, f(y_1, \ldots, y_{n+1})) = d(f(x_i, \ldots, x_i, y_i, \ldots, x_i), f(y_1, \ldots, y_n)) < d((x_i, \ldots, x_i, y_i, \ldots, x_i), (y_1, \ldots, y_{n+1})) = \mathrm{Sup}\{d(x_i, y_j) / \underset{i \neq j}{1 < j < n+1}\}$$

□

V - 2.3. ***Theorem.*** *Every hyperconvex space has a majority function; conversely if a finite space has a majority function then its balls have the 2-Helly property.*

Proof- Let E be an hyperconvex space.
Let $D = \{(x_1, x_2, x_3) / x_1 = x_2$ or $x_2 = x_3$ or $x_1 = x_3\}$. Let $h : D \longrightarrow E$ defined by setting $h(x_1, x_2, x_3) = x$ if $x_1 = x_2 = x$ or $x_1 = x_3 = x$ or $x_2 = x_3 = x$. This map is non-expansive. Since E is injective, h extends to a non-expansive mapping $\tilde{h}$ defined on E^3. This is a majority function. For the converse observe that if a space has only finitely many balls, then the 2-Helly and the finite 2-Helly properties coïncide. □

For all the examples listed in IV-1.2., IV-2.2., IV-3., IV-4., the 2-Helly property of the balls is equivalent to the hyperconvexity of the space. Thus in addition to the Rival Theorem, we have for example the following fact.

V - 2.4. ***Theorem.*** *A finite reflexive undirected graph G has a majority function if and only if G is a retract of a product of paths.*

Note that, the paths (or the fences, in the theorem above) are not necessarily finite (the graph, or the poset, needs not to be connected). For recents results about majority functions, see [62] . In this paper, the reader will find the Chineese remainder theorem in disguise of the hyperconvexity of the $\mathbb{Z}/_{p\mathbb{Z}}$.

V - 3. Partial well ordering and precompactness

Let us recall the following results.

V - 3.1. ***Theorem*** [21] *For a poset P the following are equivalent:*
(1) Every sequence of elements of P has an increasing subsequence.
(2) Every non empty subset has finitely many minimal elements.
(3) For every sequence $(x_n)_{n\in\mathbb{N}}$ *of element of E, there are* $n<m$ *such that* $x_n < x_m$.
(4) P contains no infinite antichain and no infinite descending chain.

V - 3.2. ***Theorem*** *For a metric space E (over the positive reals), the following are equivalent :*
(1) Every sequence of element of E has a Cauchy subsequence.
(2) For every $\varepsilon > 0$ *the space E can be covered by finitely many closed balls of radius* ε .
(3) There is no $\varepsilon > 0$ *and no infinite set* $X \subseteq E$ *such that* $d(x, y) > \varepsilon$ *for all* $x, y \in X$, $x \neq y$.

Posets satisfying the conditions of V-3.1. are the *partially well-ordered sets*, whereas the metric spaces satisfying those of V-3.2. are the well known *precompact spaces.*

These two theorems are two versions of the same result. In order to show that let us consider an ordered set $\underset{\sim}{V}$ with a least element 0. We select a *countable* down directed subset D of $\underset{\sim}{V}$ such that $0 \notin D$;

let $\mathfrak{U} = \{p \in \underset{\sim}{V} \ / \ u < p$ for some $u \in D\}$, thus $\mathfrak{U}$ is a filter and $0 \notin \mathfrak{U}$.

Let (E, d) be a binary space, satisfying d'1), over $\underset{\sim}{V}$. We say that a sequence $(x_n)_{n\in\mathbb{N}}$, $x_n \in E$ is a *Cauchy sequence* if for every $\varepsilon \in \mathfrak{U}$ there is an integer n such that $d(x_{n'}, x_{n''}) < \varepsilon$ for all $n < n' < n''$.

In this context, Theorem V-3.2. rephrases as follows:

V - 3.3. *Proposition* The following are equivalent:
1) Every sequence of elements of E has a Cauchy subsequence ;
2) For every $\varepsilon \in \mathfrak{U}$, every non empty subset A of E can be covered by finitely many right balls $B(x, \varepsilon) = \{y \in E \ / \ d(x, y) < \varepsilon\}$, whose centers x belong to A.
3) For every $\varepsilon \in \mathfrak{U}$ and every sequence $(x_n)_{n\in\mathbb{N}}$, there are $n < m$ such that $d(x_n, x_m) < \varepsilon$.

Proof- The standard proof of theorem V-3.2. perfectly works and gives 1) $\Rightarrow$ 3) , $\neg 2 \Rightarrow \neg 3$, 2) $\Rightarrow$ 1) . Indeed, first 1) $\Rightarrow$ 3) is trivial.

Next, if there are ε and A such that A cannot be covered by finitely many balls of radius ε , then construct inductively a sequence $(x_n)_{n\in\mathbb{N}}$ such that $x_m \in A - \cup \{B(x_n, \varepsilon) / n < m\}$. Finally in order to extract a Cauchy subsequence of a given sequence $(x_n)_{n \in \mathbb{N}}$, observe that, since D is countable, it contains a descending sequence $\varepsilon_0 > \varepsilon_1 > \varepsilon_2 > \ldots > \varepsilon_n \ldots$ such that for every $\varepsilon \in \mathcal{U}$, there is some n such that $\varepsilon_n < \varepsilon$; then define $(x_{\varphi(n)})_{n \in \mathbb{N}}$ inductively by setting $S_m = \cap \{B(x_{\varphi(m')}, \varepsilon_{m'}) / m' < m\}$ and $\varphi(m) = \mathrm{Min}\{n / x_n \in S_m$ and $\{n' / x_{n'} \in B(x_n, \varepsilon_m) \cap S_m\}$ is infinite$\}$. □

If we convert a poset P into a metric space over $\underset{\sim}{V} = \{0, a, b, 1\}$, then, for $D = \{a\}$, the Cauchy sequences are the sequences ultimately increasing. Condition 2) of Proposition V-3.3. means that every non empty subset A has finitely many element $a_1, \ldots, a_n$ such that for every $a \in A$, $a_i < a$ for some i . Thus Proposition V-3.3. gives the equivalence between 1), 2), 3) of Theorem V-3.1.
The implication 3) $\Rightarrow$ 4) is trivial. The converse requires an other application of Proposition V-3.3. . For that let $\underset{\sim}{V} = \{0, \frac{1}{2}, 1\}$, $D = \{\frac{1}{2}\}$ and let d be the distance over $\underset{\sim}{V}$ associated to the comparability graph of P . The fact that P has no infinite antichain is condition 3) of Proposition V-3.3, consequently, if $(x_n)_{n \in \mathbb{N}}$ is a sequence of element of P then it has a Cauchy subsequence. But this means that all its elements are ultimately pairwise comparable. Since P has no infinite descending chain this sequence cannot be decreasing, thus there are $n < m$ such that $x_n < x_m$.

For an arbitrary reflexive symmetric graph considered as a metric space over $\underset{\sim}{V}$, Proposition V-3.3. gives the equivalence between the following properties:
1) Every sequence ultimately consists of pairwise connected elements.
2) Every set A of vertices contains a finite subset A' such that for every $x \in A$ there is some $z \in A'$, connected to x .
3) There is no infinite independent set.

The reader has certainly recognized in this statement an equivalent version of the:

V - 3.4. ***Ramsey Theorem.*** [70]: *Every infinite graph contains either an infinite complete subgraph or an infinite independent set.*

We do not pretend that Proposition V-3.3. generalizes the Ramsey Theorem (in fact it follows from the Ramsey Theorem by an easy argument) we rather think that the Ramsey Theorem is the right abstraction of the characterization of precompactness. But we also think that this connection, even trivial, between metric spaces, posets and graphs helps to understand some developments in these areas and could suggest further reseach.

For example proofs of Ramsey Theorem with ultrafilter relate to older proofs of Theorem V-3.2. using universal nets (See J. Kelley [63], p. 198-199). Theorem V-3.1. relates to the following result of Hansell [61] .

V - 3.5. ***Theorem*** *A poset P has no infinite antichain iff every net on P has a monotone subnet.*

The theory of partially well ordered sets is rather rich, and is useful in combinatorics (see [64],[68]). Some analogs of the Higman and Kruskal theorems, in the context of metric spaces, could be interesting. The notion of cofinality of posets extends naturally to our metric spaces, in terms of density, and has some interesting instances. Finally let us mention that the classical notions of well-quasi-ordering and precompactness meet together in the study of almost periodic sequences (see [58]) . Certainly this is no accidental.

* * *

REFERENCES

[1] N. ARONSZAJN, P. PANITCHPAKDI : Extension of uniformly continuous transformations and hyperconvex metric spaces. Pacific J. Math. 6 (1956), 405-439.

[2] K. BACLAWSKI and A. BJORNER : Fixed points in partially ordered sets, Advances in Mathematics, 31 (1979), 263-287.

[3] P.D. BACSICH : Injectivity in model theory, Colloq. Math. 25 (1972), 165-176.

[4] B. BANASCHEWSKI and G. BRUNS : Categorical characterization of the MacNeille completion, Archiv. der Math. Basel 18 (1967), 369-377.

[5] L.M. BLUMENTHAL : Theory and applications of distance geometries. 1953, Clarendon Press, Oxford.

[6] L.M. BLUMENTHAL, K. MENGER : Studies in geometry 1970, W.H. Freeman and Co. San Francisco.

[7] R.E. BRUCK Jr. : A common fixed point theorem for a commuting family of non-expansive mappings. Pac. J. of Math 53 (1974) 59-79.

[8] L. CAIRE and U. CERRUTI : Fuzzy relational spaces, Rendiconti del Seminario della Facolta di Scienze dell'Universita di Cagliari, 47 (1977), 63-87.

[9] U. CERRUTI and U. HOHLE : Categorical fundations of probabilistic microgeometry. Séminaire de "Mathématique floue" LYON (1983-1984) P. 189-246.

[10] G. COHEN and M. DEZA : Distances invariantes et L-cliques sur certains demi-groupes finis, Math. Sci. hum n° 67 (1979) p. 49-69.

[11] D. DUFFUS and I. RIVAL : Retracts of partially ordered sets. J. Austral. Math. Soc. (Serie A) 27 (1979), 495-506.

[12] D. DUFFUS and I. RIVAL : A structure theory for ordered sets, J. of Discrete Math. 35 (1981), 53-118.

[13] D. DUFFUS and M. POUZET : Representing ordered sets by chains, in: Orders: Descriptions and Role (M. Pouzet and D. Richard, eds) Annals of Discrete Math. 23 (1984) p. 81-98.

[14] R. FRAISSE : Sur l'extension aux relations de quelques propriétés des ordres. Ann. Sci. Ecole Norm. Sup. (3) 71 (1954), 363-388.

[15] G. GRATZER : General Lattice theory, Acad. Press. 1978.

[16] P. HELL : Absolute retracts of graphs, Lecture notes 406 (1974) PP. 291-301.

[17] P. HELL : Graph retractions, Atti dei conveigni lincei 17, teorie combinatorie (1976) pp. 263-268.

[18] P. HELL : Rétractions de graphes. PhD. Université de Montréal. Juin 1972. 148 pages.

[19] P. HELL and I. RIVAL : Absolute retracts and varieties of reflexive graphs, preprint, 1983.

[20] D. HIGGS : Injectivity in the topos of complete Heyting algebra valued sets, Canadian J. of Math. 36 (1984) 550-568.

[21] G. HIGMAN : Ordering by divisibility in abstract algebra, Proc. London Math. Soc. (3) 2 (1952) p. 326-336.

[22] R. ISBELL : Six theorems about injective metric spaces, Comment. Math. Helv. 39 (1964), 65-76.

[23] V.I. ISTRATESCU : Fixed point theory, an introduction. Math. and its applications, Vol. 7, (1981). D. Reidel.

[24] E. JAWHARI, M. POUZET, I. RIVAL : A classification of reflexion graphs: The use of "holes". Rapport de recherche du Laboratoire d'Algèbre ordinale et algorithmique, Lyon (1983). To appear in Canadian J. of Math.

[25] E. JAWHARI : Les rétractions dans les graphes. Applications et généralisations, Thèse de 3ème cycle, N° 1318, (Juillet 1983) Lyon.

[26] W.A. KIRK : Fixed point theory for non expansive mapping, Lecture notes in Math., 886 (1981) 484-505.

[27] E.W. KISS, L. MARKI, P. PROHLE and W. THOLEN : Categorical algebraic properties. A compendium on amalgamation, congruence extension, epimorphisms, residual smallness, and injectivity, Studia Scientiarum Mathematicarum Hungarica 18 (1983), 79-141.

[28] F.W. LAWVERE, Metric spaces, generalised logic and closed categories, Rendiconti del Seminario Mathematico e Fisico di Milano 43 (1974), 135-166.

[29] H. MacNEILLE: Partially ordered sets. Trans. Amer. Math Soc. 42 (1937) 416-460.

[30] K. MENGER : Untersuchungen über allgemeine Metrik, Math. Annalen 100 (1928) 75-163.

[31] K. MENGER : Geométrie générale, Mémorial des Sciences mathématiques N° 124, (1954), Paris, Gauthiers-Villars.

[32] D. MISANE : Retracts absolus d'ensembles ordonnés et de graphes. Propriété du point fixe. Thèse de doctorat de 3ème cycle, N° 1571, (Septembre 1984), Lyon.

[33] P. NEVERMANN and I. RIVAL, Holes in ordered sets, Research paper N° 580, Nov. 1984, The University of Calgary.

[34] P. NEVERMANN and R. WILLE : The strong selection property and ordered sets of finite length, Alg. Univ. 18 (1984), 18-28.

[35] R. NOWAKOWSKI and I. RIVAL : A fixed edge theorem for graphs with loops. J. Graph theory 3 (1979) 339-350.

[36] R. NOWAKOWSKI and I. RIVAL : The smallest graph variety containing all paths, J. of Discrete Math. 43 (1983) 223-234.

[37] E. PESH : Minimal extension of graphs to absolute retracts, preprint N° 839, July 1984, Technische Hoschule Darmstadt.

[38] M. PONS VALLES : Contribucio a l'estudi d'estructures uniformes sobre conjunts ordenats, Thesis (1984), Barcelona.

[39] M. PONS VALLES : Uniform structures on boolean algebras, preprint, 10 pages (1984).

[40] M. POUZET : Contribution to the problem session on order-preserving and egde-preserving map in Graphs and Order (I. Rival, Ed) (1985) p. 574-75. D. Reidel.

[41] M. POUZET : Une approche métrique de la rétraction dans les ensembles ordonnés et les graphes. Compte-rendu des Journées infinitistes de Lyon (octobre 1984). Pub. Dépt. Math. Lyon, (1985).

[42] M. POUZET, I. RIVAL : Every countable lattice is a retract of a direct product of chains, Alg. Univ. 18 (1984) p. 295-307.

[43] M. POUZET and I.G. ROSENBERG : Embeddings and absolute retracts of relational systems, Research report n° 1265 of the CRM, Feb. 1985, University of Montréal, 23 p.

[44] A. QUILLIOT : Homomorphismes, points fixes, rétractions et jeux de poursuite dans les graphes, les ensembles ordonnés et les espaces métriques. Thèse de doctorat d'Etat, Univ. Paris VI (1983).

[45] A. QUILLIOT : An application of the Helly property to the partially ordered sets, J. Combin. Theory, serie A, 35 (1983) 185-198.

[46] I. RIVAL : A fixed point theorem for finite partially ordered sets, J. of Comb. theory (1976) 309-318.

[47] I. RIVAL, R. WILLE : The smallest order variety containing all chains, Discrete Math. 35, 203-212.

[48] Z. SEMADENI : Banach spaces of continuous functions, Vol. I, Monografie Matematyczne, Warsawa, (1971).

[49] R. SINE : On non linear contractions in Sup norm spaces. Non linear analysis, TMA, 3 (1979), 885-890.

[50] R. SINE : Fixed points and non expansive mappings (R. Sine Ed.) Contemporary Math. Vol. 18. AMS.

[51] D.R. SMART : Fixed point theorems, Cambridge tracts in Math. 66 (1974) Cambridge University press.

[52] P.M. SOARDI : Existence of fixed points of non expansive mappings in certain banach lattices, Proc. A.M.S. (1979) 25-29.

[53] A. TARSKI : A lattice theoretical fixed point theorem and its applications. Pacific J. Math. 5 (1955) - 285 - 309.

Additional References —

[54] H.J. BANDELT, A. DAHLMANN and H. SCHUTTE, Absolute Retracts of bipartite graphs, Research report. Fachbereich 6, Univ. Carl-von-Ossietzky Universität, - Oldenbourg. (51 pages) (1984).

[55] J.P. BARTHELEMY, B. LECLERC, B. MONJARDET, Ensembles ordonnés et taxonomie mathématique (M. Pouzet , D. Richard Ed.) Orders: description and roles, Annals of discrete Maths - North Holland (1984). p. 523-548.

[56] M. FRECHET, Rend. Circ. Math. Palermo, Vol. 22 (1906) p. 6 .

[57] M. FRECHET, Les espaces abstraits, Paris 1928.

[58] W.H. GOTTSCHALK and G.A. HEDLUND, Topological dynamics, Vol. 36. Colloquium pub. AMS. (1955).

[59] R. GRAHAM, P. WINKLER, Isometric embeddings of graphs Proc. Natl. Acad. Sci. USA.Mathematics Vol. 81 pp. 7259-7260 (1984).

[60] F. HAUSDORFF, Grundzüge der Mengenlehre, 1914, Leipzig.

[61] R.W. HANSELL, Monotone subnets in partially ordered sets, Proc. Amer. Math. Soc. Vol. 18 (1967) p. 854-858.

[62] K. KAARLI, Compatible function extension property, Algebra Universalis, 17 (1983) p. 200-207

[63] J.L. KELLEY, General Topology, (1955), Van Nostrand.

[64] J.B. KRUSKAL, Well quasi ordering, the tree theorem and Vazsonyi's conjecture, Trans. Amer. Math. Soc. (1960) 95, 210-225.

[65] J.B. KRUSKAL, The theory of well quasi ordering: a frequently discovered concept, J. Comb. Th. (A) 13, 197-305.

[66] B. LECLERC, Efficient and binary consensus functions on transitively valued relations, Rapport de recherche du CMS, N° 005, Paris, oct. 1983.

[67] LOTHAIRE, Combinatorics on words, Enclopaedia of Maths (G.C. Rota Ed.) Addison Weysley.

[68] E.C. MILNER, Basic wqo-and bqo-theory, Graphs and Order (I. Rival Ed.) Reidel (1985) p. 487-502.

[69] M. POUZET, Retracts: recent and old results on graphs, ordered sets and metric spaces. Circulating manuscript, 29 pages, Nov. 1983.

[70] FP. RAMSEY , On a problem of formal logic. Proc. London Math. Soc. 30, 264-286.

[71] R.L. ROTH and P.M. WINKLER, Collapse of the Metric Hierarchy for Bipartite Graphs. Preprint, Emory University (1984).

Contemporary Mathematics
Volume 57, 1986

ANTICHAINS AND CUTSETS

Mohamed H. El-Zahar and Nejib Zaguia

Abstract. *An antichain of an ordered set P intersects every maximal chain of P at most once. However there are extremal examples of ordered sets in which no antichain meets every maximal chain. Naturally we may ask for subsets of P that meet every maximal chain in P. Such subsets are called cutsets of P. A cutset for an element x in P is a subset $F \subseteq P$ of elements noncomparable to x such that $\{x\} \cup F$ is a cutset for P.*

Another context in which these concepts arise naturally is that of topological spaces associated with P. The subject has had a remarkable development over the last few years since the basic result of Bell and Ginsburg that $\mathcal{M}(P)$ is compact if and only if each element of P has a finite cutset, where $\mathcal{M}(P)$ denote the space of all maximal chains of P.

The study of cutsets , and in particular, the cutsets of restricted type such as finite, antichain, minimal, etc. yields interesting questions and results concerning the structure of the ordered set.

The purpose of this paper is to give a survey of recent developments in these directions.

1. INTRODUCTION

Chains (that is subsets in which every two elements are comparable) and *antichains* (that is subsets in which any two elements are noncomparable) have played an important role in the development of the theory of ordered sets. Perhaps the best known result in the combinatorial theory of ordered sets is the "chain decomposition theorem" of R.P. Dilworth (1950) which states that *the minimum number of chains which cover a finite ordered set equals the maximum size of an antichain.*

In particular if $C_1, C_2, \ldots, C_n$ is a family of maximal chains of minimum number whose union is an ordered set P and A is any maximal antichain of P then A meets each C_i, that is, for each i, $C_i \cap A \neq \phi$. This leads naturally to a dual question and to ask for an antichain that meets or "cuts" all maximal chains. In a finite ordered set, there is always some antichain which meets every maximal antichain, for instance the antichain of all maximal elements. However, an arbitrary maximal antichain need not meet every maximal chain. More is true. There exist ordered sets in which no antichain meets all maximal chains [see Figure 1]. Therefore, this leads to the consideration of subsets which cut all maximal chains and which may not be antichains. Restrictions on these sets yield interesting properties pertaining to antichains and the structure of ordered sets.

In the last few years, there have appeared several papers dealing with such subsets. Our aim is to give a survey of the recent developments in this direction.

Call a subset K of an ordered set P a *cutset* of P if every maximal chain C of P meets K, that is, $K \cap C \neq \phi$. If K is an antichain

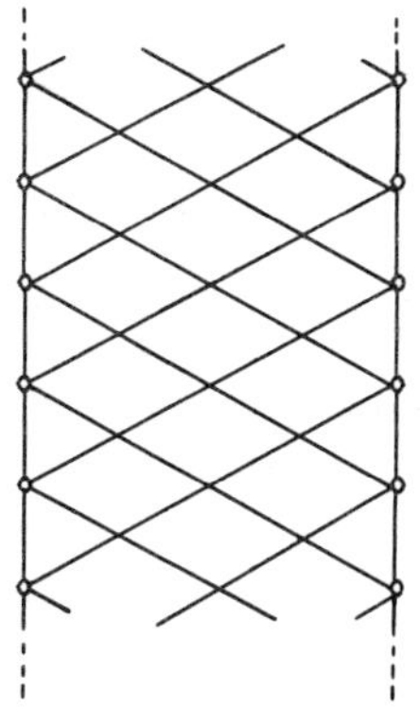

Figure 1.

then we call it an *antichain cutset* of P. The concept of cutsets was introduced by M.Bell and J.Ginsburg (1984) in a study of the compactness properties of the space $\mathcal{M}(G)$ of all maximal complete subgraphs of a graph G and, in particular, the space $\mathcal{M}(P)$ of all maximal chains of an ordered set P.

Let P be an ordered set. We denote by $\underset{\sim}{2}^P$ the set of all subsets of P. Now, if $\underset{\sim}{2} \cong \{0,1\}$ is endowed with the discrete topology then $\underset{\sim}{2}^P$ becomes a topological space which has as a subbase for the open sets all sets of the form

$$A(x) = \{Q \in \underset{\sim}{2}^P \mid x \notin Q\}$$

and

$$B(x) = \{Q \in \underset{\sim}{2}^P \mid x \in Q\}$$

where x is in P. Therefore $\mathcal{M}(P)$ is also thought of as a topological space [$\mathcal{M}(P)$ is a subspace of $\underset{\sim}{2}^P$]. The main problem which interested M.Bell and J.Ginsburg (1984) is to characterize compact topological spaces which can be represented as $\mathcal{M}(G)$ or, also, as $\mathcal{M}(P)$. It remains an open question to characterize such spaces with topological properties. However, the same authors proved that if $\mathcal{M}(P)$ is compact then P contains a finite cutset. While investigating sufficient conditions, M.Bell and J.Ginsburg (1984) were led to the definition of a cutset for an element. A subset K(x) of P is a *cutset for* x provided that each maximal chain of P contains an element of K(x) ∪ {x} and each element of K(x) is noncomparable with x. For instance the same authors have shown that *the space $\mathcal{M}(P)$ of maximal chains of an ordered set P is compact if and only if each element has a finite cutset.*

Since this result, the principal questions that have been considered bear on the order-theoretical counterparts of this compactness condition, and also on the structure of cutsets in an ordered set. Most of these studies turn on the ideas of chains and antichains.

Notice that the term "cutset" is more common in graph theory. Usually, it is used to recall a set of vertices whose removevable disconnects the graph, or also for some authors, it is used to recall a set of edges whose removable disconnects the graph.[See M.Behzad and G.Chartrand (1971).] This concept is sometimes useful in concrete situations when the emphazis is on ways of separating sets of vertices from one another. For example, in the study of flows through networks, cross sections of the network separating flow origins from destinations are considered in order to find a restrictive cross section which constitutes a bottleneck. Such bottleneck determine the flow capacity of the network as a whole.

Section 2, contains results concerning the compactness of the space $\mathcal{M}(P)$. These theorems describe either sufficient or necessary conditions under which the space $\mathcal{M}(P)$ is compact. Moreover, chain complete ordered sets will be of primary interest. An ordered set P is *chain complete* if every nonempty chain in P has a least upper bound and a greatest lower bound in P. A concept of importance in this study is the following. Let C be a chain in P and u the supremum of C. We say that u is *special* if there is an element p in C such that for every x in P and x ≥ p, we have x comparable with u. *Special infimum* is defined dually.

Actually, the basic result which led to some others is that all suprema and infima of chains of P are special whenever $\mathcal{M}(P)$ is compact. In fact, in an ordered set P, every supremum and infimum is special if and only if P does not contain a subset isomorphic to either of the ordered sets P_1 or P_2 illustrated in Figure 2, or their duals.

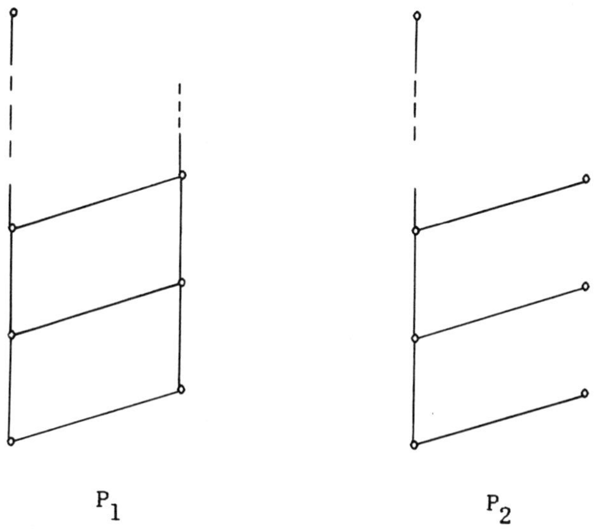

Figure 2.

Obviously the best structure of a cutset is an antichain. Section 3, contains results concerning the antichain cutsets in ordered sets. A first result is a characterization by means of "forbidden" configurations of those regular ordered sets in which every element belongs to an antichain cutset [I.Rival and N.Zaguia (1985.a)]. An ordered set is *regular* if every nonempty chain C of P has a supremum and infimum, and whenever $x < \sup C$ (respectively $x > \inf C$) then $x < c$ (respectively $x > c$) for some element c in C. We say that P *satisfies the ascending chain condition* if P does not contain a strictly decreasing infinite chain $x_0 < x_1 < \ldots$. Also, we say that P *satisfies the descending chain condition* if P does not contain a strictly increasing infinite chain $x_0 > x_1 > \ldots$. Finally, P *satisfies a chain condition* if either P satisfies the accending chain condition or P satisfies the descending chain condition.

Call a cutset K in P *minimal* if for every x in K, K-{x} is not a cutset in P. Another equivalent formulation is that for every x in K, there is a maximal chain in P such that $C \cap K = \{x\}$. I.Rival and N.Zaguia (1985.b) proved that in a regular ordered set with a chain condition every element belongs to a minimal cutset. P.A.Grillet (1969) has shown that for a regular ordered set P every maximal antichain meets every maximal chain if and only if P is $\underset{\sim}{N}$-free. [An ordered set is $\underset{\sim}{N}$-free if it contains no $\underset{\sim}{N}$, that is, a subset {a,b,c,d} suth that a<c, b<d and c covers b are the only comparabilities among these elements.] Using the same idea as in the proof of Grillet's Theorem, several authors have given a characterization of all ordered sets in which every maximal antichain is a cutset [see J.Ginsburg (1984) and I.Rival and N.Zaguia (1985.b)]. A related question is when every minimal cutset of an ordered set is an antichain. D.Higgs (1985.a) has shown that if an ordered set P contains no subset isomorphic to $\underset{\sim}{N}$ then every minimal cutset is an antichain. The converse fails in general [D.Higgs(1985.a)], however D.Higgs proved for finite ordered sets. Also I.Rival and N.Zaguia (1985.b) extended this result to regular ordered sets with a chain condition. However, the problem is still open for regular ordered sets.

Let m be a nonnegative integer. Say that an ordered set P has the *m-cutset property* if there is a cutset K(x) with at most m elements for each x in P. Also, say that an ordered set P has *the finite cutset property* if every x in P has a finite cutset. The ordered sets with the zero-cutset property are exactly the chains. Also, it is not hard to prove that the ordered sets with the one-cutset property have at most width two. J.Ginsburg, I.Rival and B.Sands (1985) have shown that an ordered set P with the finite cutset property is finite whenever all the chains of P are finite. This leads naturally to study possible relations between the length, width and m for the ordered sets with the m-cutset property. In section 4, we will discuss some aspects of this problem. As far as we know the only case where the problem is completely solved is m = 2. In fact J.Ginsburg and B.Sands(1985) and, independently, M.H.El-Zahar and N.Sauer (1985) have shown that $\omega(P) \leq \mathcal{l}(P) + 3$ for every finite ordered set P with the two-cutset property. The two proofs use different approaches. However, we will discuss the proof of M.H.El-Zahar and N.Sauer, since it uses a new concept of simple sets which might be interesting for the investigation of other questions. Section 5 deals with minimum-sized cutsets in ordered sets. For finite ordered sets, the size of minimum-sized cutsets equals the maximum number of disjoint maximal chains. This is a consequence of Menger's theorem. We will discuss the same problem for infinite ordered sets. Finally, in the last part of section 5, we will discuss briefly some problems concerning the size of minimal cutsets in the case of lattices and, in particular, the distributive lattice $\underset{\sim}{2}^n$.

2. COMPACTNESS OF THE SPACE $\mathcal{M}(P)$

Let P be an ordered set. As usual, the power set of P, $\underset{\sim}{2}^P$, is identified with the set of all functions $f\colon P \to \{0,1\}$. Now, if $\underset{\sim}{2} = \{0,1\}$ is endowed with the discrete topology then $\underset{\sim}{2}^P$ becomes a topological space which has as a subbase for the open sets all sets of the form

$$A(x) = \{Q \subseteq P \mid x \notin Q\}$$

and

$$B(x) = \{Q \subseteq P \mid x \in Q\}$$

for x in P. These sets are also closed in $\underset{\sim}{2}^P$ and form a subbase for the closed sets. Let

$$\mathcal{M}(P) = \{C \mid C \text{ is a maximal chain in } P\}.$$

$\mathcal{M}(P)$ is also thought of as a topological space, that is, a subspace of $\underset{\sim}{2}^P$. Thus all sets of the form

$$A(x) = \{C \in \mathcal{M}(P) \mid x \notin C\}$$

and

$$B(x) = \{C \in \mathcal{M}(P) \mid x \in C\}$$

for x in P, form a subbase for the open sets in $\mathcal{M}(P)$. Of course, $\underset{\sim}{2}^P$ is compact but $\mathcal{M}(P)$ need not be compact. The following theorem of M.Bell and J.Ginsburg (1984) characterizes those ordered sets P for which $\mathcal{M}(P)$ is compact.

THEOREM 2.1. *$\mathcal{M}(P)$ is compact if and only if each x in P has a finite cutset.*

Proof. Assume that $\mathcal{M}(P)$ is compact. Let x be in P, and let I(x) be the set of elements noncomparable to x. Then

$$\{ B(x) \} \cup \{ B(y) \mid y \in I(x) \}$$

is an open cover of $\mathcal{M}(P)$. Since $\mathcal{M}(P)$ is compact, this open cover contains a finite subcover. This means that there are elements $y_1, \ldots, y_n$ noncomparable to x such that $\{ B(x) \} \cup \{ B(y_i) \mid i=1, \ldots, n \}$ covers $\mathcal{M}(P)$. Therefore $\{x, y_1, \ldots, y_n\}$ meets every maximal chain in P, that is, $\{y_1, \ldots, y_n\}$ is a cutset for x.

Conversely, assume that each element in P has a finite cutset. We show that $\mathcal{M}(P)$ is closed in $\underset{\sim}{2}^P$. Let $S \in \underset{\sim}{2}^P - \mathcal{M}(P)$. If S is not a chain, then it contains two noncomparable elements x and y. Let

$$U = \{ Q \in \underset{\sim}{2}^P \mid x \in Q \text{ and } y \in Q \}.$$

U is an open set in $\underset{\sim}{2}^P$ which contains S and is disjoint from $\mathcal{M}(P)$. So, we assume that S is a chain. Since $S \notin \mathcal{M}(P)$ then there is an element z in P such that $S \cup \{z\}$ is a chain. Let K(z) be a finite cutset for z. Then the set

$$V=\{Q \in \underset{\sim}{2}^P \mid z \notin Q \text{ and } p \notin Q \text{ for all } p \in K(z)\}$$

is an open set in $\underset{\sim}{2}^P$ which contains S and is disjoint from $\mathcal{M}(P)$. Therefore $\underset{\sim}{2}^P - \mathcal{M}(P)$ is open in $\underset{\sim}{2}^P$. □

The principal direction taken in Bell and Ginsburg's paper is to characterize compact topological spaces which can be represented as $\mathcal{M}(G)$ or, particularly, as $\mathcal{M}(P)$. They show that a compact space has the form $\mathcal{M}(G)$, for some graph G, if and only if it has a binary subbase for the closed sets consisting of clopen sets (a family of sets S is called *binary* if whenever $S' \subseteq S$ and each pair of sets in S' has a nonempty intersection then $\cap S' \neq \phi$). The same authors have shown that the one-point compactification of an uncountable discrete space can be represented as $\mathcal{M}(G)$ but not as $\mathcal{M}(P)$. It remains an open question to give topological properties which characterize compact spaces of the form $\mathcal{M}(P)$. An interesting property of compact spaces of the form $\mathcal{M}(P)$, for a chain complete ordered set P, was given by J.Ginsburg, I.Rival and B.Sands (1985). They proved that such topological spaces do not contain a family of $(2^{\omega})^{+}$ disjoint open sets.

Here are some properties concerning the class $\mathfrak{F}$ of ordered sets with the finite cutset property. This class is not closed under taking subsets. However, if $P \in \mathfrak{F}$ then particular subsets of P do belong to $\mathfrak{F}$. Let a and b be elements in P. We set $(a,b)=\{x \in P \mid a<x<b \}$. A subset Q of P is *convex* if for every a, b in Q the open interval (a,b) is a subset of Q. We say that P is *up-directed* (respectively *down- directed*) if for every a, b $\in$ P, there exists c such that $a \leq c$ and $b \leq c$ (respectively $c \leq a$ and $c \leq b$). P is *directed* if it is up and down-directed.

THEOREM 2.2. [J.Ginsburg (1984)] *Let P be an ordered set in $\mathfrak{F}$.*

(i) If Q is a subset of P such that for each maximal chain C in P, $C \cap Q$ is a maximal chain in Q, then $Q \in \mathfrak{F}$.

(ii) If $Q \cong (a,b)$ an open interval in P then $Q \in \mathfrak{F}$.

(iii) If Q is a convex directed subset of P then $Q \in \mathfrak{F}$.

Proof. (i) The map $f: \mathcal{M}(P) \to \mathcal{M}(Q)$ which maps each maximal chain C in P to $C \cap Q$ is a continuous, surjective map. Hence $\mathcal{M}(Q)$ is compact.

(ii) Let $Q_0 = \{ x \in P \mid x$ is noncomparable to both a and b $\}$. Then

$$\mathcal{A}(Q_0) = \{ C \in \mathcal{A}(P) \mid a \in C \text{ and } b \in C \}$$
$$= B(a) \cap B(b).$$

Therefore $\mathcal{A}(Q_0)$ is a closed subset of $\mathcal{A}(P)$ and hence is compact. Now the open interval (a,b) considered as a subset of Q_0 satisfies the condition of (i).

(iii) Let x be in Q and let F be a finite cutset for x in P. We show that $F \cap Q$ is a cutset for x in Q. Assume it is not. Then there is a maximal chain C in Q such that $C \cap (\{x\} \cup F) = \phi$. This implies that for each element $p \in Q \cap (\{x\} \cup F)$ there is an element $c_p \in C$ noncomparable to p. Since Q is directed there are elements u, $v \in Q$ such that $v \le x \le u$ and $v \le c_p \le u$ for each p in $Q \cap (\{x\} \cup F)$. Let D be a maximal chain in P which contains the chain $\{u,v\} \cup \{c_p \mid p \in Q \cap (\{x\} \cup F)\}$. Clearly $x \notin D$ since $c_x \in D$. Therefore D contains an element p of F. Since p is noncomparable to x then $v \le p \le u$. This implies that $p \in Q$ since Q is convex. But then D contains p and c_p which is a contradiction, since p and c_p are noncomparable. □

This class $\mathcal{F}$ is also not closed under products. Even the product of two chains need not have the finite cutset property. For instance the product ZxZ ,where Z is the set of integers with the usual ordering, does not have the finite cutset property. However, $\mathcal{F}$ is closed under taking lexicographic sums.

THEOREM 2.3. [J.Ginsburg (1984)] *Let P be in* $\mathcal{F}$. *For every x in P, let* P_x *be in* $\mathcal{F}$. *Then* $\Sigma_{x \in P} P_x$ *is also in* $\mathcal{F}$.

Proof. Let $p \in \Sigma_{x \in P} P_x$, say $p \in P_{x_0}$. Let F be a finite cutset for x_0 in P and E_{x_0} be a cutset for p in P_{x_0}. For each $x \in F$ choose E_x also a finite cutset for P_x. Then $\cup \{ E_x \mid x \in \{x_0\} \cup F \}$ is a cutset for p in $\Sigma_{x \in P} P_x$. □

As a special case of Theorem 2.3, we notice that the ordinal sum

$\bigoplus_{\alpha<\kappa} P_\alpha$ of ordered sets P_α ($\alpha<\kappa$) in $\mathcal{F}$ also belongs to $\mathcal{F}$. Observe that the space $\mathcal{M}(\bigoplus_{\alpha<\kappa} P_\alpha)$ is isomorphic to the product space $\prod_{\alpha<\kappa} \mathcal{M}(P_\alpha)$. This is another way to prove that $\mathcal{M}(\bigoplus_{\alpha<\kappa} P_\alpha)$ is compact.

Before we close this section we shall briefly discuss chain complete ordered sets with the finite cutset property. M.Bell and J.Ginsburg (1984) proved the following easy but basic result.

PROPOSITION 2.4. *Let P be a chain complete ordered set with the finite cutset property. Then every supremum and infimum is special.*

Proof. Assume that $c = \sup C$ and let F be a finite cutset for c. For each element a in F, and since c is noncomparable to a, there exists an element $c_a \in C$ which is noncomparable to a. Let $d = \max \{c_a \mid a \in F\}$. Now, let $x > d$, and let D be a maximal chain containing $\{x\} \cup \{c_a \mid a \in F\}$. Since D cannot contain both a and c_a then $D \cap F = \phi$. Therefore $c \in D$ which implies that x and d are comparable. The case $c = \inf C$ is dual. □

Also it is easy to check that in a chain complete ordered set P, every supremum and infimum is special if and only if P contains no subsets isomorphic to either the ordered set P_1 or P_2 in Figure 2, or their duals. Obviously ordered sets in which every supremum and infimum is special, need not have the finite cutset property. To see that, just consider an infinite antichain.

3. MAXIMAL ANTICHAINS AND MINIMAL CUTSETS

Dilworth's theorem states that an ordered set P which contains a finite maximum-sized antichain A has a partition into $|A|$ chains. Even so, the antichain A need not meet every maximal chain in the ordered set. In other terms, A need not be a cutset. In a finite ordered set there is always some antichain which meets every maximal chain. For instance the antichain of all maximal elements. This does not mean the existence of a partition of the ordered set into such antichains. In Figure 3, we have an ordered set which cannot be expressed as a union of antichain cutsets. Indeed, x belongs to no antichain cutset at all. More is true. There exist ordered sets in which no antichain meets all maximal chains (see Figure 1). Moreover, it is simple to verify that, a finite ordered set,

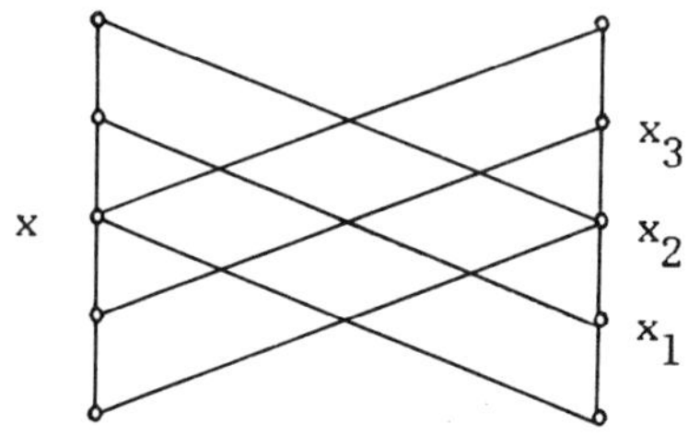

Figure 3.

in which all maximal chains have the same cardinality can be expressed as the union of antichain cutsets. For example that is the case for modular lattices. A natural way to see this is in terms of the height function h which assigns to each element x the length of a largest chain from a minimal element to x. As each maximal chain must then contain an element at each level, it follows that each level is a cutset. It is also simple to prove that if a finite ordered set P can be expressed as a disjoint union of antichain cutsets then all maximal chains in P have the same cardinality. I.Rival and N.Zaguia (1985.a) gave a characterization, by means of "forbidden" subdiagrams, of those regular ordered sets which can be expressed as a union of antichain cutsets.

THEOREM 3.1. *A regular ordered set is the union of antichain cutsets if and only if it contains no alternating-cover cycle.*

Actually, they proved a more general result. Let P be an ordered set and let x be an element in P. Call a sequence

$$x=x_0 , x_1 , x_2 , \dots , x_n=y$$

a *generalized alternating-cover path* from x to y provided that $x_i < x_{i+1}$ whenever i is even, and $x_i \succ x_{i+1}$ in P whenever i is odd. If $y = x$ and n is odd we call this sequence a *generalized alternating-cover cycle based at* x. If these are the only comparabilities among the elements $x=x_0, x_1, \dots , x_n=x$, we call it an *alternating-cover cycle based at* x (see Figure 4). An example of generalized alternating-cover cycle is illustrated in Figure 5. Notice that this ordered set contains an alternating-cover cycle

THEOREM 3.2. *In a regular ordered set, an element x is contained in an antichain cutset if and only if there is no generalized alternating-cover cycle based at x.*

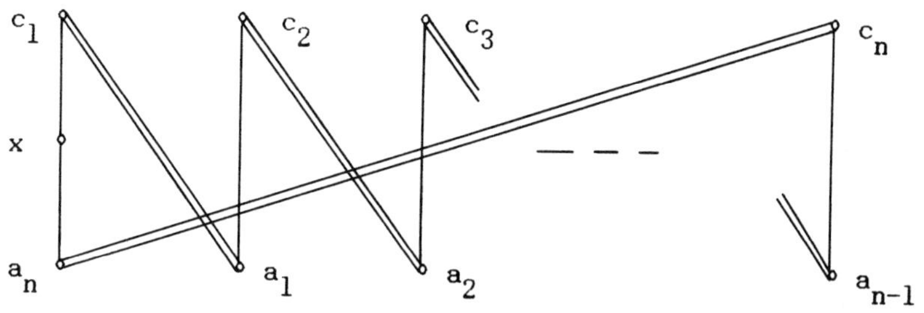

Figure 4.

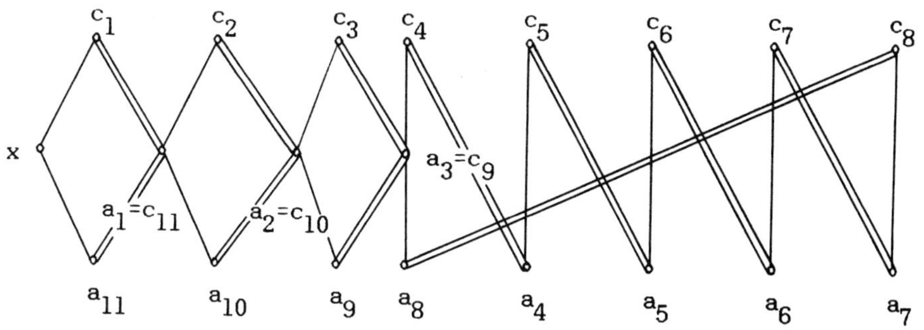

Figure 5

Without the regularity assumption neither Theorem 3.1, nor Theorem 3.2 holds. The ordered set $\underset{\sim}{2} \times (\underset{\sim}{\omega} \oplus \underset{\sim}{1})$ illustrated in Figure 6, contains no generalized alternating-cover cycle at all and yet the element x is not contained in any antichain cutset.

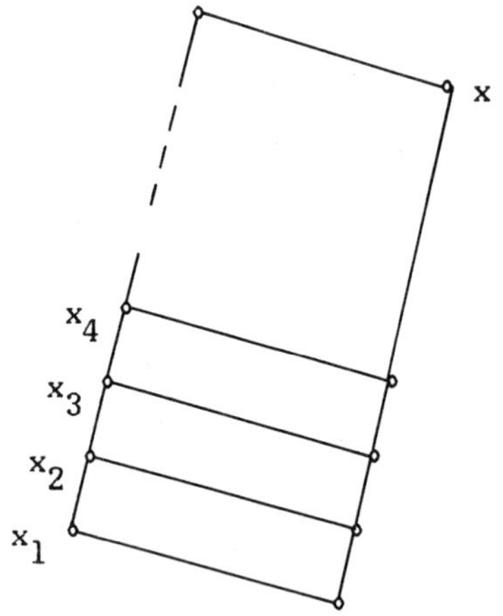

$\underset{\sim}{2} \times (\underset{\sim}{\omega} \oplus \underset{\sim}{1})$

Figure 6.

proof of Theorem 3.2. Let P be an ordered set and suppose that an element x of P is contained in a generalized alternating-cover cycle

$$\{x,\ c_1,\ a_1,\ c_2,\ a_2,\ \dots\ ,\ c_n,\ a_n,\ x\}.$$

Let us suppose, in addition, that x itself belongs to an antichain cutset A. There is a maximal antichain in P passing through the covering pair $c_1 \succ a_1$; evidently, this maximal chain must contain an element of A, say v_1, and as each element of the cutset A is noncomparable to x, $v_1 \leq a_1$. Now consider the maximal chain through the covering pair $c_2 \succ a_2$; it contains a cutset element v_2 say, and if v_1 is noncomparable to v_2 then

$v_2 \le a_2$. Continuing in this way we construct elements v_i , i=1, ... ,n-1, each belonging to A and such that $v_i \le a_i$. Finally, there is a cutset element v_n belonging to a maximal chain through $c_n \succ a_n$ which, however, is a contradiction since no cutset element can be comparable to x.

It is worth recording two remarks at this point. First notice that no regularity conditions on P are needed to establish the "necessity". The other remark is that while, a particular element of an ordered set may be contained in a generalized alternating-cover cycle, it need not be contained in an alternating-cover cycle at all (see Figure 6).

To prove the "sufficiency" fix an element x in P and assume that there is no generalized alternating-cover cycle based at x. Set

$X = \{y \in P \mid$ there is a generalized alternating-cover path from x to y$\}$

Notice that X is an up-set, and x is a minimal element in X. For if not not, then there is a minimal element y in X such that $y < x$; once we add x to the generalized alternating-cover path from x to y we obtain a generalized alternating-cover cycle based at x, which is a contradiction. Set

$$B = \{y \in P \mid y=\inf C, \text{ where } C \text{ is a maximal chain in } X\},$$

and

$$A = B \cup [(\max P) - X],$$

where max stands for the corresponding maximal elements. We claim that A is an antichain cutset of P (which clearly contains x).

Suppose t_1, $t_2 \in A$ and $t_1 < t_2$. Clearly $t_1 \notin \max P$ and so $t_1 \in B$ and there is a maximal chain C_1 in X such that $t_1 = \inf C_1$. Since P is regular, there is $y \in C_1$ such that $y < t_2$ and hence $t_2 \in X$. Since $y \in X$ it follows that $t_2 \ne \inf C_2$ for any chain C_2 in X and so $t_2 \notin A$. Therefore A is an antichain in P. Suppose that C is a maximal chain in P. If $C \subseteq X$, then $\min C \in B \subseteq A$, and if $C \subset P-X$ then $\max C \in A$ and so, in either case, $C \cap A \ne \phi$. Therefore, we may assume that $C \cap X \ne \phi$ and $C - X \ne \phi$. Let $a = \inf C \cap X$, $b = \sup C-X$. Thus $a, b \in C$. If $a \in B$, then

$A \cap C \neq \phi$ and so we can assume that $a \notin B$. Hence $C \cap X$ is not maximal in X and there is $a' \in X$ such that $a' < a$. If $a = b$, then $a' < y$ for some $y \in C-X$ (since P is regular) and this is a contradiction since $a' < y$ implies $y \in X$. Therefore $a \neq b$ and so $b -\!\!< a$. Now the alternating-cover path from x to a' may be extended to b (via a). Hence $b \in X$. But this contradicts the fact that $a = \inf X \cap C$. □

The proof of Theorem 3.1. is a direct consequence of Theorem 3.2. The key is a Lemma whose proof we omit and which provides a formal way to recognize alternating-cover cycles. In effect, it shows that every generalized alternating-cover cycle contains an alternating-cover cycle [I.Rival and N.Zaguia (1985.a)]. An algorithmically effective proof of Theorem 3.2 was given by I.Rival and N.Zaguia (1985.b). In it there is implicit an algorithm for the construction of the antichain cutset, containing a given element, in a polynomial number of steps on the number of elements of the ordered set. In particular, they gave an effective procedure to decide whether or not, a subset K of an ordered set P is a cutset.

Not every finite ordered set is a union of antichain cutsets, although every finite ordered set is a union of minimal cutsets. In fact every cutset $K(x) = \{x_1, \ldots, x_n\}$ for x in an ordered set P contains a subset M such that $M \cup \{x\}$ is a minimal cutset in P. A way to prove it, is to construct inductively a sequence of cutsets for x

$$K(x) = M_1 \supseteq \ldots \supseteq M_n = M$$

as follows. If there exists a maximal chain C in P such that $C \cap M_i = \{x_{i+1}\}$ then $M_{i+1} = M_i$, otherwise $M_{i+1} = M_i - \{x_{i+1}\}$. Now, we can easily check that $\{x\} \cup M$ is a minimal cutset in P. In general, it is not true that every element in an ordered set P, belongs to a minimal cutset. For instance in $\underset{\sim}{2} \times (\underset{\sim}{\omega} \oplus \underset{\sim}{1})$ [see Figure 6], there is no minimal cutsets which contains x. More is true. There are ordered sets which do not contain minimal cutsets at all. [See D.Higgs (1985.a) and J.Ginsburg (1984)]. However using the same idea, of a systematic reduction of the cutsets, as in the finite case, I.Rival and N.Zaguia (1985.b) have proved the following

THEOREM 3.3. *A regular ordered set with a chain condition can be expressed as a union of minimal cutsets.*

Theorem 3.3, obviously fails for arbitrary ordered sets. However I.Rival and N.Zaguia (1985.b) have conjectured that it is still true for regular ordered sets.

An antichain in an ordered set P meets every maximal chain at most once, and a cutset of P meets every maximal chain at least once. Antichains and cutsets can structurally be completely different. Still, whenever an antichain A in P is a cutset, A should be a maximal antichain. Also, whenever a cutset K of P is an antichain, K should be a minimal cutset. In the last part of this section, we shall recall characterizations of ordered sets in which maximal antichains and minimal cutsets are related. Again, most of the results are in terms of "forbidden" configurations in the ordered set. The ordered set $\underset{\sim}{N}$ plays an important role in these characterizations.

A finite ordered set is called *series-parallel* if it can be constructed from the empty set using singletons by a sequence of operations each either a linear sum ("series") or a disjoint sum ("parallel"). Finite series-parallel ordered sets form a subclass of finite $\underset{\sim}{N}$-free ordered sets. A reason for this is that a series-parallel ordered set contains no subset isomorphic to $\underset{\sim}{N}$ at all. In fact a finite ordered set is series-parallel if and only if it contains no subsets isomorphic to $\underset{\sim}{N}$. D.P.Sumner (1973) credits this result to D.J.Foulis [see also J.Valdes (1978), J.Valdes, R.E.Tarjan and E.L.Lawler (1982), and D.Kelly (1985)]. The series-parallel character of a finite ordered set P is determined by examining the direct comparability graph of P for $\underset{\sim}{N}$'s while the $\underset{\sim}{N}$-free character of P is determined by examining its directed covering graph (that is its diagram) for $\underset{\sim}{N}$'s. In a survey article in this volume, I. Rival (1985) has considered the ubiquity of the letter $\underset{\sim}{N}$ in the theory of ordered sets.

I.Rival and N.Zaguia (1985.a) have shown that in the ordered sets which contain no subsets isomorphic to $\underset{\sim}{N}$, every finite minimal cutset is an antichain. This result has recently been extended to arbitrary minimal cutsets by D.Higgs (1985.a). The converse of this theorem fails for arbitrary ordered sets. D.Higgs (1985.a) constructed an ordered set Q which is an up-tree (that is, for every $x \in Q$, $\{y \in Q \mid y \geq x\}$ is a chain), and such that Q has no minimal cutsets at all. [See also J.Ginsburg (1984).] So, we consider the ordered set $P \cong \underset{\sim}{1} \oplus (Q + \underset{\sim}{N}) \oplus \underset{\sim}{1}$, where $\underset{\sim}{1}$ is the one-element ordered set, $\oplus$ denotes the linear sum and + denotes the disjoint sum. Then obviously, every minimal cutset in P which is not either the minimum element or the maximum element of P, in turn contains a subset which is a minimal cutset in Q. This contradiction proves that the only minimal cutsets in P are the minimum element and the maximum element of P. Therefore, the ordered set P gives the desired counterexample, since P is not $\underset{\sim}{N}$-free.

After proving that the converse holds for finite ordered sets, D.Higgs (1985.a) has conjectured that this is true for all regular ordered sets. I.Rival and N.Zaguia (1985.b) gave a partial solution.

Theorem 3.4. *Let P be a regular ordered set with a chain condition, such that every minimal cutset in P is an antichain. Then P contains no subset isomorphic to* $\underset{\sim}{N}$.

The proof for the finite case and the proof for regular ordered sets with a chain condition are closely related to the fact that in both cases every element belongs to a minimal cutset. In fact, if we could generalize Theorem 3.3. to regular ordered sets, then the same proof of

Theorem 3.4 could be generalized to regular ordered sets too.

Finally, we notice that I.Rival and N.Zaguia (1985.b) adapt the proof of Grillet's Theorem, mentioned earlier, to prove a necessary and sufficient condition, that every maximal antichain is a cutset. In an ordered set P, every maximal antichain meets every maximal chain if and only if P contains no generalized $\underset{\sim}{N}$'s .

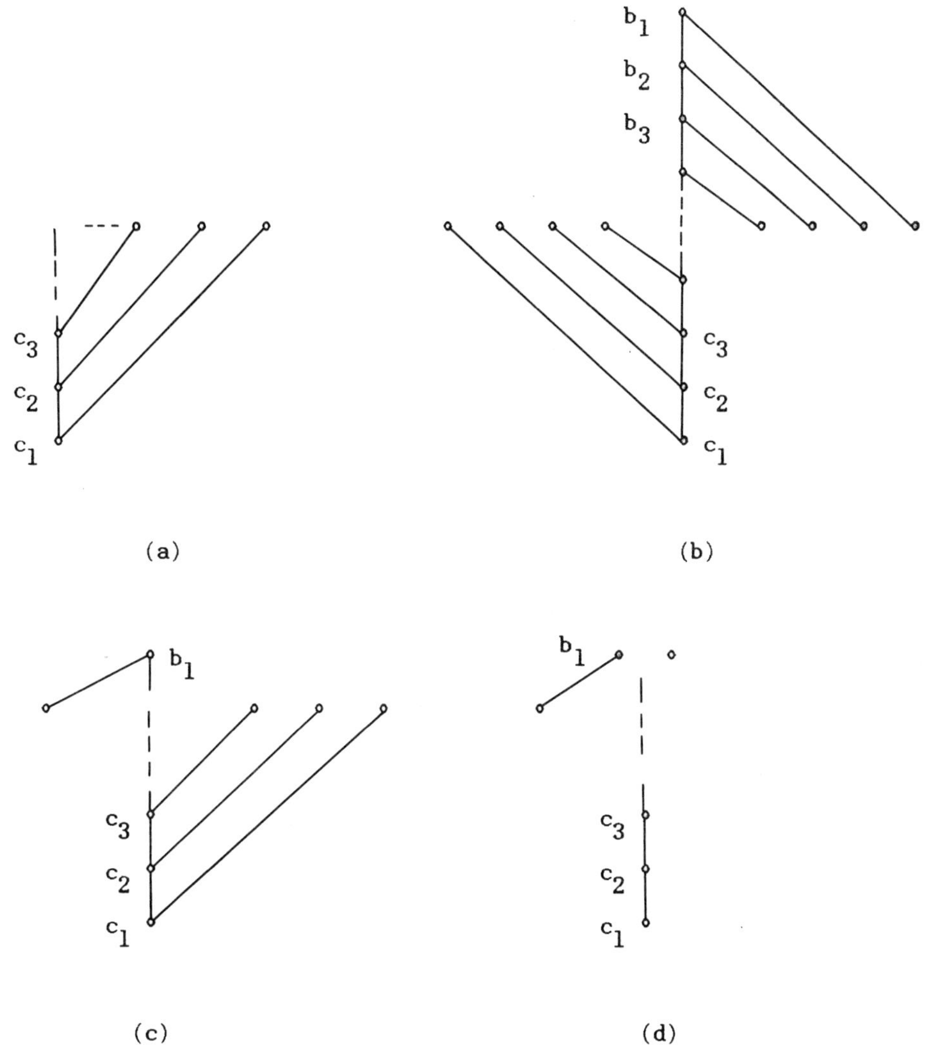

Figure 7.

A *generalized* $\underset{\sim}{N}$ is either an $\underset{\sim}{N}$ or one of the ordered sets illustrated in Figure 7 or their duals. We say, P contains a generalized $\underset{\sim}{N}$ if there is a subset of P isomorphic to a generalized $\underset{\sim}{N}$, such that the chain $\{b_1, b_2, \ldots\}$ covers the chain $\{c_1, c_2, \ldots\}$ in P itself.[Let A_1 and A_2 be subsets of an ordered set P. We write $A_1 < A_2$ if, for every $u \in A_1$ and $v \in A_2$, $u < v$. We say that A_2 covers A_1 if $A_1 < A_2$ and there is no x in P suth that $A_1 < \{x\} < A_2$. (Note that by definition $\phi < A < \phi$ for any subset $A \subseteq P$.)

Our aim in this section was to characterize the ordered sets in which every element has an antichain cutset. Naturally we may ask about cutsets with other structures. For instance, what about the ordered sets in which every element has a chain cutset? More precisely, which structural properties do these ordered sets have in common? Actually, J.Ginsburg, I.Rival and B.Sands (1985) have asked whether it is true that the width of an ordered set in which every element has an chain cutset is at most four. The ordered set illustrated in Figure 8, has width four and every element of it has a chain cutset. In the case of lattices, somewhat more is true than this conjecture. Indeed, N.Zaguia (1985) has shown that the lattices in which every element has a chain cutset are

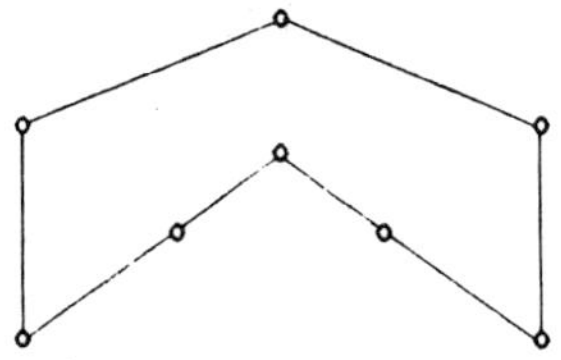

Figure 8.

exactly the lattices of width two (these are the lattices with the one-cutset property).

4. SIZE OF CUTSETS AND ANTICHAINS

J.Ginsburg, I.Rival and B.Sands (1985) proved that an ordered set with the finite cutset property which contains an infinite antichain must contain an infinite chain. They give a particularly interesting proof which uses the compactness of $\mathcal{M}(P)$.

THEOREM 4.1. *Let P be an ordered set with the finite cutset property. If all chains of P are finite then P itself is finite.*

Proof. Let C be a maximal chain in P. We have $\{C\} = \bigcap \{ B(x) \mid x \in C \}$. This is a finite intersection of open sets in $\mathcal{M}(P)$. Therefore $\{C\}$ is open in $\mathcal{M}(P)$. In other words $\mathcal{M}(P)$ is discrete. However a discrete compact space must be finite. This implies that P is finite too. □

Theorem 4.1 leads us to raise the following problem.

PROBLEM 4.2. *Determine $f(\alpha,m)$, the minimum length of an ordered set having width α and which has the m-cutset property but not the n-cutset property for $n<m$.*

It is easy to prove that an ordered set with the one-cutset property must satisfy $\omega(P) \leq 2$. To see that, let $\{a_1,a_2,a_3\}$ be an antichain in P, and let F_i ,i=1,2,3 denote the cutsets of a_i. Then, without loss of generality, we can assume that $b_1 > a_2$, a_3 and $b_2 > a_1$, a_3. Let C be a maximal chain containing b_1 and a_3. Necessarely, $b_2 \in C$. Therefore b_1 and b_2 are comparable. But $b_1 \leq b_2$ implies $a_2 < b_2$ and $b_2 \leq b_1$ implies $a_1 < b_1$ which is a contradiction.

A related result was given by J.Ginsburg, I.Rival and B.Sands (1985). They proved that a countably chain complete ordered set with the finite cutset property contains no uncountable antichain. An ordered set is countably chain complete if every countable chain has a supremum and an infimum. S.Todorcevic (1985) has shown a nice generalisation that, in a countably chain complete ordered set with the finite cutset property, every uncountable subset contains an uncountable chain. This fails for arbitrary ordered sets. The ordered set illustrated in Figure 9 has the two-cutset property and yet it contains a κ-element antichain.

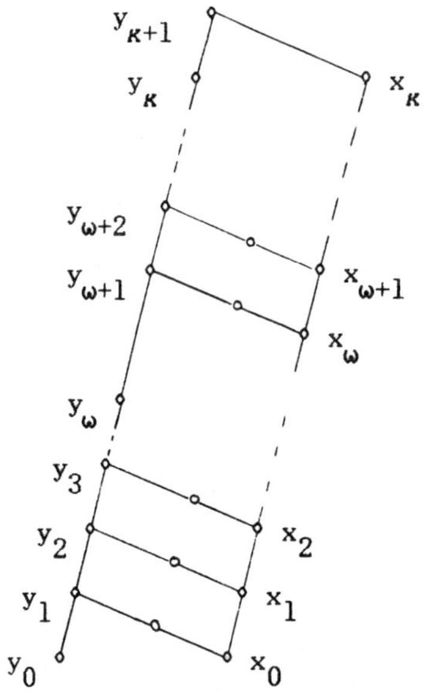

Figure 9.

Problem 4.2. is still open. In fact the only known value is $f(\alpha,2) = \alpha-3$ for a nonnegative integer $\alpha\geq 3$. This was proved by J.Ginsburg and B.Sands (1985) and, independently, by M.H.El-Zahar and N.Sauer (1985). Actually, Ginsburg and Sands used a graph theoretic approach to prove a stronger result : in a finite ordered set having the two-cutset property and width α, each element belongs to a chain with length at least α-3. El-Zahar and N.Sauer used a new concept of simple ordered sets. A finite ordered set P is said to be *simple* if the number of maximal chains of P is equal to its width. Let A be a maximum-sized antichain in P. Then P is simple if and only if each maximal chain of P meets A and every element $a \in A$ belongs to exactly one maximal chain.

These conditions can be reformulated as follows. Let

$$A^{+} = \{x \in P \mid x \geq a \text{ for some } a \in A\},$$

and

$$A^{-} = \{x \in P \mid x \leq a \text{ for some } a \in P\}.$$

Then P is simple if and only if the following two conditions are satisfied:

(i) each element $a \in A^{+}$ (respectively $a \in A^{-}$) has at most one upper cover (respectively lower cover);

(ii)if x, $y \in P$ and x covers y then either $\{x,y\} \subseteq A^{+}$ or $\{x,y\} \subseteq A^{-}$.

Still there is a third way to describe simple ordered sets. Let $C_1, C_2, \ldots, C_{\omega(P)}$ be maximal chains which cover P. Then P is simple if and only if for each pair of comparable elements x, $y \in P$, there exists i, $1 \leq i \leq \omega(P)$, such that x, $y \in C_i$.

The class of simple ordered sets is particularly interesting for the study of cutsets for at least two reasons. First of all, the number of maximal chains in a simple ordered set is minimum. Secondly, every finite ordered set with the m-cutset property contains a simple ordered set Q which also has the m-cutset property and such that $\omega(P) = \omega(Q)$. Here Q need not to be an induced subset of P. Let x be an upper cover of y in P. We denote by $P-\{y-\!\!<x\}$ the ordered set obtained from P by deleting all comparabilities of the form $u < v$ such that $u \leq x$, $y \leq v$ and there is no elements z in P noncomparable to both x and y and $u < z < v$. We say that $P-\{y-\!\!<x\}$ is obtained from P by deleting the covering relation $y -\!\!< x$.

LEMMA 4.4. (M.H.El-Zahar and N.Sauer) *Let P be a finite ordered set having the m-cutset property. Then the following subsets of P also have the m-cutset property.*

(i) $P-\{x\}$ where $x \in P$, and x has exactly one lower and one upper cover.

(ii)$P-\{y-\!\!<x\}$ where $x \in P$, x has at least two lower covers and y has at least two upper covers.

Proof. (i) Let y_1 and y_2 denote respectively the lower and upper cover of x. Let $z \in P-\{x\}$ and let K be a cutset of z in P. If $x \notin K$ then K is also a cutset for z in $P-\{y-\langle x\}$. Assume that $x \in K$. If z is noncomparable to one of y_1 and y_2 , say z is noncomparable to y_1, then $(K-\{x\}) \cup \{y_1\}$ is a cutset for z in $P-\{x\}$. Finally assume that z is comparable to both y_1 and y_2. This implies that $y_1 < z < y_2$ since x and z are noncomparable. If C is a maximal chain of P containing x then $C-\{x\}$ is not a maximal chain in $P-\{x\}$. This shows that $K-\{x\}$ is a cutset for z in $P-\{x\}$.

(ii) It is sufficient to prove that each maximal chain of $P-\{ y-\langle x \}$ is also a maximal chain in P. This is clear if neither x nor y belongs to C. Let $x \in C$. Since x has at least two lower covers in P then x is not the minimum element of C. Therefore C is also a maximal chain of P. The case $y \in C$ follows by duality. □

THEOREM 4.5. (M.H.El-Zahar and N.Sauer) *Let P be a finite ordered set with the m-cutset property. Then P contains a simple ordered set Q which also has the m-cutset property and* $\omega(P) = \omega(Q)$.

Proof. Let $Q \subseteq P$ be minimal, with respect to the size and the number of comparalities, such that Q has the m-cutset property and $\omega(P) = \omega(Q)$. We shall prove that Q is simple. Let A be a maximum-sized antichain in Q and let

$$A^{+} = \{x \in Q \mid x \geq a \text{ for some } a \in A\},$$

and

$$A^{-} = \{x \in Q \mid x \leq a \text{ for some } a \in A\}.$$

Assume that Q is not simple and that there is an element $x \in A^{+}$ which has two upper covers. First we suppose that each $y \in Q$, $y > x$, y has exactly one lower cover. Let $Q' = Q-\{y \in Q \mid y > x\}$. Clearly $A \subseteq Q'$ and, hence, $\omega(Q) = \omega(Q)$. Consider an element $z \in Q'$ and let K denote its cutset in Q. If $K \subseteq Q'$ then K is also a cutset for z in Q'. Otherwise, K contains an element y, $y > x$. Then $(K \cap Q') \cup \{x\}$ is a cutset for z in Q'. This shows that Q' has the m-cutset property which contradicts the minimality of Q. Next we assume that there is an element $y \in Q$, $y > x$, such that y has at least two lower covers. We may assume that x and y are chosen such that the number of elements $z \in Q$ for which $z < z < y$ is minimum. This implies that if $x < z < y$ then z has exactly one lower cover and one upper cover. But then Lemma 4.4 (i), implies that $Q-\{z\}$ has the m-cutset property, which contradicts the minimality of Q. Therefore y covers x in Q. From Lemma 4.4 (ii), $Q-\{x -\!< y\}$ has the m-cutset property, which is again a contradiction. Thus each $x \in A^+$ has at most one upper cover and, by duality, each $x \in A^-$ has at most one lower cover in Q.

Assume that $x \in A^+-A^-$ and $y \in A^--A^+$ where $y -\!< x$. This implies that there are elements a and a' in A such that $a < x$ and $y < a'$. In other terms, x has two lower covers and y has two upper covers in Q. In view of Lemma 4.4 (ii), this contradicts the minimality of Q. □

Reducing the problem to simple ordered sets is the key to the proof of the principal result.

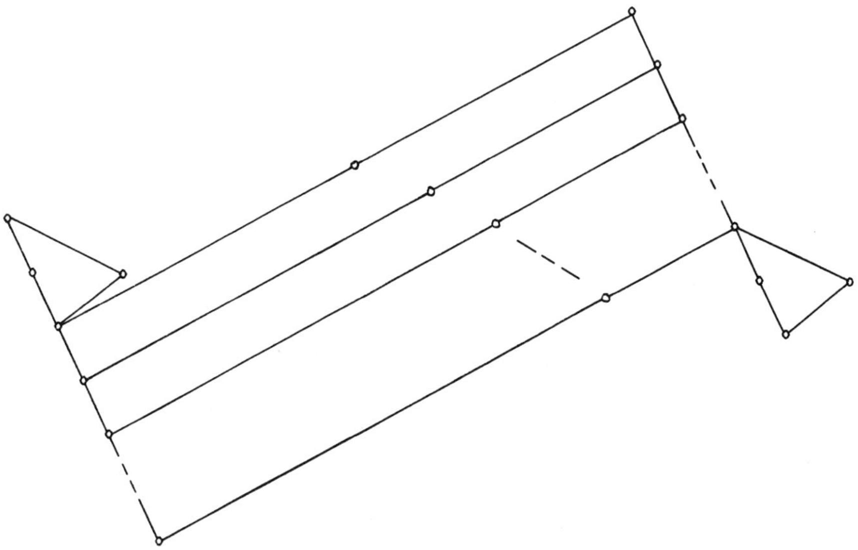

Figure 10

THEOREM 4.6. *Let P be a finite ordered set with the two-cutset property. Then* $\omega(P) \le l(P) + 3$.

The ordered sets illustrated in Figure 10 and Figure 11 are optimal examples for Theorem 4.6.

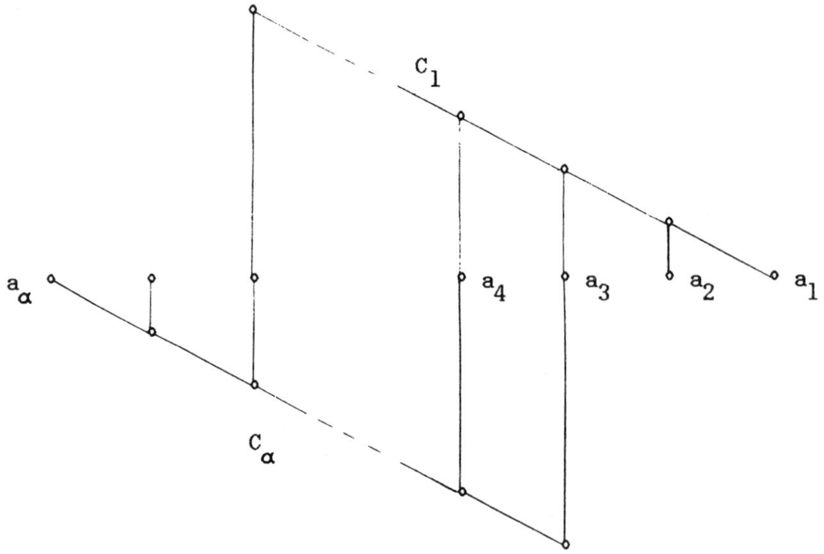

Figure 11

An interesting result related to Problem 4.2 was proved by N.Sauer and R.Woodrow (1984). In a chain complete ordered set with the finite cutset property, every element belongs to a finite maximal chain. This fails for arbitrary ordered sets (see Figure 12). The following more general question is still open.

PROBLEM 4.7. *Let* κ *be an infinite cardinal. Let P be a chain complete ordered set and assume that each element x in P has a cutset* $K(x)$ *where* $|K(x)| < \kappa$. *Is it then true that every x in P is contained in a maximal antichain of size less than* κ.

For this problem the assumption that κ is infinite is essential. The ordered set illustrated in Figure 9 has the two-cutset property yet, the element x belongs to no maximal antichain of size less than κ (for an infinite cardinal κ). Note that this ordered set is not chain

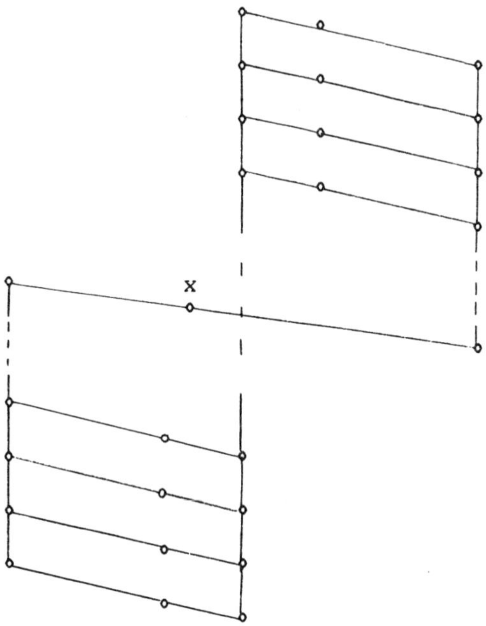

Figure 12

complete. The ordered set in Figure 13 is finite (hence chain complete), has the three-cutset property but the element x does not belong to a maximal antichain of size less than (k+2). Both examples were given by Sauer and Woodrow.

However the same authors have shown that in an ordered set with the two-cutset property, every element belongs to a maximal antichain of size at most four. The ordered set illustrated in Figure 14 ,which is due to N.Sauer and R.Woodrow (1984), shows that this result is best possible.

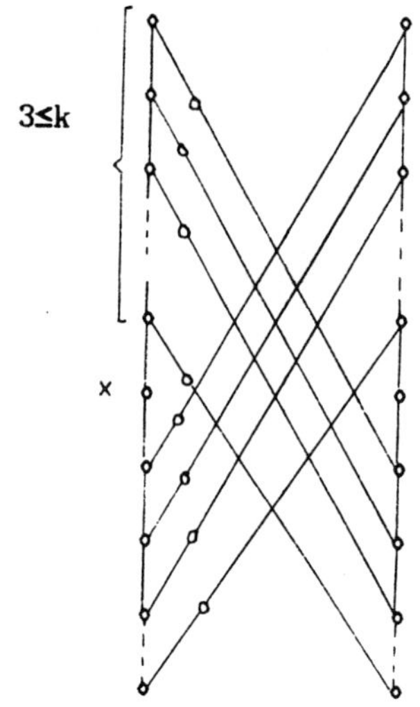

Figure 13

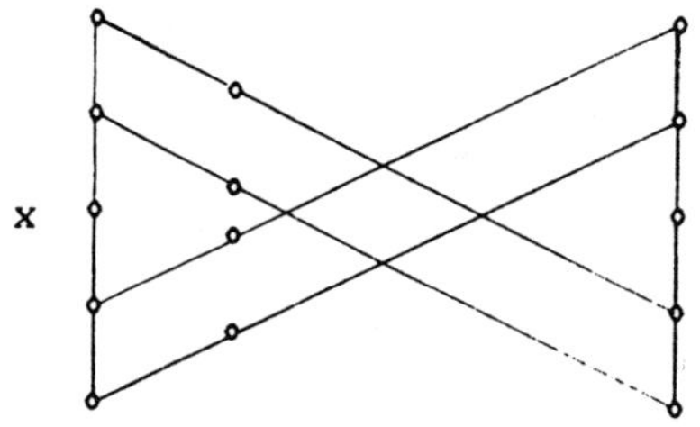

Figure 14

5. MINIMUM-SIZED CUTSETS

Dilworth's theorem can be rephrased as a minimax theorem, (that is theorem of the form : the maximum value of x is equal to the minimum value of y). It gives a structural property of some ordered sets as a function of the size of the maximum-sized antichain. A natural question is this.

What about the size of the minimum-sized cutset?

When the ordered set P is finite, the size of the minimum-sized cutset in P equals the size of the maximum-sized disjoint family of maximal chains in P. This fact can be viewed as a consequence of one of the most important theorems in graph theory, which is Menger's theorem : *Let X and Y be two sets of vertices in a finite graph (or digraph) G. Then the minimum size of an XY-separating set is equal to the maximum number of disjoint XY-paths (or directed paths) in G.* [K.Menger (1927)]

An XY-*path* is a path (or directed path) from some x in X to some y in Y passing through no other vertex of X or Y. (If $v \in X \cap Y$ then v itself is an XY-path.) A set of vertices Z is said to *separate* Y from X, or is an XY-*separating set*, if every XY-path contains a vertex of Z.

Now to prove the previous statement about the size of minimum-sized cutsets in a finite ordered set P, denote by G the directed covering graph of P, that is, the diagram of P. If X is the set of minimal elements in P and Y is the set of maximal elements in P, then the XY-paths in G are exactly the maximal chains in P. Also, the XY-separating sets in G are the cutsets in P. Therefore, from Menger's theorem it follows that the size of the minimum-sized cutset in P equals the maximum number of disjoint maximal chains in P.

For infinite graphs, P.Erdös proved a generalization of Menger's theorem for sets of vertices X and Y which have a finite separating set, [cf. D.König (1933)]. For infinite ordered sets, the problem seems much more difficult, even if we suppose that the size of minimum-sized cutsets is finite. Actually it is false for arbitrary ordered sets. In Figure 15, we have an ordered set in which every cutset has at least two elements and yet all the maximal chains intersect each other.

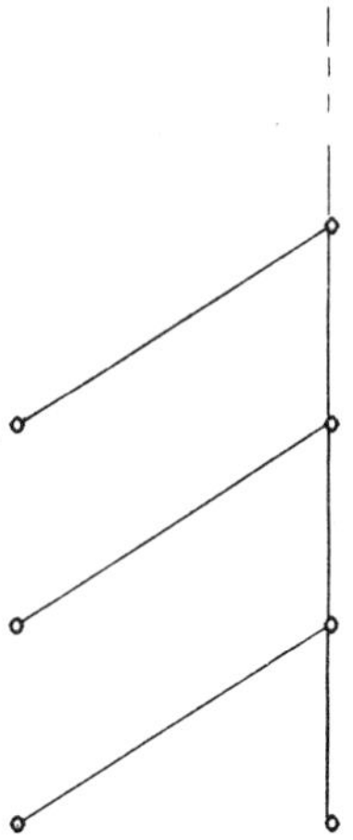

Figure 15

However, we expect that a completeness condition on the ordered set might be enough.

CONJECTURE [cf. N.Zaguia (1985)] *In a chain complete ordered set P, there is a family F of disjoint maximal chains and a cutset A that are in one-to-one correspondence. Each element of A is in a unique maximal chain of F and each maximal chain of F contains a unique element of A.*

For instance, B.Sands proved this conjecture in the very special case when all the maximal chains in the ordered set intersect each other. Finally, notice that a similar conjecture for infinite graphs has been proposed by P.Erdös [cf. D.R.Woodall (1978)].

We end this section with some results concerning the minimum-sized cutsets in lattices. We already know that for arbitrary ordered sets with the m-cutset property, there is no obvious relation between the width of the ordered set and m. However, R.Woodrow has verified that in a lattice with the two-cutset property every element belongs to a maximal antichain of size three. In fact, this antichain can be chosen as a cutset in the lattice [I.Rival and N.Zaguia (1985.a)]. However, for m $\geq$ 3 the situation is almost the same as in the case of arbitrary ordered sets. I.Rival and N.Zaguia (1985.a) have shown that for every positive integer k, there is a positive integer m and a finite lattice with the m-cutset property in which every nontrivial antichain has at least k.m elements (see Figure 16). By trivial antichain, we mean the $\underset{\sim}{0}$ or $\underset{\sim}{1}$ of the lattice.

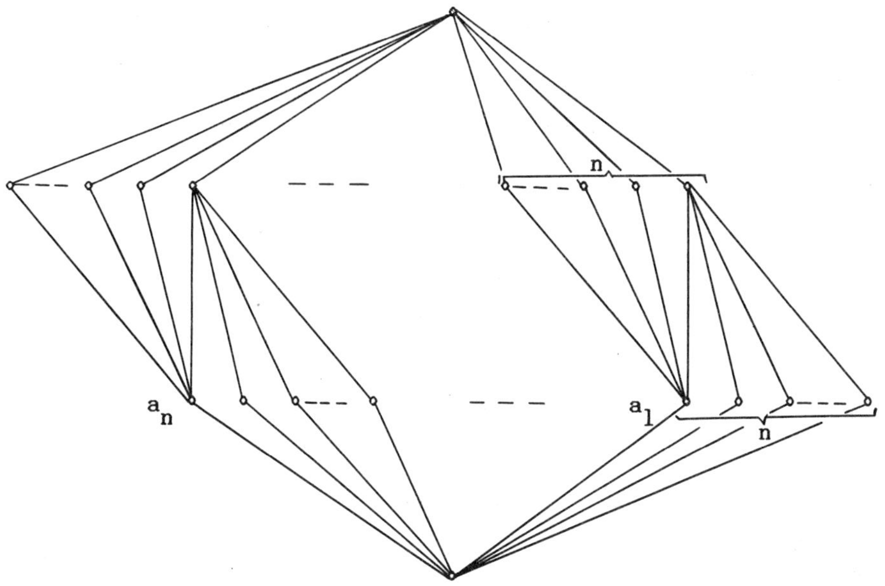

Figure 16

In the case of distributive lattices, the situation seems to be different, since the size of the maximal antichains imposes some constraints on the size of distributive lattice. In fact, J.Kahn and M.Saks have announced that they have a proof of the conjecture of B.Sands [see U.Faigle and B.Sands (1985)] which states that the ratio of the size of a finite distributive lattice and the width of the lattice approaches infinity when the size of the distributive lattice approaches infinity. It was conjectured that in a distributive lattice, if the minimum-sized cutset for an element x is m then x belongs to a maximal antichain of size m+1. [See N.Zaguia (1985).] However, when such an antichain exists, it need not be a cutset. To see this, we recall an example constructed by R.Nowakowski (1985). Let n be a positive integer and let $\underset{\sim}{2}^n$ be the lattice of all subsets of an n-element set ordered by inclusion. For each integer m such that $0 < m < n$, we denote by L_m the set of all elements in $\underset{\sim}{2}^n$ with height m, that is, the level m in $\underset{\sim}{2}^n$. In $\underset{\sim}{2}^n$ there are not many antichain cutsets at all. In fact I.Rival and N.Zaguia (1985.a) have shown that the only antichain cutsets in $\underset{\sim}{2}^n$ are the levels. Now, consider an element x in $\underset{\sim}{2}^n$ such that $h(x) = m < [n/2]$, and set $K(x) = \{y \in \underset{\sim}{2}^n \mid y \nleq x \text{ and } y \succ z \text{ for some } z < x\}$. It is simple to prove that $|K(x)| = (n-m)\ (2^n-1)$ and that K(x) is a cutset for x. Now, x has only one antichain cutset of size $\left[{n \atop m} \right]-1$ (where $\left[{j \atop i} \right]$ stands for the number of i-element subsets of a j-element set), and it is easy to check that if $n > 7$, $(n-m)\ (2^n-1) < \left[{n \atop m} \right]-1$.

Finally, notice that the last conjecture is true for the distributive lattice $\underset{\sim}{2}^n$. This is an easy consequence of a result announced by R.Nowakowski (1985) which states that if x is an element in $\underset{\sim}{2}^n$, for $n \geq 7$, such that $h(x) = m$ then the size of a minimum-sized cutset for x is $\min\ \{\ (2^m - 1)\ (n - m)\ ,\ (2^{n-m} - 1)\ m\ \}$.

REFERENCES

M.Behzad and Chartrand (1971) *Introduction to the Theory of Graphs*, Allyn and Bacon, Inc., Boston.

M.Bell (1982) The space of complete subgraphs of a graph, *Comm. Math. Univ. Carol.* 23 , 3, 525-536.

M.Bell and J.Ginsburg (1984) Compact spaces and spaces of maximal complete subgraphs, *Trans. Amer. Math. Soc.*, 283, 329-338.

R.P.Dilworth (1950) A decomposition theorem for partially ordered sets, *Ann. Math.* 51, 161-166.

M.H.El-Zahar and N.Sauer (1985) The length, the width and the cutset-number of finite partially ordered sets, *Order*, to appear.

D.R.Escalante (1972) Schnittverbände in Graphen, *Abh. Math. Sem. Humburg* 38, 199-220.

U.Faigle and B.Sands (1985) A size-width inequality for distributive lattices, preprint.

P.C.Gilmore and A.J.Hoffman (1964) A characterization of comparability graphs and of interval graphs, *Can. J. Math.* 16, 539-548.

J.Ginsburg (1984) Compactness and subsets of ordered sets that meet all maximal chains, *Order* 1, 147-157.

J.Ginsburg, I.Rival and B.Sands (1985) Antichains and finite sets that meet all maximal chains, preprint

J.Ginsburg and B.Sands (1985) Chains and antichains in partially ordered sets with the 3-cutset property, preprint

P.A.Grillet (1969) Maximal chains and antichains, *Fund. Math.* 15, 157-167.

D.Higgs (1985.a) A companion to Grillet's theorem on maximal chains and antichains, *Order* 1,

D.Higgs (1985.b) Lattices of cross-cuts I, preprint.

D.Kelly (1985) Comparability graphs, in *Graphs and Order* (I.Rival, ed.), D.Reidel, Dordrecht, pp. 3-40.

K.M.Koh (1983) On the lattices of maximum-sized antichains of a finite poset, *Algebra Universalis* 17, 73-86.

D.König (1933) Über Trennende Knotenpunkte in Graphen (nebst Anwendungen auf Determinanten und Matrizen), *Acta Litt. Sci. Szeged* 6, 155-179.

B.Leclerc and B.Monjardet (1973) Orders "C.A.C.", *Fund. Math.* 79, 11-22.

K.Menger (1927) Zur allgemeinen Kurventheorie, *Fund. Math.* 10, 96-115.

B.Meyer (1982) On the lattices of cutsets in finite graphs, *Europ. J. Comb.* 3, 153-157.

R.Nowakowski (1985) Cutsets of Boolean Lattices, preprint.

I.Rival (1985) Stories about the letter $\underset{\sim}{N}$, this volume.

I.Rival and N.Zaguia (1985.a) Antichain cutsets, *Order* 1, 235-247.

I.Rival and N.Zaguia (1985.b) Effective constructions of cutsets for finite and infinite ordered sets, preprint.

B.Sands (1985) Personal communication.

N.Sauer and R.E.Woodrow (1984) Finite cutsets and finite antichains, *Order* 1, 35-46.

D.P.Sumner (1973) Graphs indecomposable with respect to the X-join, *Disc. Math.* 6, 281-298.

S.Todorcevic (1985) Posets with finite cutsets, preprint.

J.Valdes (1978) Parsing flowcharts and series-parallel graphs, *Technical Report* STAN-CS-78-682, Stanford.

J.Valdes.,R.E.Tarjan and E.L.Lawler (1982) The recognition of series-parallel diagraphs, *SIAM J. Comp.* 11, 298-313.

D.R.Woodall (1978) Minimax theorems in graph theory, in *Selected Topics in Graph Theory*, (L.W.Beineke and R.J.Wilson, eds.), Acad. Press, 237-269.

R.E.Woodrow (1984) Personal communication.

N.Zaguia (1985) *Ph. D. dissertation*, The University of Calgary.

Contemporary Mathematics
Volume 57, 1986

STORIES ABOUT ORDER AND THE LETTER **N** (*en*)

by

Ivan Rival

Abstract. Because of its similarity in shape to the upper case, fourteenth letter of the modern English alphabet, this common four-element ordered set has come to be known as an '**N**' (*en*). It plays a heuristic role in the theory or ordered sets, for instance, in problems about the structure of subsets that meet each maximal chain -- cutsets. It figures, too, in combinatorial optimization problems as a convenient description of classes of orders for which effective solutions can be found -- **N**-free ordered sets. The recent 'stories' about the letter '**N**' feature subdiagrams of the diagram of an ordered set.

INTRODUCTION

The letter **N** (*en*) is the fourteenth of the modern English alphabet. Historically it represents the Greek *nu* (ν) and the Semitic *nun* (נ). It has preserved its original form, with little change from the earlier Greek forms ᴎ and ͷ, corresponding to the Phoenician ɣ. The letter is common in English and is silent only infrequently (e.g. 'damn'). Of course, too, there are systems of reference for this letter, as for them all, which, in the absence of 'pen and ink' have exploited what resources were at hand (cf. Figure 1).

0271-4132/86 $1.00 + $.25 per page

□ ■

Flag code   Semaphore   Morse code

Figure 1

In mathematics, the simplest and first conception of 'order' pertains to the natural numbers $0 < 1 < 2 < \ldots$. The lower case **n** (*en*) is almost universally used to denote a *natural number*[1]. Its upper case **N** (and often **ℕ**) usually stands for the ordered set of natural numbers; pictorially it is found represented horizontally,

$$\underset{0\quad\ 1\quad\ 2\qquad n\quad\ n+1}{\circ\!\!-\!\!-\!\!\circ\!\!-\!\!-\!\!\circ\!\!-\cdots-\!\!\circ\!\!-\!\!-\!\!\circ\!\!-\ \ \cdots}$$

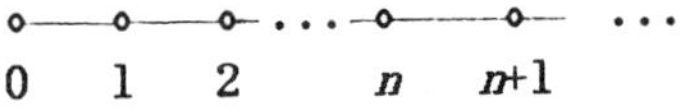

as well as vertically (see Figure 2). Though it disrupts the horizontal flow of text this vertical pictorial representation is more common today.

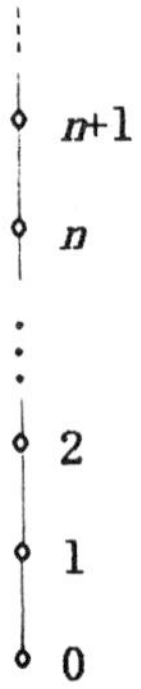

Figure 2

The letter **N** goes beyond such simple convention and usage in mathematics, and especially so in the theory of ordered sets. There is actually much more to say. For one thing, there are other orders on the set of natural numbers which are, in shape, closely related to the letter **N**. Take for instance the divisibility relation on the set $\{2,3,8,12\}$ of natural numbers. Untangling a vertical

---

[1]This is a credible convention also for both the French (*entier naturel*) as well as the German (*naturliche Zahl*).

pictorial representation of it leaves a distinct resemblence to this letter's shape (see Figure 3). The inclusion relation too is pictorially associated with the letter – take the elements {1}, {3}, {1,2,3}, {1,2,4} from among the set of all subsets of {1,2,3,4} ordered by inclusion (see Figure 4). This four-element ordered set has come to be known as an '**N**'. As we shall see it plays a heuristic role in the theory of ordered sets. In recent years this association between *order* and the *letter* **N** bears particularly on a further feature of the order's 'diagonal' comparability, and the letter's 'diagonal' slash: the diagonal pair of vertices of a subset, identified as an **N**, should be in the covering relation. Thus, in Figure 5 {2,3,6,8} is an **N** but, since 2 < 6 < 12, {2,3,8,12} is not **N**. Similarly, in Figure 6, {{1},{3},{1,3},{1,2,4}} is an **N**, but {{1},{3},{1,2,3},{1,2,4}} is not. In the usual pictorial schemes this may be emphasized by a convention of double lines.

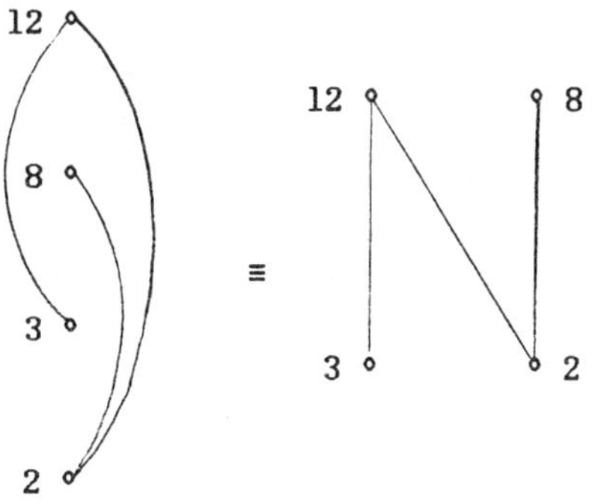

Figure 3

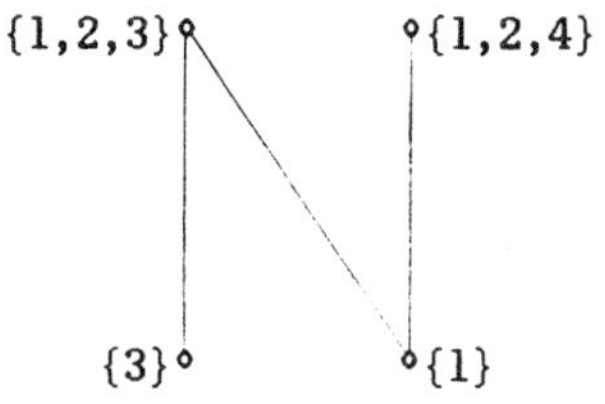

Figure 4

The feature seems to be a quite important one!

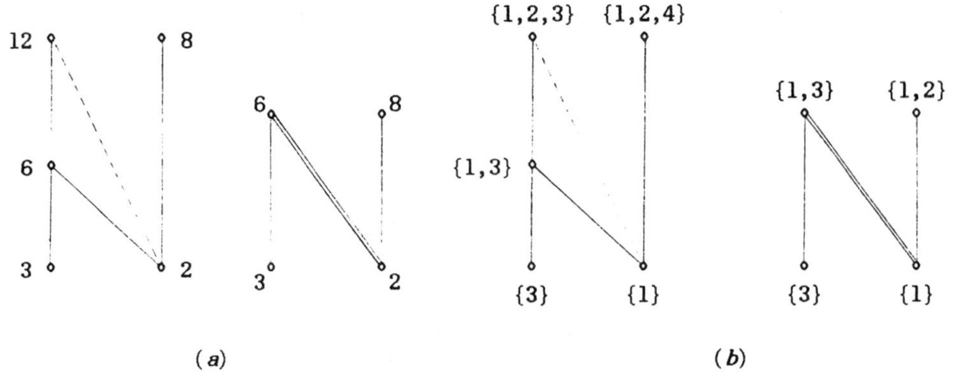

Figure 5

SERIES-PARALLEL ORDERED SETS

The letter **N** in combinatorial optimization problems is a suggestive device to describe ordered sets for which effective solutions can be found. The first intimation of the importance of the letter occurs in the study of 'series-parallel ordered sets'. Call an ordered set *P series-parallel* if it can be constructed from singletons using the operations of disjoint sum (+) and linear sum (⊕) (see Figure 6). In other words, *P* can be decomposed into singletons using only disjoint sum and linear sum. All 'trees', for example, are series-parallel.

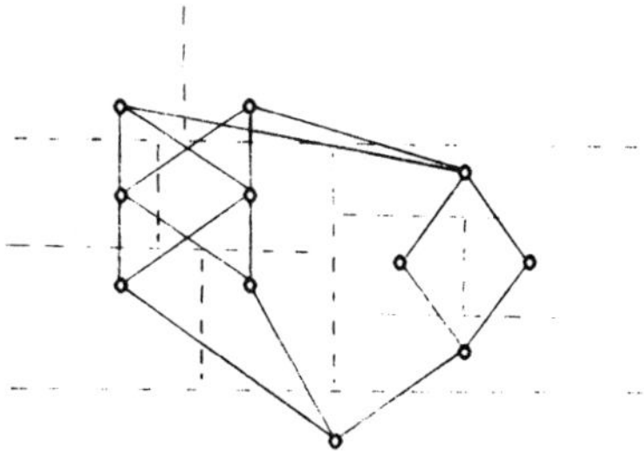

$P \cong (1 \oplus (((1 + 1) \oplus (1 + 1)) + (1 \oplus (1 + 1) \oplus 1)) \oplus (1 + 1)$

Figure 6

SERIES-PARALLEL-**N** THEOREM. *A finite ordered set is series-parallel if and only if it contains no subset isomorphic to* **N**. [29], [33].

Proof. Let P be series-parallel. Then P cannot contain a subset isomorphic to **N** for the elements of such a subset could not be decomposed using + and ⊕ into singletons.

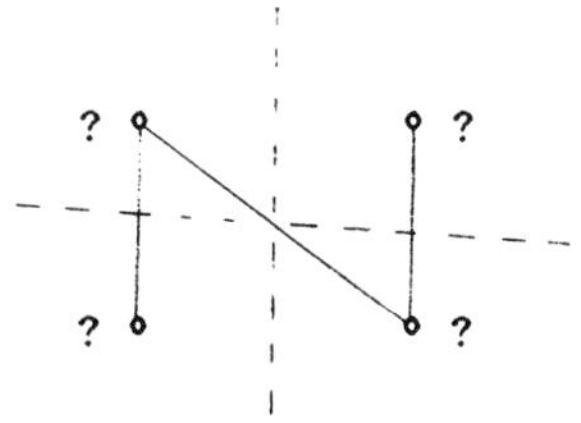

Figure 7

Let P be a finite ordered set which contains no subset isomorphic to **N**. We show, by induction on $|P|$ that, P can be decomposed into singletons using + and ⊕.

If P is not connected then $P = P_1 + P_2$.

Let P be connected. Let x be a maximal element in P and let y be a minimal element in P. Then there is a shortest 'zig-zag' $x = z_0 > z_1 < z_2 > \dots < z_{k-1} > z_k = y$, $k \geq 1$, connecting x and y. As P contains no subset isomorphic to **N**, $k = 1$, that is, $x > y$. In other words, every maximal element is above every minimal element.

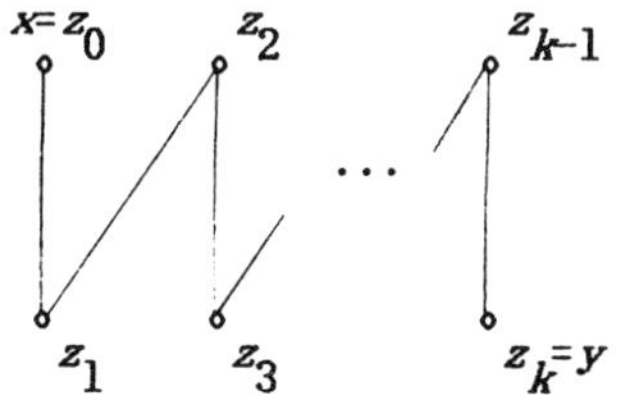

Figure 8

Let $M = \min P$ the set of all minimal elements of P and, for $m \in M$, let U_m denote the up set of all elements $x \geq m$. Set

$$I = \bigcap_{m \in M} U_m .$$

Now, $I \neq \phi$; in fact, every maximal element of P belongs to I. Also, $I \cap M = \phi$, otherwise $|M| = 1$ and $P \cong M \oplus (P - M)$ so we could apply the induction hypothesis. In particular, $P - I \neq \phi$ too.

Finally, we show that

$$P \cong (P - I) \oplus I.$$

Suppose there is $x \in P - I$ and $y \in I$ such that $x \nleq y$. Evidently, $x \notin M$ so choose a minimal element m satisfying $m \leq x$. Of course, $m \leq y$ too. Moreover, from $x \notin I$, it follows that there is another minimal element m' satisfying $m' \nleq x$ although $m' < y$. Then, $\{m', y, m, x\}$ is a subset isomorphic to **N**. □

This construction scheme renders full service in several classical scheduling problems. For instance, suppose that a set P of jobs subject to precedence constraints, is processed on a single machine and, in addition to a processing time p_j each job x_j in P has a specified weight w_j. A classical problem is to find a schedule which minimizes the weighted sum $\sum_j w_j p_j$ of the job completion times. In the special case that the ordered set P is an antichain there is a famous result in scheduling theory called the *ratio rule* [27] according to which it is possible to find an 'optimal' schedule 'effectively' by sequencing the jobs in order of nondecreasing ratios

$$\frac{p_{j_1}}{w_{j_1}} \leq \frac{p_{j_2}}{w_{j_2}} \leq \dots .$$

If the (finite) ordered set P is arbitrary then the problem becomes *NP*-hard [15], [18]. An important exception is the class of series-parallel ordered sets. In fact, the ratio rule has been generalized to apply to series-parallel ordered sets [15] and the algorithm has even been generalized to apply to a variety of other sequencing problems too [16], [20]. The literature is rich in results, with worn epithets, well-known to the operations research cognoscenti. Among these are the single machine problems of minimizing *total weighted discounted completion time* [17], *expected cost of fault detection* [11], [20], *minimum initial resource requirement* [1], [20] or the two machine *permutation flow shop problem with time lags* [26].

The algorithmic character of the series-parallel idea arises even in work on infinite ordered sets. Some aspects of recursion theory follow the lines of the mathematics of the finite especially where it concerns (infinite) ordered sets which can be mechanically determined or whose 'elementary sentences' can be described by an algorithm.

Loosely speaking, an ordered set is *recursive* provided that both its underlying set and its order are algorithmically recognizable. For a class $\mathcal{X}$ of ordered sets, the *elementary theory* Th$\mathcal{X}$ of $\mathcal{X}$ is the set of all elementary sentences true in every member of $\mathcal{X}$. An elementary theory is *decidable* provided that it is recursive, that is, there is an algorithm to recognize the elementary sentences true in this theory (cf. [19]).

Here is a striking result from the realm of infinite, recursive ordered sets. For a finite ordered set P let $\mathcal{X}_P$ denote the class of all ordered sets which have no subset isomorphic to P.

Th$\mathcal{X}_P$ *is decidable if and only if* P *is series-parallel.* [25].

N-FREE ORDERED SETS

The contemporary stories about the letter are concerned with the 'covering relation'. Say that *a covers b* (*b is covered by a*, *a* is an *upper cover* of *b*, or *b* is a *lower cover* of *a*) if, for any element c, $a > c \geq b$ implies $c = b$. We write $a \succ b$. An ordered set is **N**-*free* if it contains no cover-preserving subset isomorphic to **N**. The distinction with series-parallel is important and it can be neatly cast in the language ot the graphical schemes used to pictorially represent a (finite) ordered set: the comparability graph and the diagram. Whereas the series-parallel character

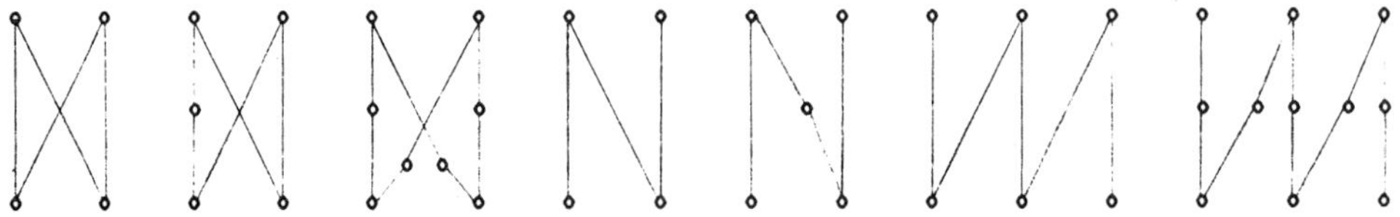

(*a*) (*b*) (*c*) (*d*) (*e*) (*f*) (*g*)

Each of the ordered sets (*a*), (*c*), (*e*), (*g*) is **N**-free; each of (*b*), (*d*), (*f*) contains a subdiagram isomorphic to **N**.

Figure 9

of an ordered set is determined by the non-containment of **N** in its directed comparability graph (as a subset (order induced) isomorphic to **N**), its **N**-free character is determined by the non-containment of **N** in its diagram (as a cover-preserving subset isomorphic to **N** or, equivalently, 'subdiagram' isomorphic to **N**). Every series-parallel ordered set is **N**-free yet, every finite ordered set can be embedded as a subset in an **N**-free ordered set -- a 'subdivision' of every edge in the diagram of an ordered set, produces an **N**-free one.

An elegant early result in the theory of **N**-free ordered sets is this structure theorem.

CHAIN-MEETS-ANTICHAIN THEOREM. *In a finite ordered set every maximal chain meets every maximal antichain if and only if it is* **N**-*free.* [13].

Proof. Suppose P contains a subdiagram $\{a < c \succ b < d\}$ isomorphic to **N**. Extend $\{a, d\}$ to a maximal antichain A and extend $\{b, c\}$ to a maximal chain C. Then $C \cap A = \phi$.

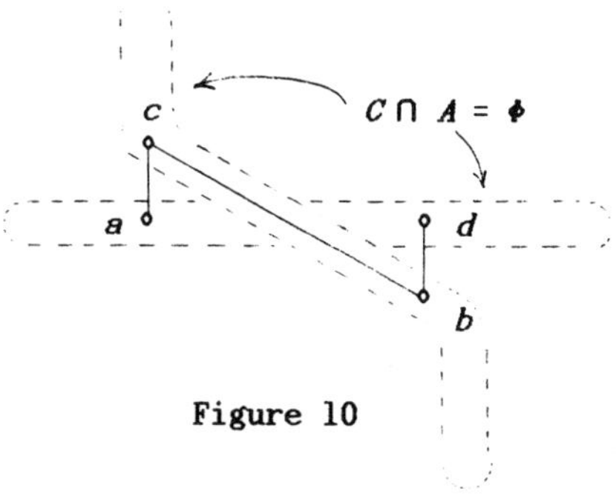

Figure 10

Suppose that there is a maximal antichain A which is disjoint from some maximal chain C. Let $C = \{c_1 < c_2 < \ldots\}$. Each c_i is comparable to some $a_i \in A$. Since C is maximal it contains elements 'above' A and also elements 'below'. Let

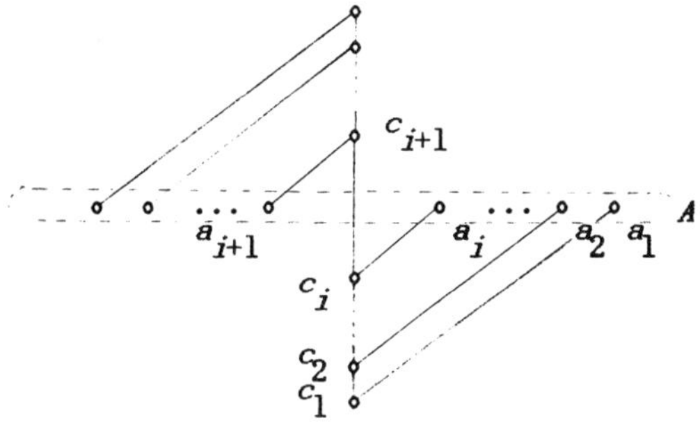

Figure 11

$$c_i = \max\{c_j \in C \mid c_j \leq a \text{ for some } a \in A\}.$$

Then $\{a_{i+1}, c_{i+1}, c_i, a_i\}$ is isomorphic to **N**. Evidently, c_{i+1} covers c_i and, if we choose a and b satisfying $c_{i+1} \succ a'_{i+1} \geq a_{i+1}$ and $a_i \geq a'_i \succ c_i$ then $\{a'_{i+1}, c_{i+1}, c_i, a'_i\}$ is a subdiagram isomorphic to **N**. □

This structure theory has evolved considerably in recent years [2], (cf. [9]). Call a subset A of P a *cutset* if $A \cap C \neq \phi$ for each maximal chain C in P. If A is, in addition, an antichain then we call it an *antichain cutset.* Thus, min$P$ is an antichain cutset for every finite ordered set. In an **N**-free ordered set every maximal antichain is an antichain cutset. Thus, an **N**-free ordered set can be expressed as the union of (not necessarily disjoint) antichain cutsets. This is not true for every (finite) ordered set. There are even ordered sets - albeit infinite - which contain no antichain cutset at all! (See Figure 13.) A subset $\{x, a_1, c_1, a_2, c_2, \ldots, a_n, c_n\}$, $n \geq 2$, of P is an *alternating cover cycle of P* provided that it contains only the comparability relations indicated in Figure 14 and, $c_1 \succ a_1$, $c_2 \succ a_2$, $\ldots$, $c_n \succ a_n$ are covering relations. Thus, each of the ordered sets in Figure 11 and Figure 12 contains an alternating cover cycle $\{x, a_1, c_1, a_2, c_2\}$. The principal result is this 'Antichain Cutset Union Theorem'.

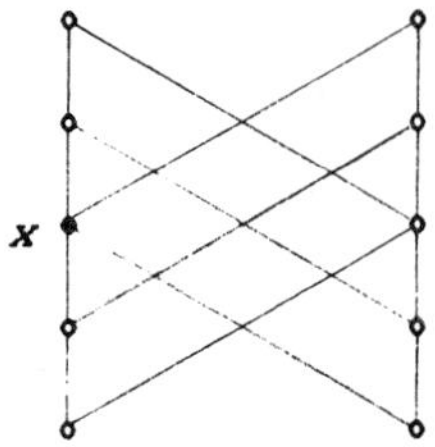

There is no antichain cutset containing x.

Figure 12

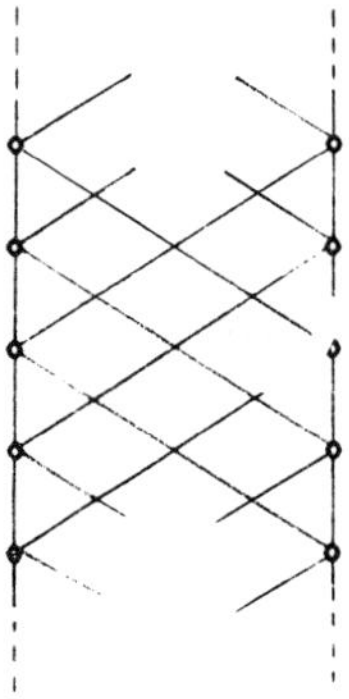

Figure 13

A finite ordered set is the union of antichain cutsets if and only if it contains no alternating-cover cycle [22].

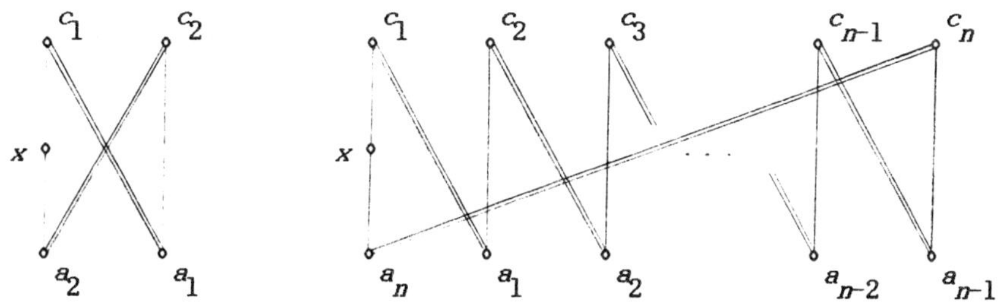

Figure 14

Insofar as **N**-free ordered sets constitute a much larger class than series-parallel it is interesting to turn to the combinatorial optimization issues, for which series-parallel ordered sets are so well-known and so well-suited. One of these problems is especially closely linked to the current interest in **N**-free ordered sets: the 'jump number problem'.

We are to schedule a set of jobs for processing, one at a time, by a single machine. Precedence constraints, due perhaps to technological limitations, prohibit the start of certain jobs until certain others are already completed. A job which is performed immediately after one which is not constrained to precede it requires some additional cost -- a 'setup' or a 'jump'. The simplest variation is already difficult enough: *schedule the jobs to minimize the number of jumps.*

In the language of ordered sets this is commonly rendered as follows. Let P be a finite ordered set and let $\mathcal{L}(P)$ stand for all of its linear extensions. For $L \in \mathcal{L}(P)$ let $\text{jump}(P, L)$ count the number of pairs (a, b) of elements of P such that $a \succ b$ *in* L and $a \not\geq b$ *in* P. Each such pair (a, b) is a *jump*. Put

$$\text{jump}(P) = \min\{s(P, L) \mid L \in \mathcal{L}(P)\}$$

the *jump number* of P. The problem is to *construct* $L \in \mathcal{L}(P)$ *such that* jump(P, L) = jump(P). It is the theme of a long list of articles over the last few years (e.g. [4], [21], [30], [8], [12], [32], [5], [6], [10], [24], [23], [7]).

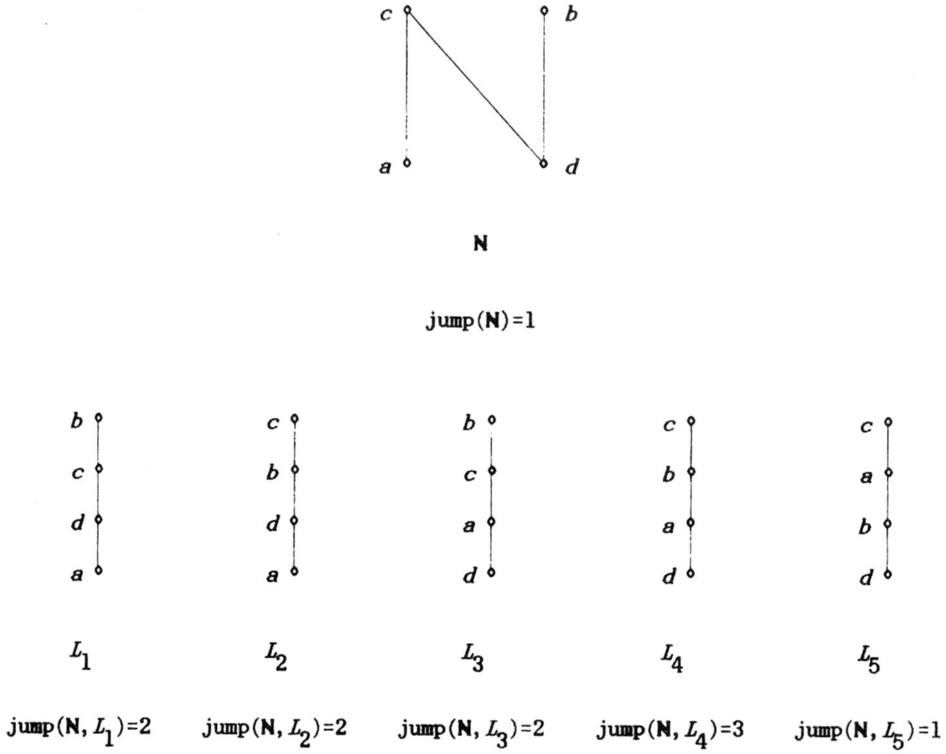

All of the linear extensions of **N**.

Figure 15

The algorithmic character of series-parallel ordered sets renders then particularly amenable. Actually, for series-parallel ordered sets there is a natural 'greedy' linear extension which is simple to construct and which is always *jump-optimal*, that is, jump(P, L) = jump(P). A *greedy linear extension* is a linear extension $x_1 < x_2 < \ldots$ of P such that $x_1 \in \min P$ and, for $i \geq 1$, $x_{i+1} \in \min(P - \{x_1, x_2, \ldots, x_i\})$ and, if possible, $x_{i+1} > x_i$. Thus, 'climb as high as you can'. For instance, in Figure 15, L_1, L_2 and L_5 are greedy linear extensions of **N** and L_3, L_4 are not. In fact, *for any series-parallel ordered set any greedy linear extension is jump-optimal* [4].

What about **N**-free ordered sets?

Indeed, this is now a fundamental result in the theory, the '**N**-Free Jump Number Theorem':

For a finite **N**-*free ordered set every greedy linear extension is jump-optimal. Indeed, in this case, every jump optimal linear extension is greedy too* [21].

There are already many quite different proofs (e.g. [21], [30], [32], [10])! The original proof uses a transformation technique which, in turn, is of independent interest and has wider application. Here is the heart it.

N-LEMMA [21]. *Let* P *be a finite ordered set, let* a, b $\in$ P *and let* L $\in \mathcal{L}$(P) *satisfy* a $\succ$ b(L) *but* a $\not\geq$ b(P). *Then either there is an* L' $\in \mathcal{L}$(P) *satisfying* a $<$ b(L') *and* jump(P,L') $\leq$ jump(P,L) *or,* {a' $<$ c $\succ$ d $<$ b'} *is a subdiagram of* P *isomorphic to* **N** *for some* a $\leq$ a' $\prec$ c *and* b $\geq$ b' $\succ$ d.

Proof. Let A be a chain in $P \cap L$ with bottom element a and maximal with respect to

$$x \in A \quad \text{implies} \quad x \not\geq b(P).$$

As $a \in A$, $A \neq \emptyset$. Let B be a chain in $P \cap L$ with top element b and maximal with respect to

$$y \in B \text{ implies } y \nleq a(P).$$

Write $L = C \oplus B \oplus A \oplus D$ and put $L' = C \oplus A \oplus B \oplus D$.

If $L \notin \mathcal{L}(P)$ then there is $x \in A$ and $y \in B$ such that $x > y(P)$. In this case let

$$c = \inf_P\{x \in A \mid x > y(P) \text{ some } y \in B\}$$

and

$$d = \sup_P\{y \in B \mid y < c(P).$$

According to the construction of A and B, $c > a(P)$ and $d < b(P)$. If $c > z \geq d(P)$ then, according to the construction of c, $z \notin A$. Then $z \in B$ and $z = d$. This shows that $c \succ d(P)$. Let $a' \in A$ satisfy $a' \prec c(L)$ and let $b' \in B$ satisfy $d \prec b'(L)$. Then $\{a', c, d, b'\}$ is a subdiagram isomorphic to **N**.

Thus, let $L \in \mathcal{L}(P)$.

Is $\text{jump}(P, L') \leq \text{jump}(P, L)$? Let $a' \succ \sup_L A(L)$ and let $b' \prec \inf_L B(L)$. Apart from the jump $(\inf_L B, \sup_L A)$ in L', the only possible jump in L', different from those in L, are (a', b) and (a, b'). Suppose $a' \ngtr b(P)$ but $a' > \sup_L A(P)$. From the construction of A it must be that $a' > b(P)$. It follows that if (a', b) is a jump in L' then $(a', \sup_L A)$ is a jump in L. Similarly if (a, b') is a jump in L' then $(\inf_L B, b')$ is a jump in L. Therefore $\text{jump}(P, L') \leq \text{jump}(P, L)$. □

The **N**-Lemma gives rise to a simple procedure to transform one linear extension into another. Let $L \in \mathcal{L}(P)$, let $a \succ b(L)$ such that $a \ngeq b(P)$, let A be a chain in $P \cap L$ with bottom element a and maximal with respect to, $x \in A$ implies $x \ngeq b(P)$, and let B be a chain in $P \cap L$ with top element b and maximal with respect to, $y \in B$ implies $y \nleq a(P)$. Then L can be expressed in the form

$$L = C \oplus B \oplus A \oplus D.$$

We say that

$$L(a/b) = C \oplus A \oplus B \oplus D$$

is *obtained from* L *by interchanging chains* or, $L(a/b)$ is a *chain interchange* of L. According to the **N**-Lemma it is a linear extension with no larger jump number than L, as long as $A \cup B$ is **N**-free in P. It may be applied, for instance, in an **N**-free ordered set to transform any greedy linear extension, by a sequence of chain interchanges, to any jump-optimal linear extension [21].

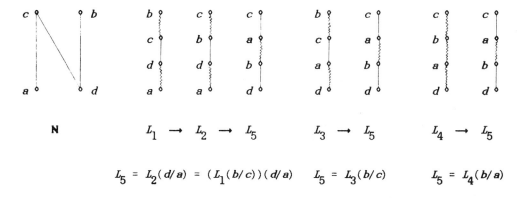

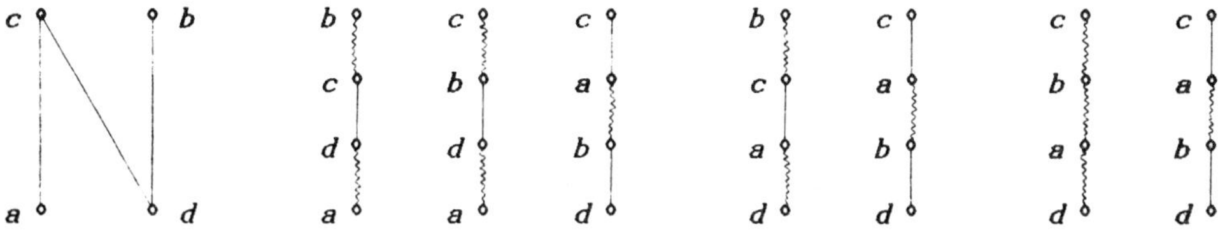

N $\quad L_1 \to L_2 \to L_5 \quad L_3 \to L_5 \quad L_4 \to L_5$

$L_5 = L_2(d/a) = (L_1(b/c))(d/a) \quad L_5 = L_3(b/c) \quad L_5 = L_4(b/a)$

Chain interchanges transform each of the linear extensions of **N** to the jump-optimal one. (A jump is illustrated as a 'squiggle'.)

Figure 16

The jump number problem has been uncommonly fruitful in advancing and, indeed, in unifying our understanding of ordered sets. Hitherto, for example, there has been no reason even to consider any common aspect of 'chain decompositions' and 'linear extensions'. Yet, there seems to be such a common ground. Let $C_1, C_2, \ldots, C_m$ be any sequence of disjoint chains of a finite ordered set P whose set union is all of P,

$$C_1 \cup C_2 \cup \ldots \cup C_m = P.$$

The linear sum of these chains

$$C_1 \oplus C_2 \oplus \ldots \oplus C_m$$

is a total order, although it need not be a linear extension of P. Is there a ***minimum*** chain decomposition whose corresponding linear sum is a *linear extension*? On the other hand, any linear extension L of P can be expressed as the linear sum of chains $C_1, C_2, \ldots, C_k$ in P,

$$L = C_1 \oplus C_2 \oplus \ldots \oplus C_k$$

so chosen that top $C_i \not\le$ bottom $C_{i+1}(P)$, for each $i = 1,2,\ldots k-1$. Of course, the union

$$C_1 \cup C_2 \cup \ldots \cup C_k = P.$$

Is there always a *linear extension* whose sequence of chains is a ***minimum*** chain decomposition? The answer is no. Actually, the minimum number of a sequence of disjoint chains whose linear sum is a linear extension is precisely

$$\text{jump}(P) + 1$$

and

$$\text{jump}(P) + 1 \geq \text{width}(P),$$

where width(P) is the maximum number of pairwise noncomparable elements of P.

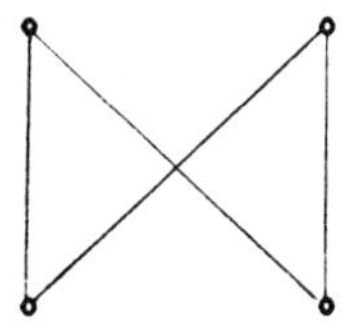

jump(P) + 1 = 3 > 2 = width(P)

Figure 17

In at least one important case these two fundamental constructions for ordered sets can be reconciled. According to the 'Decomposition Extension Theorem',

in a finite, cycle-free ordered set there is a ***minimum*** *chain decomposition which corresponds to a linear extension* [6].

In symbols, in a finite, cycle-free ordered set P,

$$\text{jump}(P) + 1 = \text{width}(P).$$

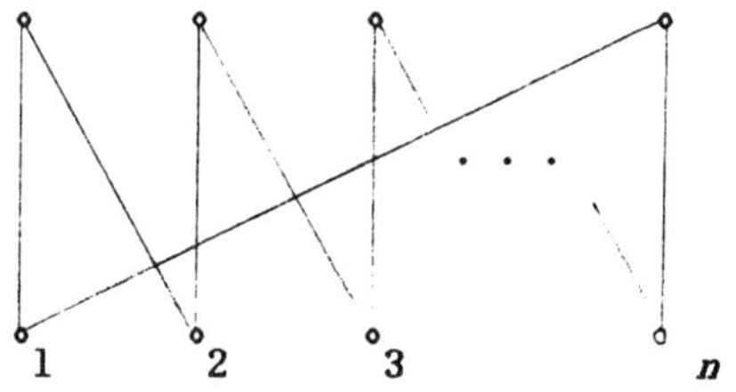

A cycle, $n \geq 2$

Figure 18

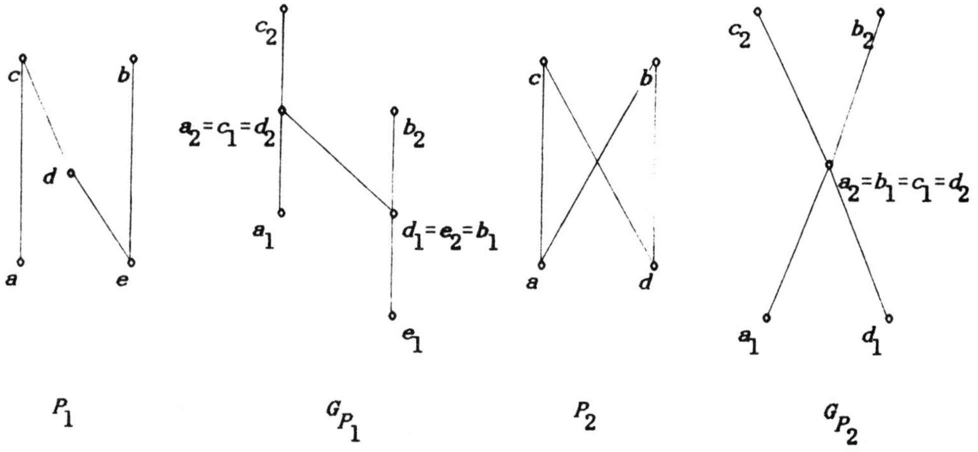

P_i is the line digraph of G_{P_i}. The vertices of P_i correspond to the edges of G_{P_i} and $x < y(P_i)$ if $x_2 = y_1(G_i)$.

Figure 19

Another proof of the **N**-Free Jump Number Theorem exploits the graph theoretical fact that *an ordered set is* **N**-free *if and only if its diagram is a line digraph* (cf. [30]). (See Figure 19.) In fact,

$$\text{jump}(P) = |P| - |G_P| + 1$$

as long as P *is* **N**-free [30]. Yet another proof has uncovered, in a subtle way, that there is a natural matroid structure corresponding to an **N**-free ordered set [32]. Indeed, the collection of bottom elements

$$\mathfrak{B}(P) = \{\{\text{bottom}(C_i) \mid i = 1,2,\ldots\} \text{ where } L = C_1 \oplus C_2 \oplus \ldots \text{ is a greedy linear extension of } P\}$$

is a family of bases of a matroid.

This enthusiasm notwithstanding, the class of **N**-free ordered sets is not always an appropriate generalization of series-parallel ordered sets -- at least from the standpoint of computational complexity. For instance, while the jump number problem is 'polynomially solvable' for any **N**-free ordered set, it is *NP*-complete for the ***minimum weighted sum completion time problem***, although it is polynomially solvable for any series-parallel ordered set [14]. The point here seems to be that a series-parallel ordered set is inductively defined and, in contrast, any ordered set can be embedded in an **N**-free one.

THE LETTERS **M** (em) AND **W** (double yoo)

Our thesis is that the study of diagrams, especially of conditions formulated in terms of subdiagrams, is leading to intriguing advances in the theory of ordered sets. Small subdiagrams may remind us of familiar shapes, as of letters. With an open mind to such mnemonics we have established several further results especially concerning greedy linear extensions [22], [26].

Let a, b be elements of an ordered set P and suppose that $a \ngeq b$. It is well known that there is a linear extension L of P in which $a < b$ [31]. Indeed, there is even a greedy linear extension in which $a < b$ (cf. [7]). What if we take three elements a, b, c in P such that $a \ngeq b$, $b \ngeq c$, and $a \ngeq c$? Again, there is a linear extension L of P in which $a < b < c$. Nevertheless, it is not always possible to choose L greedy. The simplest example is the **W**, illustrated in Figure 20. If P has no subdiagram isomorphic to **W** we call P **W**-*free.*

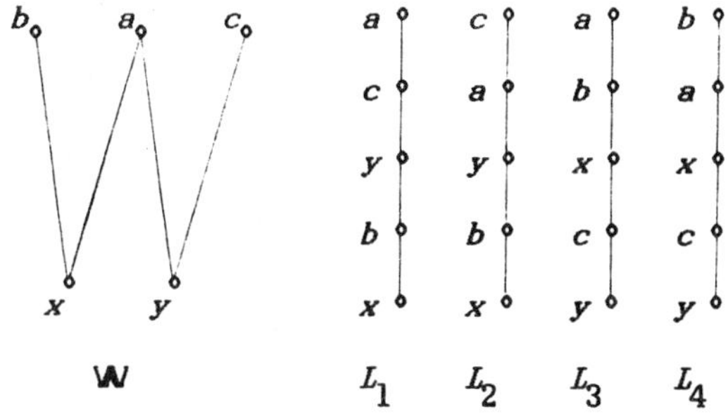

L_1, L_2, L_3, L_4 are all of the greedy linear extensions of **W**.
None satisfies $a < b < c$.

Figure 20

Let P *be* **W**-*free. For every antichain* $a_1, a_2, \ldots, a_n$, $n \geq 2$, *there is a greedy linear extension* L *of* P *such that* $a_1 < a_2 < \ldots < a_n(L)$. [7].

M

Figure 21

The converse cannot hold for the ordered set **M** illustrated in Figure 21 has **W** as subdiagram yet, for every antichain $\{a, b, c\}$ there is a greedy linear extension in which $a < b < c$. If P is **N**-free, rather more can be said. *For any antichain* $a_1, a_2, \ldots, a_n$,

$n \geq 2$, *there is a greedy linear extension* $L = C_1 \oplus C_2 \oplus \ldots$ *in which the sequence* $a_1, a_2, \ldots, a_n$ *is distributed consecutively in a sequence of these chains*, that is, there is $j \geq 1$ and $a_i \in C_{j+i}$ for each $i = 1,2,\ldots,n$. [7].

The techniques needed to prove these results can be turned to good account to derive an unexpected application to the 'dimension' of an ordered set. The ***dimension*** $\dim P$ of P is the least number of linear extensions of P whose intersection is P. The ***greedy dimension*** $\dim_g P$ of P is the least number of greedy linear extensions whose intersection is P [3].

If P *is* **W**-*free then* $\dim_g P = \dim P$. [24].

All of these ideas are related to this apparently fundamental and difficult problem. [23].

GREEDY LINEAR EXTENSION PROBLEM. *Let* P *be an ordered set and let* P' *be a partial extension. Under which conditions is there a linear extension of* P' *which is a greedy linear extension of* P?

Call an ordered set ***greedy*** if every one of its greedy linear extensions is jump-optimal. **N**-free ordered sets are, for instance, greedy. (Although **N** is not greedy **W** is!)

If **N** *is a subdiagram of a greedy ordered set then either* **W** *or* **X** *is too.* [24].

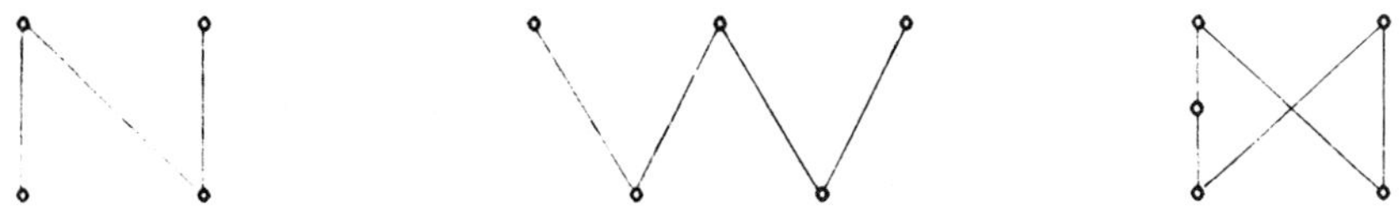

N **W** **X**

Figure 22

The problem to characterize greedy ordered sets is open. We have this solution in the case that it has length one (bipartite).

An ordered set of length one is greedy if and only if every **N** = {a < c ≻ d < b} *can be extended to a* **W** = {x > a < c ≻ d < b} *but which, in turn, cannot be extended to an* **M** = {y < x > a < c ≻ d < b}. [24].

Pictorially, we might express this condition by

for every 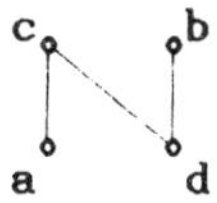 *there is* 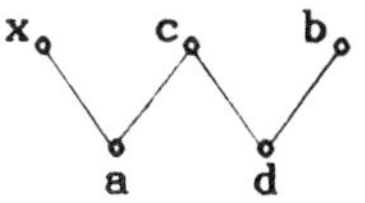 *and no* 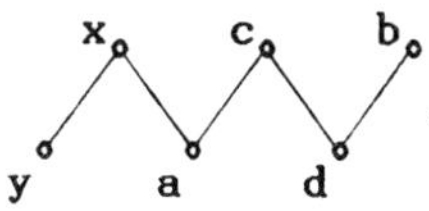.

There is little doubt that the letter **N** is serving us uncommonly well. Will there be such a fuller heuristic alphabet for the theory of ordered sets? Perhaps the letters **M** and **W** suggest so? On the other hand, the physical anatomy of both of these letters might suggest otherwise: for **W**, in English we say "double yoo"; in French, "double vay". From our enlightened vantage point we might now say "double en"!

REFERENCES

1. H.M Abdel-Wahab and T. Kameda (1978) Scheduling to minimize maximum cumulative cost subject to series-parallel procedence constraints, *Oper. Res.* 26, 151-158.

2. M. Bell and J. Ginsburg (1984) Compact spaces and spaces of maximal complete subgraphs, *Trans. Amer. Math. Soc.* 283, 329-338.

3. V. Bouchitté, M. Habib and R. Jégou (1985) On the greedy dimension of a partial roder, *Order* 1, 219-224.

4. O. Cogis and M. Habib (1979) Nombre de sauts et graphes série-parallèles, *RAIRO Inform. Théor.* 13, 13-18.

5. C.J. Colbourn and W.R. Pulleyblank (1985) Minimizing setups of ordered sets of fixed width, *Order* 1, 225-229.

6. D. Duffus, I. Rival and P. Winkler (1982) Minimizing setups for cycle-free ordered sets, *Proc. Amer. Math. Soc.* 85, 509-513.

7. M.H. El-Zahar and I. Rival (1985) Greedy linear extensions to minimize jumps, *Discrete Appl. Math.* 11, 143-156.

8. M.H. El-Zahar and J. Schmerl (1984) On the size of jump-critical ordered sets, *Order* 1, 3-5.

9. M.H. El-Zahar and N. Zaguia (1986) Antichains and cutsets, in *Combinatorics and Ordered Sets* (ed. I. Rival), *Contemporary Math.*, *Amer. Math. Soc.* (this volume).

10. U. Faigle and R. Schrader (1984) Minimizing completion time for a class of scheduling problems, *Inform. Proc. Letters* 19, 27-29.

11. M.R. Garey (1973) Optimal task sequencing with precedence constraints, *Discrete Math.* 4, 37-56.

12. G. Gierz and W. Poguntke (1983) Minimizing setups for ordered sets: a linear algebraic approach, *SIAM J. Algebraic Discrete Math.* 4, 132-144.

13. P.A. Grillet (1969) Maximal chains and antichains, *Fund. Math.* 15, 157-167.

14. M. Habib and R. Möhring (1986) *Discrete Math.* (to appear).

15. E.L. Lawler (1978) Sequencing jobs to minimize total weighted completion time subject to precedence constraints, *Ann. Discrete Math.* 2, 75-90.

16. E.L. Lawler (1978) Sequencing problems with series parallel precedence constraints, *Proc. Summer School on Combinatorial Optimization*, Urbino, Italy.

17. E.L. Lawler and B.D. Sivazlian (1978) Minimization of time varying costs in single machine sequencing, *Oper. Res.* 26, 563-569.

18. J.K. Lenstra and A.H.G. Rinnooy Kan (1978) Complexity of scheduling under precedence constraints, *Oper. Res.* 26, 22-35.

19. G.F. McNulty (1982) Infinite ordered sets, a recursive perspective, in *Ordered Sets* (ed. I. Rival), D. Reidel Publ. Co, Dordrecht, pp.299-330.

20. C.L. Monma and J.B. Sidney (1979) Sequencing with series-parallel precedence constraints, *Math. Oper. Res.* 4, 215-224.

21. I. Rival (1983) Optimal linear extensions by interchanging chains, *Proc. Amer. Math. Soc.* 89, 387-394.

22. I. Rival and N. Zaguia (1985) Antichain cutsets, *Order* 1, 235-247.

23. I. Rival and N. Zaguia (1986) Greedy linear extensions with constraints, *Discrete Math.* (to appear).

24. I. Rival and N. Zaguia (1986) Constructing greedy linear extensions by interchanging chains, *Order* (to appear).

25. J. Schmerl (1980) Decidability and $\aleph_0$ categoricity of theories of partially ordered sets, *J. Symbolic Logic* 45, 585-611.

26. J.B. Sidney (1979) The two-machine maximum flow time problem with series parallel precedence relations, *Oper. Res.* 27, 782-791.

27. W.E. Smith (1956) Various optimizers for single-stage production, *Naval Res. Logist. Quart.* 3, 59-66.

28. G. Steiner (1985) On finding the jump number of a partial order by substitution decomposition, *Order* 2, 9-23.

29. D.P. Sumner (1973) Graphs indecomposable with respect to the X-join, *Discrete Math.* 6, 281-298.

30. M.M. Syslo (1984) Minimizing the jump-number for partially ordered sets: a graph-theoretic approach, *Order*, 7-19.

31. E. Szpilrajn (1930) Sur l'extension de l'ordre partiel, *Fund. Math.* 16, 386-389.

32. M. Truszczynski (1985) Jump number problem: the role of matroids, *Order* 2, 1-8.

33. J. Valdes, R.E. Tarjan and E.L. Lawler (1982) The recognition of series-parallel digraphs, *SIAM J. Computing*, 11, 298-313.

Department of Mathematics and Statistics
The University of Calgary
Calgary, Canada T2N 1N4

ABCDEFGHIJ – 89876

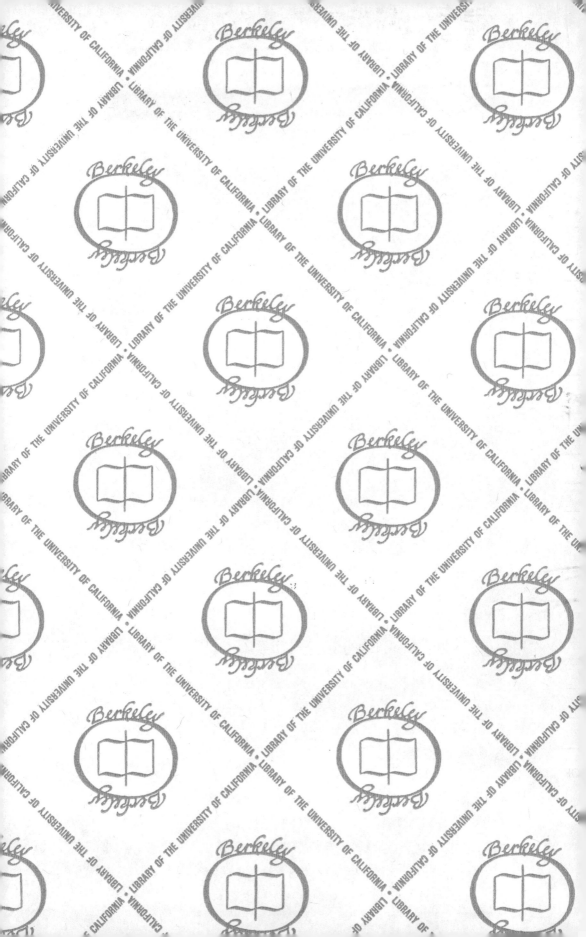